MEMBRANE SEPARATIONS IN BIOTECHNOLOGY

BIOTECHNOLOGY AND BIOPROCESSING SERIES

Series Editor

W. Courtney McGregor

XOMA Corporation
Berkeley, California

ADDITIONAL VOLUMES IN PREPARATION

MEMBRANE SEPARATIONS IN BIOTECHNOLOGY

Second Edition, Revised and Expanded

edited by
William K. Wang

GlaxoSmithKline Pharmaceuticals
King of Prussia, Pennsylvania

CRC Press is an imprint of the
Taylor & Francis Group, an **informa** business

CRC Press
Taylor & Francis Group
6000 Broken Sound Parkway NW, Suite 300
Boca Raton, FL 33487-2742

First issued in paperback 2019

ISBN-13: 978-0-8247-0248-9 (hbk)
ISBN-13: 978-0-367-39743-2 (pbk)

Visit the Taylor & Francis Web site at
http://www.taylorandfrancis.com

and the CRC Press Web site at
http://www.crcpress.com

Series Introduction

Biotechnology encompasses all the basic and applied sciences as well as the engineering disciplines required to fully exploit our growing knowledge of living systems and bring new or better products to the marketplace. In the era of biotechnology that began with recombinant DNA and cell fusion techniques, methods and processes have developed mostly in service of protein production. That development is documented in this series, which was originally called Bioprocess Technology. Many protein products that are derived from the technology are already marketed and more are on the way.

With the rapid expansion of genomics, many new biological targets will likely be identified, paving the way for the development of an even wider array of products, mostly proteins. As knowledge of the targets develop, so will rational drug design, which in turn may lead to development of small molecules as healthcare products. Rational genetic manipulation of cells as factories for growing products is also developing. Other examples of the application of genomics in health care include the development of gene therapy by insertion of genes into cells and the blocking of gene expression with antisense nucleotides. In such new directions, nucleotides and other small molecules as well as protein products will evolve. Technologies will develop in parallel.

Transgenic technology, in which the genome of an organism is altered by inclusion of foreign genetic material, is also just beginning to develop. Recombinant protein products can already be made, for example, in the milk of transgenic animals, as an alternative to conventional bioreactors. Newer applications for transgenic technology in agriculture may take time to develop, however. Questions continue to be raised about the long-term environmental consequences of such manipulation.

As technology develops in newer as well as established areas, and as knowledge of it becomes available for publishing, it will be documented in this continuing series under the more general series name of Biotechnology and Bioprocessing.

W. Courtney McGregor

Preface

When the first edition of *Membrane Separations in Biotechnology* was published in 1986, biotechnology was just beginning an era of rapid growth that continues today. As the biotechnology industry has matured, increased competition and regulatory requirements have placed a premium on the development of more productive and efficient biotechnology processes. As a result, the membrane technologies described in the first edition have been refined and optimized. Further, new membrane applications have been developed to overcome the challenges faced by the biotechnology industry. This second edition documents some of these developments by updating and supplementing the material published in the first edition. This edition also places more emphasis on pharmaceutical applications, although much of the technology can be used in other industries.

The focus of this volume is not to provide an all-inclusive, theoretical treatment of each of the subjects presented. Rather, discussions regarding actual applications, in industry and academia, have been collected. The authors describe their efforts to apply membrane technology to improve the performance of their bioprocesses. The case studies presented in this volume cover a wide variety of applications including cell culture harvesting, product purification, and process validation. Some of the contributions discuss applications for novel membrane products while others represent advances using established technology. In two of the chapters discussing more mature applications—one describing hollow-fiber membrane bioreactors and one summarizing harvesting of bioreactors—extensive background material is provided for reference. For other applications, such as those on virus removal and clearance, the strategies presented are broad but the data are case-specific. While the results presented here may not be universal to all applications, the solutions to the problems faced by each of the contributors may provide inspiration for other users as well as spark new ideas for further technological developments. As in the first volume, contributions from membrane manufacturers and suppliers have been excluded to avoid the possibility of discussing material in an excessively optimistic or promotional form.

This edition is intended to describe the approaches used by the individual authors to improve the performance and productivity of their bioprocesses. They discuss their experiences, both positive and negative, when using membrane technology for their applications. A discussion of the various issues the authors encountered is provided so that the reader can fully appreciate the scope of their work. The authors provide the rationale for the specific membrane technology chosen as well as a bioprocess description. Finally, they conclude with their thoughts regarding the success of membrane technology in their application. Again, this does not imply that the technology presented is universal, but rather demonstrates how membrane technology may be another tool available to the development scientist as new bioprocesses are designed.

The case studies presented here will be useful to a wide range of process development scientists and engineers who frequently encounter challenges similar to those that were faced by the authors. Since the contributions in this volume are real-world applications, the ideas that provide the basis for each example may be useful in a variety of situations. As with the first edition, it is hoped that this volume will be useful to the experienced user as well as the novice. The approaches used in these applications may also assist membrane suppliers in the improvement of current products and in the development of future bioprocess technologies. Finally, the student may discover that these contributions provide insights into the nature of the problems faced and solutions developed outside the classroom.

I wish to thank all the contributing authors for their time in preparing manuscripts and their willingness to share their ideas, results, and observations. Their efforts made this book possible.

William K. Wang

Contents

Contributors

Jenifer L. Ayala   Associate Scientist, CMC and BRD, Berlex Biosciences, Richmond, California

Mahmood R. Azari   Director of Project Implementation, Recombinant Business Unit/Manufacturing, Hyland Immuno Division, Baxter Healthcare Corporation, Thousand Oaks, California

Bernard P. Bautista   Process Development Scientist, Process Sciences, Biotechnology Unit, Pharmaceutical Division, Bayer Corporation, Berkeley, California

Bruce D. Bowen   Department of Chemical and Bio-Resource Engineering, University of British Columbia, Vancouver, British Columbia, Canada

Francisco J. Castillo   Scientific Director, Fermentation and Cell Culture Development, Berlex Biosciences, Richmond, California

G. Michael Connell   Senior Associate Process Scientist, Process Sciences, Biotechnology Unit, Pharmaceutical Division, Bayer Corporation, Berkeley, California

Alfred C. Dadson Jr.   Principal Scientist I, Process Sciences, Biotechnology Unit, Pharmaceutical Division, Bayer Corporation, Berkeley, California

Robert H. Davis   Patten Professor, Department of Chemical Engineering, University of Colorado, Boulder, Colorado

Igor Yu. Galaev   Associate Professor, Department of Biotechnology, Center for Chemistry and Chemical Engineering, Lund University, Lund, Sweden

Kent E. Göklen   Director, BioProcess Research and Development, Merck Research Laboratories, Rahway, New Jersey

Joseph I. Horwitz*   Fermentation and Cell Culture Development, Berlex Biosciences, Richmond, California

Ping Yu Huang   Senior Process Engineer, Manufacturing Services, Protein Design Labs, Inc., Fremont, California

Brian D. Kelley   Associate Director, Purification Process Development, Genetics Institute, Andover, Massachusetts

Marek Labecki   Department of Chemical and Biological Engineering, The Biotechnology Laboratory, University of British Columbia, Vancouver, British Columbia, Canada

Bo Mattiasson   Professor, Department of Biotechnology, Center for Chemistry and Chemical Engineering, Lund University, Lund, Sweden

W. Courtney McGregor   Vice President, Technical Development and Santa Monica Operations, XOMA Corporation, Santa Monica, California

Michael Meagher   Assistant Professor, Food Science Department, University of Nebraska, Lincoln, Nebraska

Gautam Mitra   Manufacturing Department, SmithKline Beecham Biologicals, Rixensart, Belgium

Harold G. Monbouquette   Associate Professor, Chemical Engineering Department, University of California, Los Angeles, California

Thomas J. Monica   Scientist, Fermentation and Cell Culture Development, Berlex Biosciences, Richmond, California

Tana J. Montgomery   Associate Scientist, Clinical Production, Berlex Biosciences, Richmond, California

Paul K. Ng   Principal Research Scientist II, Purification Development, Process Sciences, Biotechnology Unit, Pharmaceutical Division, Bayer Corporation, Berkeley, California

Kristen F. Ogle   Senior Research Scientist, Renal Division, Baxter Healthcare Corporation, McGaw Park, Illinois

---

* *Current affiliation*: Principal Research Scientist, Department of Mammalian Cell Culture Development, Centocor, Inc., Malvern, Pennsylvania

John Peterson   Associate Scientist, Protein Design Labs, Inc., Fremont, California

Jon T. Petrone   Principal Engineer, Drug Substance Development, Genetics Institute, Andover, Massachusetts

James M. Piret   Professor, Department of Chemical and Biological Engineering and Biotechnology Laboratory, University of British Columbia, Vancouver, British Columbia, Canada

Georg Roth   Senior Scientist, Fermentation and Cell Culture Development, Berlex Biosciences, Richmond, California

Gregory Russotti   Research Fellow, BioProcess Research and Development, Merck Research Laboratories, Rahway, New Jersey

Vicki L. Schlegel   Assistant Professor, Food Science Department, University of Nebraska, Lincoln, Nebraska

Gary M. Schoofs   Clinical Production Manager, Clinical Production, Berlex Biosciences, Richmond, California

Peter Schu   Director, Vaccines Bulk Manufacturing, Manufacturing Department, SmithKline Beecham Biologicals, Rixensart, Belgium

Clayton E. Smith*   Berlex Biosciences, Richmond, California

Robert van Reis   Distinguished Engineer, Department of Recovery Sciences, Genentech, Inc., South San Francisco, California

William K. Wang   Assistant Director, Biopharmaceutical Development, Downstream Process Development, SmithKline Beecham Pharmaceuticals, King of Prussia, Pennsylvania

Andrew L. Zydney   Professor, Department of Chemical Engineering, University of Delaware, Newark, Delaware

---

* *Current affiliation*: Consultant, Biotechnology Transfer, Bio-Marin, Berkeley, California

1
Protein Transport in Ultrafiltration Hollow-Fiber Bioreactors for Mammalian Cell Culture

Marek Labecki, Bruce D. Bowen, and James M. Piret
University of British Columbia, Vancouver, British Columbia, Canada

I. INTRODUCTION

Artificial organ devices, blood purification, production of therapeutic and diagnostic proteins, separation of complex solute mixtures, desalination and purification of water—these are just a few example applications involving the use of hollow-fiber membranes. In spite of some application-dependent differences, for instance in the permeability of the membranes for solutes and water, most hollow-fiber modules share a number of design and performance features. These include a shell-and-tube arrangement and high packing density of the fibers, possible occurrence of various concentration polarization phenomena, and the importance of flow configuration and membrane properties. Consequently, although this chapter focuses on cell culture bioreactors, the significance and implications of many of the analyses presented here are not necessarily limited to this single application of hollow-fiber devices.

The first uses of hollow-fiber reactors to immobilize enzymes and whole cells were reported in the early 1970s [1,2]. Since then, certain advantages of whole-cell immobilization over enzyme immobilization have become more apparent, such as the elimination of an expensive enzyme purification step, greater stability of intracellular enzymes, and the ability of whole cells to catalyze complete metabolic pathways [3]. Important advantages of perfused immobilized-cell culture over traditional suspension culture can include protection of cells from shear stresses, higher cell densities, higher productivities, reduced medium

1

requirements, and increased product concentrations [4]. Combined with a large membrane surface area per unit reactor volume, availability, and ease of production of hollow-fiber systems, these features can ultimately translate into lower costs for hollow-fiber culture.

Current interest in commercial cell culture systems stems mainly from the increasing demand for various therapeutic and diagnostic proteins, such as erythropoietin (stimulator of erythrocyte production), human growth hormone, factor VIII (a blood coagulation factor), interferons, or monoclonal antibodies (MAbs) [4,5]. Recombinant DNA technology has emerged as an attractive tool for obtaining many of these products from bacterial, plant, insect, or yeast cells. Although these cultures have comparatively simple medium requirements and high growth rates, it has been found that most of the proteins can be expressed in fully functional form only by mammalian cells [6]. Potential problems associated with the use of genetically engineered bacteria include also the production of endotoxins, which can be a source of product contamination; intracellular location of the product, which would require its extraction from the cell lysate [5]; or cell overgrowth leading to fiber damage in hollow-fiber cultures [7]. Another concern is that orders-of-magnitude higher oxygen uptake rates of bacterial versus mammalian cells can lead to severe oxygen depletion in reactors utilizing microorganisms [4]. The traditional way of producing MAbs from ascites tumors in mice, although competitive to cell culture in terms of cost and product concentration, suffers from several disadvantages, including ethical issues, practical problems with handling large numbers of animals (in large-scale production), high risk of product contamination by viruses and extraneous antibodies, and limitations in obtaining human antibodies from rodents [8,9]. Thus, recent decades have seen a rapid expansion of research concerning the use of mammalian cell bioreactors for the production of many useful proteins.

Numerous designs of hollow-fiber bioreactors (HFBRs) have been proposed [10], but the most conventional option involves entrapment of the mammalian cells in the extracapillary space (ECS) or shell side of a cylindrical ultrafiltration membrane cartridge with medium recirculated through the fiber lumina or intracapillary space (ICS), as illustrated in Figure 1. This configuration mimics to a certain extent the capillary-tissue system in living organisms. The ICS recycle flow, analogous to blood flow through capillaries, exchanges low-molecular-weight* (low-MW) nutrients and metabolites with the ECS, thus providing the cells with a stable growth environment. Higher-MW proteins required for cell growth (e.g., transferrin), which do not permeate easily through the membranes, are periodically supplied with fresh medium pumped into the shell side through one of the ECS ports while the product protein is harvested through the other

---

* Strictly speaking, a more appropriate term is "molar mass." The term molecular weight is used here in accordance with the general convention.

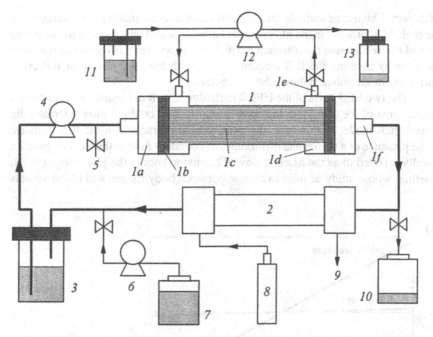

Figure 1 Conventional axial-flow HFBR system configuration (not to scale): *1*, HFBR cartridge; *1a*, upstream ICS manifold; *1b*, epoxy support for the fibers; *1c*, hollow fibers; *1d*, downstream ECS manifold; *1e*, downstream ECS port; *1f*, downstream ICS port; *2*, oxygenator; *3*, ICS recycle medium reservoir; *4*, ICS recycle pump; *5*, ICS sampling port; *6*, ICS feed pump; *7*, fresh ICS medium reservoir; *8*, air cylinder; *9*, waste gas; *10*, waste ICS medium reservoir; *11*, ECS medium reservoir; *12*, harvesting pump; *13*, harvest reservoir. The terms upstream and downstream are used with reference to the main ICS flow. The ICS medium is recirculated through the cartridge *1* and the oxygenator *2* using the recycle pump *4* and is periodically replenished using the feed pump *6*, while the waste medium is collected in the reservoir *10*. The oxygenator *2* is placed downstream of the bioreactor to avoid air release in the cartridge due to the ICS axial pressure drop and to allow air bubbles possibly carried with the recycle stream to be trapped in the medium reservoir *3*. The sampling port *5* enables monitoring of the ICS nutrient and metabolite levels. Fresh ECS medium is intermittently supplied using the harvesting pump *12*, while the harvest is collected in the reservoir *13*. Oxygen, pH, and temperature probes are omitted in this diagram.

ECS port. The use of suitable ultrafiltration membranes enables concentration of the high-MW product in the ECS prior to harvesting and thus reduces the time and cost of the subsequent downstream purification processes. Oxygen is provided to the cells by aerating the ICS medium (often with 5% CO_2 added for buffering purposes) in an online hollow-fiber oxygenator.

The two ECS ports of the HFBR cartridge shown in Figure 1 remain closed during most of the operation; the corresponding flow configuration is termed the closed-shell mode. In this case, the axial pressure gradient in the ICS induces, in the presence of a permeable membrane, a secondary flow in the ECS (Figure 2), usually referred to as the Starling flow. This name honors the physiologist E. H. Starling, whose study of fluid exchange between body tissues and blood vessels

a)

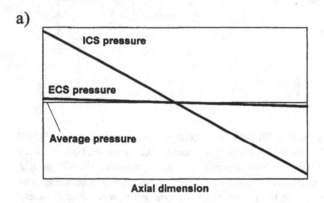

b)

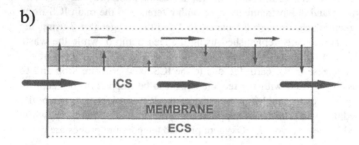

Figure 2 ICS and ECS pressure distributions (a) and flow paths (b) for a representative hollow fiber in a closed-shell HFBR (not to scale). Fluid passes from the ICS to the ECS in the upstream part and from the ECS to the ICS in the downstream part of the reactor. The secondary flow induced in the ECS (Starling flow) has a maximum at or near the half-length of the ECS. The exact pressure profiles and flow patterns existing in the HFBR will depend on the membrane permeability, cell density, solute distributions, and other factors (see Section II).

dates back to the turn of the past century [11]. Starling flow can cause convective accumulation of the cells and high-MW proteins in the downstream part of the reactor [12,13]. This phenomenon may have serious implications for HFBR performance and will be discussed later. Examples of open-shell operations are inoculation of the ECS with cell-containing medium and product harvesting from the ECS; they are typically carried out countercurrently (inoculation) or cocurrently (harvesting) with respect to the ICS flow, although other configurations are also possible. Usually, at all stages of cell culture, the HFBR cartridge is inclined at about 45° to the horizontal, with the ICS recycle flow directed upward; a configuration whose intent is to reduce the effects of downstream polarization and cell sedimentation.

The promises and challenges of cell cultivation in HFBRs have stimulated a great deal of modeling work. Most models include substrate consumption kinetics and neglect convective transport in the membranes and in the ECS [1,3,14–18]. The importance of ECS convective transport in HFBR operation was recognized in many experimental [12,13,19,20] as well as theoretical studies [21–27]. Models that include the effect of osmotic pressure on ECS protein distribution [25–29], hindered transmembrane protein transport [29], and the dynamics of cell growth in HFBRs [30] have recently been proposed.

The following section of this chapter provides background for understanding various aspects of hollow-fiber cell culture and discusses the complex spectrum of related factors and challenges. Its contents should appeal both to general HFBR users and to engineers who wish to strengthen their underlying knowledge and thus benefit their design or modeling endeavours. Theoretical analyses of HFBR fluid flow and protein transport are described in more detail in Section III. This section will likely be of greatest interest to engineers, although the theoretical results presented therein do have a broader relevance and may provide useful insights to experimentalists. Section IV links the preceding parts of the chapter and concludes with a discussion of alternatives and a prognosis for the future of the field. For additional information on hollow-fiber bioreactors, the reader is encouraged to consult one or more of the following good review papers on the subject: Piret and Cooney [4], Tharakan et al. [10], Heath and Belfort [31], and Brotherton and Chau [32].

II. HOLLOW-FIBER MAMMALIAN CELL CULTURE: FACTORS AND CHALLENGES

Efforts to enhance the performance of immobilized-cell cultures are focused on two major objectives: providing the cells with optimum growth conditions, and obtaining large quantities of highly concentrated product at low cost. In most cases, the optimal conditions for cell growth and for product formation are different; i.e., product generation is usually not growth associated [10,33,34]. Ideally, the first objective should have priority during the initial (growth) phase of the culture, while the second one should take precedence at the later stages and may involve subjecting the cells to environmental or nutritional stress [35]. On the other hand, an increase in the specific (i.e., per-cell) product generation rate resulting from such stress may not always improve the overall product yield because of the concomitant decrease in the cell density* [36]. The complication of operating procedures and a lack of reproducible optimization data are further reasons why the distinction between optimal growth and optimal production conditions is often ignored in practice.

The extent to which these major objectives of hollow-fiber cell culture can be accomplished is influenced by a complex variety of interdependent design and operating parameters as well as physical, biological, and biochemical phenomena. The most important of these factors are summarized below. This systematic presentation should be helpful in understanding the complexity of HFBR systems, related operational and modeling challenges, and the role of protein transport in HFBR operation.

A. Cell Types

Animal cells used in culture can be anchorage-dependent, i.e., require a surface substratum for attachment and growth; or anchorage-independent, i.e., able to grow in suspension [5]. The two most commonly used types of anchorage-dependent cells are fibroblast and epithelial cells, while the most important anchorage-independent cells are lymphoblasts and hematopoietic cells [37]. Some of the frequently studied epithelial cell lines are CHO-K1 (Chinese hamster ovary cells),† used extensively for the production of recombinant proteins; and HeLa (human cervical cancer cells), used mostly in research. Well-known examples of fibroblast cells include the BHK-21 cell line (baby hamster kidney cells), used for the production of clotting factors and veterinary vaccines; Vero cells (African green monkey kidney cells), used for the production of polio vaccine; and a family of cell lines (e.g., MRC-5) isolated from embryonic human lung and used for the production of various human vaccines [37]. Available surface area normally represents a physical growth limitation to anchorage-dependent cells, which grow in an adherent monolayer [38]. Therefore, systems having large ratios of surface to volume, such as hollow fibers, appear ideal for cultivating these cells, provided a suitable membrane material has been selected or the fibers have

---

* Cell density is used here to mean the number of cells per unit volume of the ECS.
† Many of the epithelial and fibroblast cell lines, including CHO-K1, can readily grow in suspension. This is a result of transformation, which usually leads to a partial or complete loss of anchorage dependence (see next paragraph).

been precoated in order to facilitate cell attachment [39,40]. Under certain culture conditions, anchorage-dependent cells can also grow in multilayers [41] or form aggregates that allow them to grow in suspension [36].

Most types of mammalian cells used in culture are transformed cell lines; i.e., they are either genetically modified or derived from neoplastic (tumor) tissues. Compared with normal cells, the transformed cells are usually immortal and tumorigenic, exhibit higher growth rates, have less sophisticated medium demands, and lose their anchorage dependence [5,38]. Genetic modification has as its objective the enhancement of the very low protein production yields typical of normal human cells, and can be carried out using recombinant DNA technology, leading to recombinant cell lines; or by hybridization, which gives rise to hybridomas [36]. The first hybridoma cell line capable of continuous secretion of a single type of antibody to a specific antigen was produced in 1975 by Köhler and Milstein [42], who fused short-living antibody-producing B-lymphocytes (from mouse spleen) with immortal myelomas (tumor cells). This technique has since been improved to obtain more stable and more productive cell lines [34,43,44], and the cultivation of hybridomas for the production of monoclonal antibodies is currently one of the leading applications of hollow-fiber bioreactors.

B. Extracellular Matrix

Many types of mammalian cells in culture synthesize and deposit some sort of extracellular matrix (ECM). Transformation usually decreases ECM production, leading to reduced anchorage dependence [45]. The major constituents of ECM are collagens, forming insoluble protein fibrils; glycoproteins (e.g., fibronectin or laminin), containing a protein moiety with carbohydrates attached to it; and proteoglycans, composed predominantly of a polysaccharide known as glycosaminoglycan, bound to a core protein [45]. Owing to their net negative charge, the glycosaminoglycan molecules are able to bind cations as well as large amounts of solvation water and thus act as shock absorbers in connective tissues in vivo [46]. Flow through several types of solid tissues was found to obey Darcy's law (i.e., to exhibit a linear dependence between flow rate and applied pressure gradient), and the tissue hydraulic permeabilities correlated well with the glycosaminoglycan contents [47]. A comprehensive study by Levick [48] demonstrated, however, that none of the ECM components alone could account for the low hydraulic permeabilities of most tissues, and that the main resistance the ECM offers to fluid flow arises from the combined effects of collagen fibrils, glycosaminoglycan, and the proteoglycan core protein. A dramatic effect of ECM on convective flow in hollow-fiber bioreactors was observed by Ryll et al. [49], who reported extreme difficulties in product removal from an ECS packed with BHK cells. Although high densities similar to those of solid tissues in vivo are typically observed in hollow-fiber cultures [38,50,51], problems with product recovery have not been encountered in the case of hybridomas or other cell lines which do not produce ECM [49]. This seems to agree with an observation that the drag effect of the cells, estimated from the Carman-Kozeny equation by treating the tissue as a porous bed in which the cells are impervious obstacles, contributes only negligibly to the interstitial hydraulic resistance [48].

C. Inoculation

After initial propagation of a cell line in batch culture (e.g., in T-flasks, shake flasks, or spinner flasks), the inoculum suspended in growth medium is introduced into the ECS of the HFBR. The inoculation can be carried out using one of several possible flow configurations (Figure 3). In the countercurrent arrangement (Figure 3a), the cells are added to the upper (downstream) part of the reactor, while the ECS is continuously perfused via lumen recycle flow. The dead-end configuration (Figure 3b) can be useful when concentration of cells and high-MW growth factors is of importance. Attention is required to ensure that a sufficient number of cells is present in the ECS at the end of inoculation; typically, cell concentrations greater than 10^6 cells/mL are used to effect a successful startup of the culture. It has been reported that a higher cell number inoculum can shorten the time needed to reach the maximum level of antibody production [52]. A major challenge of the inoculation procedure and of the initial stages of HFBR operation is to achieve and maintain maximally uniform distribution and growth of the cells in the extracapillary space. This is not an easy task, considering that some

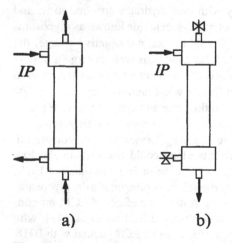

a)   b)

Figure 3 Examples of flow configurations used for HFBR inoculation: (a) countercurrent; (b) dead-end. *IP*, inoculation port.

of the cells usually settle by gravity to the bottom of the reactor, often filling up fiber-free spaces like the ECS manifolds where they are in danger of oxygen starvation; while some cells may be obstructed by the more densely packed regions of the fiber bundle, never leaving the vicinity of the inoculation port [53]. These problems may be alleviated, for example, by adjusting the inclination angle of the reactor cartridge or by periodically rotating the HFBR about its longitudinal axis. Nevertheless, system-specific optimization of the inoculation procedure, including a proper choice of flow configuration, seems very important from the point of view of an efficient use of the reactor space and therefore requires further study.

D. Cell Growth in HFBRs

After inoculation, a lag phase of several days can occur during which the cells adapt to HFBR culture conditions. Within the 2 to 4 weeks of the subsequent growth phase, the cells essentially fill the ECS of the reactor. The lack of physical space will usually limit the duration of the growth phase and mark the onset of the stationary (or steady-state) period, in which the rates of cell growth and cell death are approximately equal [5,38,54]. The ECS cell density at steady state can reach as high as 10^9 cells/mL, which is equivalent to a bed of spheres having a diameter of 10 μm each (i.e., small mammalian cells [46]) and packed to about 50% porosity. The hydraulic permeability of a cell-packed ECS can be decreased by as many as 7 or 8 orders of magnitude, compared with that at cell-free conditions [27,47], while the effective solute diffusivities can be reduced by a factor ranging from about 2 to 3 for small solutes like oxygen or glucose to about 25 or more for proteins like bovine serum albumin (BSA) [47]. At high cell densities, the HFBR system can be very sensitive to variations in the culture parameters, thus necessitating a particularly strict control of culture conditions [55]. This steady-state phase should theoretically continue indefinitely; in practice, an irreversible decline in overall cell viability is usually observed after 1 to 3 months of the culture. The shift to the death phase is likely due to the accumulation of dead cells and cell debris, as well as perhaps high-MW inhibitory factors [56], which are very difficult to remove from the ECS in a steady and effective fashion.

E. Effects of Nutrients and Metabolites

One of the most important factors which influence cell growth in the bioreactor is the availability of nutrients to the cells. The nutrients should be supplied at rates at least as high as the rates of their consumption, and there often exists a nutrient level optimal for cell growth, which may be different from the level optimal for product generation. For example, the approximate range of oxygen demand for mammalian cells is $0.053-0.59 \times 10^{-9}$ mmol/cell/h [57], the optimal dissolved oxygen tension (DOT) for hybridoma growth has been reported to be 50% to 60% of air saturation, while MAb secretion can be maximal at DOT equal to 25% of air saturation [58,59]. Adequate oxygen delivery has raised special concern owing to the constraints imposed by its low solubility in water (0.22 mmol/mL at 37°C, in equilibrium with atmospheric air) and the toxicity of hyperbaric oxygen concentrations [60]. Unlike that of any other low-MW nutrient, the supply of oxygen cannot then be enhanced by simply adjusting its concentration in the medium to a higher level. At the high cell densities typically encountered in hollow-fiber cultures, oxygen depletion can easily become a problem if the system is not properly designed; for example, if the fibers in the cartridge are packed too sparsely, or if the ICS flow rate is too low. This conclusion agrees with results of several theoretical [19,31] as well as experimental [61] studies identifying oxygen as the most important limiting substrate in hollow-fiber cultures. Currently, continuous oxygenation through an external unit, often another hollow-fiber module, which is installed online within the ICS recycle loop (Figure 1), is the standard means of oxygen supply in closed-shell HFBR systems.

Other nutrients required in a mammalian cell culture include carbohydrates, amino acids, lipids, fatty acids, vitamins, salts, and trace elements [62]. Glucose and glutamine are of particular importance because of their high demands by the cells. Glucose is consumed by mammalian cells at rates comparable to that of oxygen, while glutamine is utilized at rates three to four times lower than glucose [63,64] and 5 to 10 times higher than other amino acids [65]. The type of metabolic endproducts and their potential inhibitory effect on mammalian cell growth depend strongly on the cell line, medium composition, and other culture conditions [4]. The most common metabolites are lactic acid, ammonium, and carbon dioxide, but cells can also produce alanine, glutamic acid, and other amino acids [65]. Chresand et al. [66] reported that a sodium lactate concentration of 10 mM inhibited by about 50% the growth of Ehrlich ascites tumor cells, and that lactate buildup in the hollow-fiber culture of these cells became inhibitory to growth even before oxygen was limiting. On the other hand, growth and antibody production of CRL-1606 hybridomas cultivated in HFBRs by Piret et al. [61] were virtually unaffected by metabolite accumulation (up to 40 mM lactate [67]) or depletion of nutrients other than oxygen. Other studies confirmed that glucose and glutamine should not be limiting to mammalian cell growth as long as their concentrations are maintained at sufficiently high levels dependent on the cell line and culture conditions [67–69]. Increased MAb production rates in HFBRs have been observed for those hybridoma cell lines that have higher demands for glucose and glutamine [69].

Concern about diffusional limitations in hollow-fiber bioreactors due to insufficient rates of nutrient delivery and metabolite removal stimulated work on the feasibility of convective enhancement of these transport processes [32]. To that end, alternative HFBR designs such as a dual-circuit system with two fiber bundles operated at different pressures [70,71], a cross-flow flat-bed reactor [72], or a radial-flow cartridge with a central feed distributor [41] were proposed (see also Section IV). Convective flow was predicted to have a marked influence on the behavior of low-MW solutes in an experimental bioreactor composed of two concentric hollow-fiber membranes, which was investigated by Salmon et al. [23]. Schonberg and Belfort [22] also concluded from their model studies that radial convection could enhance bioreactor performance. Unfortunately, their analysis is inapplicable to closed-shell systems because of the assumption that fluid passes across the membrane only in one direction (from the ICS to the ECS). For conventional ultrafiltration closed-shell HFBRs such as Amicon P30 [16] or Gambro GFE-15 [29], the role of convection in the transport of low-MW solutes like oxygen across the membrane should be negligible (transmembrane Peclet numbers $<10^{-2}$ at an ICS flow rate of 200 mL/min). In fact, even if the convective transmembrane flow were significant, it would likely increase nutrient maldistribution in closed-shell HFBRs by acting against diffusion in the downstream part of the reactor, where the potential nutrient depletion is expected to be most severe [4]. A theoretical study by Pillarella and Zydney [73] demonstrated that convection played a significant role in establishing a desired insulin response in a hollow-fiber bioartificial pancreas, where insulin-producing beta cells were cultured as a monolayer on the fiber surface. In such devices, however, nutrient consumption is relatively small because of the low cell numbers involved.

F. Serum Supplementation

Most mammalian cells have traditionally been cultivated in media containing fetal bovine serum (FBS), which is a rich source of nutrients, carrier proteins (e.g., albumins or transferrin), attachment factors (needed by anchorage-dependent cells), growth factors (e.g., interleukin-6), and hormones (e.g., insulin or hydrocortisone) [74]. The positive effects of using serum include protection against pH fluctuations (buffering capacity), neutralization of some toxins (such as heavy metals), protease inhibition, and protection against shear forces [36]. The use of serum also has several drawbacks, such as its high cost, undefined composition, lot-to-lot variation, potential cytotoxicity, risk of contamination by viruses or mycoplasma, and complication of downstream processing [38,75,76]. A vast amount of research has been conducted on cell adaptation to specially formulated serum-free and protein-free media [68,75–78]. Serum-free media have also been successfully used in the production of MAbs in hollow-fiber bioreactors [19,49,64,79–81]. Although the use of serum in hollow-fiber culture, if required, can largely be confined to the cell compartment, i.e., the ECS of the bioreactor, addition of serum to the ICS medium is often necessary to compensate for partial leakage of some serum components through the membranes [82]. Various aspects of serum- and protein-free animal cell culture have been discussed, for example, by Freshney [38], Jäger [74], and Maurer [76].

G. Effects of pH, Osmolarity and Temperature

The optimal pH range for mammalian cells is about 7.0 to 7.4; a pH below 6.8 is usually inhibitory to cell growth [40], although it may cause an increase in specific antibody production [35,69]. Some hybridoma cells have been able to adapt to slightly elevated pH (7.7), eventually reaching growth and antibody production rates similar to those at optimum pH levels [35]. Variations in pH are expected to occur naturally in the culture as a result of cell metabolic activity, such as glucose consumption leading to the formation of lactic acid. To avoid accumulation of acidic metabolites, it has been recommended that glucose concentrations in the medium be maintained at sufficiently low levels (<2 g/L) [40]. Selection of a proper buffering system is another important factor affecting the pH of the culture. For example, pH gradients existing in a hollow-fiber culture of hybridoma CRL-1606 cells were shown to decrease when HEPES (N-2-hydroxyethylpiperazine-N'-2-ethane sulphonic acid) buffer was used instead of the CO_2-bicarbonate system [61]. Three major strategies are used to control pH in HFBRs: adjustment of the amount of CO_2 delivered with the air through the oxygenator; addition of a base (NaHCO$_3$ or NaOH) to the culture medium; and variation of the medium supply rate [55,69]. The main challenge lies, however, not as much in the regulation of pH level in the medium delivered to the culture as in the prediction, measurement, and ultimately minimization of local pH gradients which may arise in the heterogeneous environment of a hollow-fiber reactor, especially in the presence of mass transfer limitations at high cell densities. The magnitude of these local gradients can be roughly estimated from the difference between the measured pH values of the ECS and ICS samples [61], but the actual pH distribution within the extracapillary space is much more difficult to determine.

The osmolarity range optimal for most cells is approximately 260 to 320 mOsm/L [38], which corresponds to the osmotic pressure range of 700 to 850 kPa. Partial or complete rejection of large molecules by the membranes may lead to the development of osmotic pressure gradients in the ECS of the HFBR, but this will likely have a negligible effect on the cells. For instance, assuming that ECS convective flow has caused downstream polarization of BSA to the concentration of 70 g/L, the resulting local increase in osmolarity will be only about 1 mOsm/L. Caution should nonetheless be exercised when adjusting medium composition, e.g., by adding a base for the purpose of pH control, as this may cause unacceptable changes in osmolarity [38,62]. Both hypo-osmotic and hyperosmotic stresses have been found to supress hybridoma growth [33,83,84], while their effect on monoclonal antibody production can be positive, especially at elevated osmolarity [33,83–85].

The bioreactor system is usually thermostatted at 36 to 37°C, which is the optimal cultivation temperature for most mammalian cells [38]. It is especially important to avoid hyperthermic conditions, as temperatures in excess of 2°C above optimal will rapidly result in cell death. While primary cell metabolism and growth rates are negatively affected by changes in temperature, secondary metabolism including protein production may be enhanced under conditions of a thermal shock (within a reasonable range) [67,86,87]. In some cases, thermal degradation of product proteins may be attenuated at lower temperatures, leading to improved product characteristics [62]. The temperature, as well as the pH and the medium composition, can also affect the stability of some nutrients; for example, glutamine has been known to undergo a relatively rapid (half-life of several days) chemical decomposition at 37°C [88]. It has been recommended that the optimal temperature for a given cell culture be determined experimentally, preferably after a period of cell adaptation [62]. Minor temperature variations, which can be expected to occur in the system as a result of nonuniform heating or fluctuations within the control range, can also influence transport phenomena in the HFBR. About 2% changes in fluid viscosity and solute diffusivities will result from each 1°C of temperature variation near 37°C.

H. Starling Flow and Downstream Polarization

The ICS flow rate, Q_L, is an important operating parameter which controls the rate of nutrient delivery to the cells and determines the magnitude of the ECS Starling flow in a closed-shell HFBR. Excessively rapid ICS flows can lead to fiber breakage or to the denaturation of ICS proteins. Starling flow has been known to cause convective downstream polarization of cells and growth factor proteins (Figure 4), leading to nonuniform cell growth and thus inefficient utilization of reactor space [12,13]. In addition to its dependence on Q_L, the magnitude of the Starling flow is a function of the membrane and ECS hydraulic permeabilities, L_p and k_S, which themselves depend on other factors. For example, L_p can be influenced by membrane fouling, while k_S is dependent on the cell density, presence of ECM, as well as on geometric parameters such as the length and spacing of the fibers. The importance of Starling flow to ECS protein redistribution can be assessed by inspecting the axial ECS Peclet number, which can be readily estimated using the simplifying assumptions that the axial ICS pressure drop is linear and the ECS pressure is constant and equal to the average ICS pressure. For the Amicon P30 polysulphone membrane HFBR used by Piret and Cooney [16], $Q_L = 100$ mL/min, $L_p = 3.5 \times 10^{-13}$ m, the fiber lumen radius $R_L = 10^{-4}$ m, the effective fiber length $L = 0.17$ m, and the ECS cross-sectional area $A_S \approx 1.9 \times 10^{-4}$ m^2, which results in an average axial ECS velocity $u_S = 2Q_L L_p L^2/(A_S R_L^3) \approx 1.8 \times 10^{-4}$ m/s. Thus, for a typical protein diffusivity $D = 10^{-10}$ m^2/s and a length scale $d > 2.5 \times 10^{-4}$ m (i.e., assuming conservatively that d is not less than the average interfiber distance), the axial Peclet number Pe

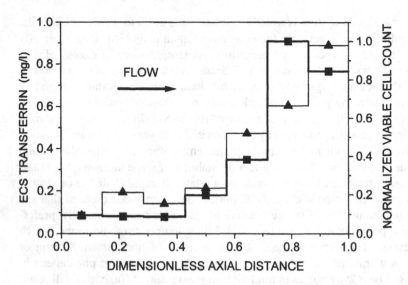

Figure 4 Downstream polarization of growth factor protein (transferrin; triangles) and mammalian cells (squares) in a hollow-fiber bioreactor after 8 days of closed-shell operation. Steps correspond to axial segments into which the reactor was cut upon freezing in liquid nitrogen. (Adapted from Ref. 13.)

$= u_S d/D > 450$. A similar analysis for a low-permeability Cuprophan membrane Gambro cartridge [29], using $Q_L = 300$ mL/min, $L_p = 7 \times 10^{-15}$ m, $R_L = 1.1 \times 10^{-4}$ m, $L \approx 0.24$ m, $A_S \approx 3.1 \times 10^{-4}$ m², and $d > 8.4 \times 10^{-5}$ m, yields $u_S \approx 10^{-5}$ m/s and $Pe > 8$. In either case, the relatively large Peclet number indicates a dominance of convection over diffusion in the axial transport of ECS proteins.

Dimensionless analysis can also be used to determine which mechanisms are important in the transport of cells in the ECS. Table 1, which has been adapted from Zeman and Zydney [89], compares the characteristic velocities associated with different forces acting on the cells suspended in the ECS of closed-shell HFBRs (in the absence of surface interactions). The calculations were made using $u_S = 10^{-5}$ m/s, length scale $d = 10^{-4}$ m (average interfiber distance), gravitational constant $g = 9.81$ m/s², fluid viscosity $\mu \approx 7 \times 10^{-4}$ Pa · s (water at 37°C), fluid density $\rho \approx 1000$ kg/m³, average density of a mammalian cell $\rho_c \approx 1060$ kg/m³ [90], and cell radius $r_c \approx 7 \times 10^{-6}$ m. The wall shear rate was estimated using eq. (10) in Section IIIA.1 to be $\gamma_w = (\partial u_S/\partial r)_{r=R_F} \approx 0.4$ s⁻¹ for the parameters of the Gambro cartridge [29]; virtually the same result could be obtained by assuming Hagen-Poiseuille flow and a parabolic velocity profile in the ECS, which leads to the expression $\gamma_w \approx 4 u_S/d$. The cells were assumed to be impermeable to flow, and the cell diffusion coefficient was calculated from the Stokes-

Table 1 Characteristic Velocities Associated with Different Mechanisms of Cell
Transport in Closed-Shell HFBRs

Transport mechanism	Characteristic velocity	
	Formula	Magnitude (m/s)
Convection	$U_{conv} = u_S$	10^{-5}
Sedimentation	$U_{sed} = 2r_c^2(\rho_c - \rho)g/(9\mu)$	9×10^{-6}
Shear-induced diffusion	$U_{shear} = D_{shear}/d = 0.1\gamma_w r_c^2/d$	2×10^{-8}
Brownian diffusion	$U_{diff} = D_c/d = kT/(6\pi\mu r_c d)$	5×10^{-10}
Inertial lift	$U_{lift} = \gamma_w^2 r_c^3\rho/\mu$	8×10^{-11}

Source: Ref. 89.

Einstein equation as $D_c = kT/(6\pi\mu r_c) \approx 5 \times 10^{-14}$ m²/s, where $k = 1.38 \times 10^{-23}$
J/K is the Boltzmann constant, and $T = 310$ K is the temperature. The estimated
convective and sedimentation velocities are of the same order of magnitude and
much larger than the other characteristic velocities (Table 1), which clearly demonstrates that Starling flow and sedimentation may play the most significant roles
in cell redistribution in the ECS.

Convective polarization of ECS proteins results in the local elevation of
the osmotic pressure, fluid viscosity, and fluid density in the ECS. The osmotic
pressure increases until the ECS total pressure locally reaches the level of the
pressure on the lumen side. At this point, since there is negligible transmembrane
flow, the axial fluid flow in this protein-rich region is practically shut down.
Consequently, a stagnant zone develops in the downstream part of the reactor,
which leads to a reduction in the effective ECS length available to Starling flow
and thereby to a decrease in its magnitude [25]. Protein polarization can also be
engendered by natural convection induced by fluid density gradients under the
action of gravitational forces. In this case, proteins can polarize at the bottom
of the reactor (similarly to cell sedimentation), which has been both observed
experimentally and predicted by dimensionless analysis (Grashof number $Gr =
O(1)$, which indicates a similar importance of viscous and buoyant forces acting
on a concentrated protein solution) [13]. As a result of convective or buoyancy-
induced polarization, accumulation of the proteins in the fiber-free ECS manifolds usually takes place, thus effectively lowering the protein concentration in
the fiber bundle and leading to the formation of additional stagnant zones in the
closed-shell HFBR. Unless the outlet ECS port is positioned downward during
harvesting, the existence of such zones may diminish product recovery. The understanding and control of the effects of the manifold regions and gravity on the
distributions of proteins and cells in the ECS represents a considerable operational
and modeling challenge. In vertical or inclined cartridges with an upward ICS

recycle flow, the natural-convective protein polarization due to density gradients could be partly or entirely neutralized by the opposite action of forced convection due to Starling flow. Inclination of the cartridge at about 45° to the horizontal is often preferred to the vertical orientation, in which the cells might completely fill up the bottom ECS manifold. Since the fluid viscosity usually shows a fairly strong dependence on protein concentration, e.g., BSA at 50 g/L produces about a 25% increase in viscosity [91], the effect of locally elevated viscosity of the ECS fluid can also contribute to the development of stagnant zones in the reactor. Nonuniformities of Starling flow and their relation to fiber maldistribution have been investigated using magnetic resonance imaging (MRI) [92,93]. It should be noted that the above-described effects of Starling flow may be substantially diminished in some of the alternative HFBR designs (see Section IV). For example, in the expansion chamber system of Endotronics (currently Cellex Biosciences) [94], either the ICS or the ECS is pressurized in order to force the fluid across the membrane along the entire length of the fiber (see also Section III.L).

I. Membrane Material

The material and properties of the membranes can significantly influence protein distribution and therefore play an important role in the operation of a hollow-fiber system. Membrane materials used in HFBR cell culture include cellulose and its derivatives [9,49,50,55,64,69,79,82,95–97], polysulfone [13,16,41,50, 61,72,98], poly(vinyl chloride)-acrylic copolymer [50,72], poly(methylmethacrylate) (PMMA) [100], and polyacrylonitrile (PAN) [72]. Cellulose is a highly hydrophilic and mechanically strong material and it is used to manufacture fairly thin (5 to 10 μm) isotropic membranes [79,99]. Cellulose-based membranes have relatively small pores and low hydraulic permeabilities, and usually exhibit low affinity for protein adsorption and low cytotoxicity, which makes them suitable for applications such as mammalian cell culture and hemodialysis [79,99]. Cuprophan, or regenerated cellulose, exhibits a relatively high crystallinity (~50%) in its dry form, but can become amorphous and swollen upon sorption of large quantities of water [100]. A manifestation of this morphological change is radial and axial expansion of Cuprophan hollow fibers under wet conditions. Since the expanded fibers are fixed at both ends of the ECS, they assume a wavy appearance, which increases the heterogeneity of their distribution [26,101]. Cellulose acetate (CA) (more exactly, diacetate) is obtained by acetylating (i.e., substituting H– with CH_3CO-) approximately 80% of the hydroxyl groups in cellulose, while cellulose triacetate (CTA) is completely acetylated. Hemophan is another cellulose derivative, in which a fraction (<1%) of the hydroxyl groups are replaced with $(CH_3CH_2)_2N-CH_2CH_2-O-$ (diethylaminoethoxy) [102]. Compared with Cuprophan, the Hemophan, CA, and CTA membranes are less hydrophilic, and usually have slightly larger pores and higher permeabilities [99]. In contrast to isotropic membranes, asymmetric (or anisotropic) membranes consist of a thin (1 to 2 μm) ultrafiltration skin and a thicker (50 to 75 μm) spongy support layer; they usually have higher hydraulic permeabilities and are made of polysulfone or other hydrophobic polymers [79,102]. To ensure successful growth of anchorage-dependent cells, many types of membranes, especially those based on cellulose, may require precoating with materials promoting adhesion, such as poly-D-lysine [41,95,96]. PMMA membrane HFBRs have been reported to support both high hybridoma MAb productivity and high growth rates of anchorage-dependent Vero cells [95]. Although some membranes may undergo chemical or mechanical changes (e.g., compaction or chemical decomposition) under inappropriate pressure, temperature, or pH conditions [89], these can generally be avoided as long as manufacturer's recommendations are followed.

J. Membrane Fouling and Mass Transfer Limitations

The phenomena of membrane fouling and formation of a concentration polarization boundary layer are often of concern in ultrafiltration processes as they usually cause a temporal decline of transmembrane fluid and solute fluxes. The extent of concentration polarization on the lumen and shell sides of hollow-fiber membranes (not to be confused with the downstream and vertical polarization phenomena described earlier) can be assessed using the stagnant film theory [103], modified for dialysis-type systems by Zydney [104]. Using this theory, it has been estimated that protein polarization on low-permeability membranes in closed-shell HFBRs under typical operating conditions should be negligible [29]. However, the same conclusion may not be true in the presence of large transmembrane fluid fluxes, such as in some open-shell operations or in systems with highly permeable membranes. The stagnant film theory represents a fairly idealized picture of mass transfer in a hollow-fiber reactor; a more rigorous approach should include a two- or three-dimensional fluid flow and protein transport analysis for both the ICS and ECS as well as hindered protein transport through the membranes. An additional modeling challenge is associated with flow heterogeneities and hence variations in mass transfer conditions from fiber to fiber, which can be expected to exist in real HFBR systems.

For hollow-fiber cell-culture bioreactors, two factors playing a major role in membrane fouling are the adsorption of proteins, often accompanied by their denaturation, and pore blockage by cell debris. Proteins can adsorb to the surface of fully retentive membranes or on the pore walls of semipermeable membranes [89,105]. Substantial amounts of proteins and cell debris can often accumulate in the spongy support matrix of asymmetric membranes. The adsorption of proteins is usually entropically driven and exhibits a maximum near the protein isoelectric point. The latter effect can be explained by noting that the protein molecule is more globular and therefore needs fewer sites for adsorption when its net charge is smaller, and that the repulsive interactions between the adsorbed molecules are minimal when the protein is electrically neutral [106]. The increased protein adsorption observed at higher ionic strengths is a result of the counterion shielding of charged protein molecules, which reduces their repulsive interactions and promotes a more globular conformation [106]. In some cases, charged protein molecules can adsorb on surfaces having charges of the same sign by adopting an orientation in which the charge-carrying groups are away from the surface [89]. It has been noted by many investigators that the extent of protein adsorption is greater on more hydrophobic surfaces [106]. This is related to the energetically favorable displacement of water molecules from the hydrophobic surface to the bulk of the solution, where they can participate to a greater extent in hydrogen bonding and dipole interactions; and to the attractive Van der Waals interactions between the membrane surface and the hydrophobic groups of the protein molecule [107]. About an order of magnitude greater amounts of BSA were found to adsorb on polysulfone and polyamide membranes, compared with less hydrophobic CA membranes [108], while another study reported no detectable adsorption of BSA and transferrin on hydrophilic regenerated-cellulose membranes [26]. The extent of protein adsorption can be very different on membranes differing in morphology (e.g., in surface roughness), even if they are made of the same polymer [108,109].

Adsorption of proteins usually leads to a reduction of the membrane hydraulic and diffusive permeabilities, and hence to a decline in the magnitude of transmembrane fluxes and ECS Starling flow. A fourfold decrease in the L_p value of hollow-fiber polysulfone membranes as a result of BSA adsorption has been observed [26]. Another important implication of protein adsorption in hollow-fiber cell culture may be growth reduction as a result of the depletion of growth-factor proteins from the ECS medium. This is a particular concern for serum-free cultures, where the proteins are present at low concentrations. For example, >99% of the transferrin introduced to the ECS in the inoculation medium was found to adsorb to polysulfone hollow fibers within 2 h [26]. This problem was partly overcome by adding to the ECS an excessive amount of a relatively inexpensive protein (e.g., BSA) to competitively block the nonspecific adsorption sites [20,26]. Theoretical studies of protein adsorption on membranes have been carried out, for example, by Zydney et al. [110,111].

K. Membrane Molecular Weight Cutoff

Another important parameter affecting protein distribution in HFBRs is the membrane molecular weight cutoff (MWCO), which is intended to indicate the critical MW of a solute that will be nearly completely (95%) rejected by the membrane. Unfortunately, the concept of MWCO can be misleading, especially when the solute in question is a protein. Firstly, a single precise definition of MWCO has not been established; secondly, it is the size of the molecule rather than its mass that determines the extent of solute rejection [89]; and thirdly, MWCO values for most membranes are determined using rejection data for dextrans, whose molecules (random uncharged coils) differ in shape and properties from protein molecules (mostly globular polyelectrolytes) [89]. In fact, leakage of proteins through membranes having an MWCO significantly lower than the protein MW has been observed; for instance, transferrin (MW = 77 kDa) was found to pass through 30-kDa cutoff membranes [26], and immunoglobulin G (IgG; MW = 155 kDa) through 70-kDa cutoff membranes [52]. The performance of a hollow-fiber cell culture can strongly depend on the membrane MWCO (or pore size). It has been observed that cell lines which grow poorly in microporous membrane cartridges (pore size > 0.1 μm) can grow well in ultrafiltration modules (e.g., MWCO $\leq$ 10 kDa) and vice versa [56].

   This behavior is related to the type and size of positive and negative regulatory proteins provided by the medium or secreted by the cells. On the one hand, accumulation in the ECS of inhibitory factors, such as transforming growth factor-beta (TGF-β; MW = 30 kDa), is undesirable because it can lead to a decline in cell growth and antibody production [56]. On the other hand, the accumulation of macromolecular factors that stimulate cell growth or prevent cell death can improve bioreactor performance. In the former case, the use of large-pore (or high-MWCO) membranes is thus recommended to maximize the leakage of growth inhibitors to the ICS; in the latter case, ultrafiltration membranes having sufficiently small pores (or low-MWCO) should be selected to retain the growth factors in the ECS [56]. Improved IgG productivities in HFBRs with higher-MWCO membranes were observed, for example, by Altshuler et al. [98]. (Evans and Miller [52] reported similar trends, although caution should be exercised when interpreting their results, because they used different cell lines with different membranes.) There are, however, some drawbacks associated with the use of high-MWCO membranes, including: (1) growth-factor and product protein denaturation caused by oxidation at the air-liquid interface (in the oxygenator and ICS recycle reservoir) [29] and by the shearing action of the ICS pump [112]; (2) increased cost of the medium because of the dilution of serum or expensive protein additives in the ICS recycle volume (which can be 1 to 2 orders of magnitude larger than the ECS volume); and (3) lower product concentrations because of its dilution in the ICS recycle volume. If a choice of a higher-MWCO membrane is preferred for a particular cell line, then the use of protein-free media may help alleviate some of these problems. The vast majority of hollow-fiber cartridges used for cell culture at the present time, however, utilize ultrafiltration membranes having MWCOs of approximately 10 to 30 kDa [9,49,69,82,94,95].

L. Harvesting Strategy

A well-designed harvesting strategy is important for an efficient and productive bioreactor operation, as it directly influences the ECS protein and cell distributions and determines the product concentration in the harvest. Unless significant quantities of the product are present in the lumen side as a result of leakage from the ECS through high-MWCO membranes, product recovery from the ICS recycle fluid is not usually undertaken and is therefore not considered here. Some example harvesting configurations are shown in Figure 5. Most commonly, the product is harvested from the ECS in a periodic (or batch) fashion, every 1 to 3 days, with closed-shell operation maintained during the non-harvesting periods. The downstream and downward convective polarization phenomena can be used to help concentrate the product near the harvest port. To that end, measures such as a change in the HFBR cartridge orientation or a temporary increase in the ICS flow rate can be taken just before harvesting. Piret and Cooney [13] suggested that the direction of the ICS flow might be alternated during the exponential phase of cell growth in order to obtain more homogeneous distributions of growth factors and cells in the ECS, while unidirectional flow and a downward position of the ECS downstream (harvest) port should be maintained prior to harvesting or when steady-state production has been established. However, it should be noted that the time needed to convectively polarize the ECS proteins may be considerably longer for cell-packed HFBRs, compared with reactors operated at low cell densities [28].

The cocurrent flow configuration (Figure 5a) is the preferred variant for batch harvesting (see also Section IIIB2), because it (1) takes advantage of the downstream polarization of the product; (2) enables a simultaneous replenishment

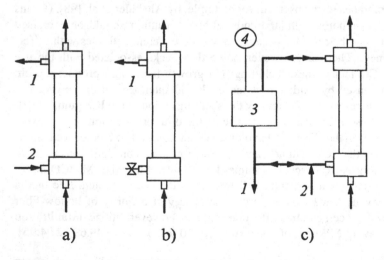

Figure 5   Examples of HFBR harvesting configurations: (a) cocurrent, (b) cross-flow, (c) harvesting with ECS fluid cycling through an expansion chamber. *1*, harvest outlet; *2*, fresh medium intake; *3*, expansion chamber; *4*, pressure regulator. (Adapted from Ref. 94.)

of the ECS medium (unlike the cross-flow harvesting mode; Figure 5b); and (3) ensures a more complete product recovery at high cell densities (i.e., at reduced ECS hydraulic permeabilities), owing to smaller transmembrane pressure differences and hence less significant short-circuiting of ECS flow through fiber lumina, compared with the countercurrent configuration. Cross-flow harvesting (Figure 5b) can be performed in a batch or continuous mode, although it is generally less practical because of potential medium degassing, if a pump is used to withdraw the ECS fluid; or a decline in the ECS outlet flow rate with increasing cell density, if the harvest is allowed to bleed out of the ECS. An example of continuous product harvesting and continuous addition of fresh medium to the ECS is a protocol developed by Endotronics (now Cellex Biosciences) and shown schematically in Figure 5c [94]. Here, the medium is cycled between the ICS and the ECS, passing through an external expansion chamber installed within the ECS loop. Both the expansion chamber (Figure 5c) and the ICS recycle reservoir (shown in Figure 1) are equipped with pressure regulators used to control the magnitude and direction of transmembrane fluid flow. Some investigators [81,113] have pointed out the advantage of continuous harvesting in minimizing product exposure to the potentially degradative environment of the bioreactor (e.g., the elevated temperature or the action of proteolytic enzymes). The rate and configuration of the harvesting flow, the ECS cell density, the degree of product polarization, and the harvest volume and frequency in batch harvesting are all important factors affecting the efficiency of product recovery from the ECS. Many of these operating parameters may require variation over time in order to accommodate the effects of increasing cell density and changing growth conditions in the ECS. Time dependence of the ECS hydraulic permeability, flow heterogeneities existing in the ECS, complex influences of the polarization phenomena, and a largely unknown product distribution prior to harvesting make the modeling of harvesting procedures a fairly challenging task.

III. MODELS OF PROTEIN TRANSPORT IN HOLLOW-FIBER BIOREACTORS

Although the analysis of hydrodynamics in HFBRs is important, one must remember that fluid flow influences bioreactor productivity in an indirect fashion. Conclusions derived from hydrodynamic analyses are therefore meaningful only when discussed in conjunction with their effect on solute transport or cell growth. Most of the early theoretical studies of HFBRs concentrated on modeling the distribution and consumption of low-MW substrates perceived as critical to cell growth and product generation [1,3,14–16]. Since diffusion is the dominant mode of transport of such low-MW solutes in the membranes and ECS, these models usually neglected Starling flow and ECS convection [16,18] or included only a radial convective component [21,22]. A growing interest in understanding the role of fluid flow in HFBRs was stimulated by anticipation that convection might help alleviate potential problems of substrate limitation, metabolite inhibition, or growth heterogeneities. More recently, many investigators have realized that the distribution of proteins such as growth factors, growth inhibitors, or products in the hollow-fiber system can be as important to efficient bioreactor operation as the transport of low-MW solutes. It has since become evident that convective flow plays a significant role in protein redistribution throughout the reactor and that a meaningful model of this process must, in most cases, also take into account the influence of osmotic pressure on HFBR hydrodynamics. This section discusses the major modeling efforts relevant to protein transport in HFBRs.

In reactors equipped with highly permeable membranes, e.g., microporous or ultrafiltration membranes having very high MWCOs, proteins can pass unhindered between the ECS and the ICS, and hence their distribution can be described using models similar to those developed for low-MW substrates. For axial-flow HFBRs, such models should include axial convection in the ECS, as do the formulations by Salmon et al. [23] or by Pillarella and Zydney [73]. The former study, unfortunately, is only of limited interest because of the unconventional geometry of the system under investigation. In the models of radial-flow HFBRs, such as those proposed by Chau et al. [114,115], axial ECS convection can usually be neglected. These radial-flow designs are also considered unconventional and therefore are not a major focus of this chapter. None of the above-mentioned models has included, often justifiably, the influence of osmotic pressure or gravity on protein distribution. Such effects can, however, become very important in the presence of significant concentration gradients which may develop in low-MWCO ultrafiltration membrane HFBRs, where the proteins are normally retained in the ECS. Modeling of protein transport in this type of bioreactor is considerably more complex and will be described here in some detail. Hollow-fiber membranes that are semipermeable, or leaky, present an even greater modeling challenge; in this case, the theory needs to be extended to include an analysis of hindered transmembrane protein transport. The model formulations presented below generally neglect protein adsorption and the effects of local density and viscosity gradients on convective flow in the bioreactor.

A. Model Formulations

1. Krogh Cylinder Models (KCMs)

Most models of hollow-fiber systems have been based on the assumption that the fluid and solute behaviors in the vicinity of each fiber are the same, such that

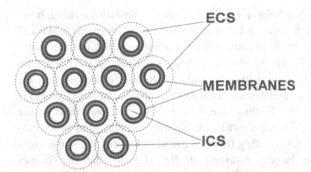

Figure 6 Cross section through a bundle of uniformly spaced hollow fibers in the Krogh cylinder approximation (not to scale); dotted line marks the boundary of each fiber unit (Krogh cylinder).

a multifiber reactor can be represented by a single fiber surrounded by an annular fluid envelope. This approximation was introduced in 1919 by Krogh [116], who investigated physiological fluid flow through capillaries in tissue; the term "Krogh cylinder" has since been used to refer to such a single capillary unit. The Krogh cylinders are assumed to contain parallel, uniformly spaced fibers (Figure 6), with no fluid or solute exchange between them. Since the reactor volume is entirely and equally divided among all fibers, partial overlapping of the neighboring Krogh cylinders is necessary to properly account for the void volumes between them. A nonoverlapping arrangement, assumed by some investigators [22,24,32,73], neglects these void spaces, even though they constitute about 10% of the total reactor volume (and often more than 20% of the ECS volume) and hence can be responsible for proportionally large mass imbalances in the system. For the fiber arrangement shown in Figure 6, the radius of each Krogh cylinder can be calculated as:

$$R_K = R_{HFBR} \sqrt{\frac{L}{nL_F}} \qquad (1)$$

where R_{HFBR} is the radius of the cylindrical bioreactor cartridge, L is the ECS length, n is the number of fibers, and L_F is the permeable length of the fiber. For straight, parallel fibers, L_F equals L, but the two lengths will differ if, for instance, the fibers expand axially as a result of absorbing water. It should be noted that the Krogh cylinder concept cannot rigorously account for the presence of the fiber-free ECS manifolds, which can occupy as much as 25% to 30% of the ECS volume [29].

If the fluid flow in the HFBR is independent of protein transport and time, then the hydrodynamic analysis is greatly simplified and can, by itself, be a valuable means of predicting solute behavior in the bioreactor. Strictly speaking, however, the assumption of such independence is justified only when high-MWCO membranes are employed, the ECS protein concentrations are low, and the membrane permeability and cell density do not change significantly over the time periods of interest. This stipulation may be satisfied to a sufficient degree, for example, in hollow-fiber artificial organs such as the bioartificial pancreas [117] or liver support devices [118]. In these cases, the islets of Langerhans or the hepatocytes grown in the ECS are steadily exchanging proteins and other biochemicals with the bloodstream passing through the fiber lumina, while the high-MW plasma proteins are largely retained in the ICS by the membranes [73,119,120]. Pillarella and Zydney [73] modeled insulin (MW = 5.2 kDa) transport in a closed-shell hollow-fiber system of this type by using the Krogh cylinder approximation and solving the following transient, spatially two-dimensional, convective-diffusion equation for the protein present in the lumen, membrane matrix, and shell:

$$\frac{\partial C}{\partial t} + u\frac{\partial C}{\partial z} + v\frac{\partial C}{\partial r} = \frac{D}{r}\frac{\partial}{\partial r}\left(r\frac{\partial C}{\partial r}\right) + \chi_{ins} \qquad (2)$$

where C is the local insulin concentration; u and v are the axial and radial velocities; D is the insulin diffusivity, assumed constant; and χ_{ins}, the rate of insulin production, is a function of the local cell density, local glucose concentration, and time. Essentially cell-free conditions in the ECS were considered, and insulin was assumed to pass unhindered through the membrane. Osmotic effects, axial convection in the matrix, and axial diffusion in all three regions were neglected. Boundary conditions for eq. (2) include the symmetry of the radial concentration profile at $r = 0$ and $r = R_K$, continuity of protein flux at the lumen-matrix ($r = R_L$) and matrix-shell ($r = R_F$) interfaces, radially uniform concentration at the lumen inlet, and no flux through the end boundaries of the shell. The initial condition at $t = 0$ was the known, uniform protein concentration in the lumen, matrix, and shell.

Since the flow equations in the above approach were decoupled from the solute mass balance equations, the velocities in eq. (2) could be evaluated analytically using the hydrodynamic theory developed by Kelsey et al. [24], which was based in part on an earlier analysis by Apelblat et al. [121]. This theory assumed laminar flow in all regions of the bioreactor, fully developed flow in the lumen, as well as negligible axial velocity gradients and inertial terms in the momentum equation. Many of these assumptions are justified by the low membrane permeabilities and small aspect ratios (both R_L/R_F and $(R_K - R_F)/L_F$ are of order 10^{-3}) and hence very small radial Reynolds numbers in typical HFBRs. Under these conditions, the continuity and Navier-Stokes equations for steady-state flow of a newtonian, incompressible fluid in the ICS and the essentially cell-free ECS take the forms

$$\frac{\partial u}{\partial z} + \frac{v}{r} + \frac{\partial v}{\partial r} = 0 \tag{3}$$

and

$$\frac{1}{r}\frac{\partial}{\partial r}\left(r\frac{\partial u}{\partial r}\right) = \frac{1}{\mu}\frac{dP}{dz} \tag{4}$$

where μ is the fluid viscosity and P is the hydrostatic pressure, which is a function of z only. The combination of Darcy's law with fluid continuity for flow in the membrane matrix, together with the assumptions that the axial velocity in the membrane is negligible and that the pressure is continuous across the lumen-matrix and matrix-shell interfaces, leads to the following equation for the radial superficial velocity in the membrane:

$$v_m(z, r) = \frac{L_p}{\mu}\frac{R_L}{r}[P_L(z) - P_S(z)] \tag{5}$$

Upon integrating eq. (4) twice with respect to r, making use of symmetry boundary conditions ($\partial u/\partial r = 0$ and $v = 0$) at $r = 0$ and $r = R_K$ and a no-slip condition ($u = 0$) at $r = R_L$ and $r = R_F$, combining the result with eq. (3), and finally substituting the obtained radial velocities $v_L(R_L, z)$ and $v_S(R_F, z)$ into eq. (5), one arrives at the following set of coupled ordinary differential equations for the lumen and shell pressures:

$$\frac{d^2P_L}{dz^2} = \frac{16L_p}{R_L^3}[P_L(z) - P_S(z)] \tag{6}$$

and

$$\frac{d^2P_S}{dz^2} = -\frac{16L_p}{\gamma R_L^3}[P_L(z) - P_S(z)] \tag{7}$$

where

$$\gamma = [4R_K^4 \ln(R_K/R_F) + 4R_K^2R_F^2 - 3R_K^4 - R_F^4]/R_L^4 \tag{8}$$

and R_L, R_F, and R_K are the lumen radius, fiber outer radius, and Krogh cylinder radius, respectively. Labecki et al. [101] proposed that the z coordinate in the above equations be replaced by $\zeta = zL_F/L$ in order to account for any axial expansion of the wet fibers. An analytical solution to eqs. (6) and (7) is readily obtained and leads eventually, after applying the appropriate known-pressure, known-flow (known pressure-derivative), or no-flux (zero pressure-derivative) boundary conditions, to expressions for the ICS and ECS fluid velocities [24].

For example, the algebraic forms of the velocities for a closed-shell HFBR are as follows:

$$u_L(z, r) = u_0 \frac{1 - (r/R_L)^2}{1 + 1/\gamma} \left\{ \frac{\cosh[\lambda(z/L - 1/2)]}{\cosh(\lambda/2)} + \frac{1}{\gamma} \right\} \tag{9}$$

$$u_S(z, r) = u_0 \frac{r^2 - R_F^2 - 2R_K^2 \ln (r/R_F)}{R_L^2(1 + \gamma)} \left\{ \frac{\cosh[\lambda(z/L - 1/2)]}{\cosh(\lambda/2)} - 1 \right\} \tag{10}$$

$$v_L(z, r) = -u_0 r[2 - (r/R_L)^2] \sqrt{\frac{L_p}{R_L^3(1 + 1/\gamma)}} \frac{\sinh[\lambda(z/L - 1/2)]}{\cosh(\lambda/2)} \tag{11}$$

$$v_S(z, r) = -u_0 r \lambda \Phi(r) \sqrt{\frac{L_p}{R_L^3 \gamma(1 + \gamma)}} \frac{\sinh[\lambda(z/L - 1/2)]}{\cosh(\lambda/2)} \tag{12}$$

where μ_0 is the inlet centerline lumen velocity,

$$\lambda = 4L \sqrt{\frac{L_p(1 + 1/\gamma)}{R_L^3}} \tag{13}$$

and

$$\Phi(r) = \left[2R_K^2 - 2R_F^2 + r^2 - 3\frac{R_K^4}{r^2} + 2\frac{R_K^2 R_F^2}{r^2} \right.$$
$$\left. + 4\frac{R_K^4}{r^2} \ln(R_K/R_F) - 4R_K^2 \ln(r/R_F) \right] \Big/ R_L^2 \tag{14}$$

It should be noted that the Krogh cylinder approach assumes that the fluid in open-shell operations enters or exits the ECS axially at either end of the annular space; in reality, the inflow or outflow occurs radially through the ECS manifold, which extends over a finite distance along the outer circumference of the module and typically offers a much larger surface area for the incoming or outgoing flow. Therefore, the KCM formulations of the v_S or P_S boundary conditions for an open-shell reactor are rather crude approximations. Another frequently challenged boundary condition is the assumption of zero tangential velocity (no slip) at the surface of a permeable material. Based on an empirical study performed over three decades ago [122] and later also supported theoretically [123], the fractional increase in flow rate due to slip along the surface of a hollow-fiber membrane can be estimated to be of the order of $\sqrt{L_p/R_L}$, i.e., evidently negligible for hollow-fiber systems of any practical interest.

Taylor et al. [25] developed a similar transport theory for protein that is completely retained in the ECS of an essentially cell-free HFBR. In this case, however, osmotic effects were taken into account. Accordingly, the radial superficial velocity in the membrane was calculated as

$$v_m(z, r) = \frac{L_p}{\mu} \frac{R_L}{r} [P_L(z) - P_S(z) + \Pi_S(z, R_F)] \tag{15}$$

with the osmotic pressure, Π_S, evaluated as a function of the local protein concentration using the formula derived by Vilker et al. [125]:

$$\Pi_S(C_S) = \frac{R_g T}{M_p} [\sqrt{(Z_p C_S)^2 + (2m_s M_p)^2} - 2m_s M_p + C_S + A_1 C_S^2 + A_2 C_S^3] \tag{16}$$

where R_g is the gas constant, T is the absolute temperature, M_p is the molecular weight of the protein, Z_p is the protein charge number (which depends on pH), m_s is the molar salt concentration, and the coefficients A_1 and A_2 are functions of Z_p. A simpler, virial form of $\Pi_S(C_S)$ has also been used in other studies [26–29]. Taylor et al. [25] described the time-dependent protein distribution within the ECS using the following equation:

$$\frac{\partial C_S}{\partial t} = D \left[\frac{\partial^2 C_S}{\partial z^2} + \frac{1}{r} \frac{\partial}{\partial r} \left(r \frac{\partial C_S}{\partial r} \right) \right] - u_S \frac{\partial C_S}{\partial z} - v_S \frac{\partial C_S}{\partial r} \tag{17}$$

which differs from eq. (2) by inclusion of the axial diffusive term and by neglect of any sink or source terms due, for instance, to protein uptake or production by the cells. The initial condition for eq. (17) is the known uniform concentration at $t = 0$, while the boundary conditions include zero concentration derivatives at the solid and Krogh cylinder boundaries and a quasi-steady balance between protein diffusion and convection at the membrane surface. Unlike in the formulation of Kelsey et al. [24], the axial pressure derivatives needed to calculate the local fluid velocities in the ECS were expressed in terms of the radially averaged axial velocity in the lumen, $\bar{u}_L$. Substitution of these pressure-derivative terms into the differentiated form of eq. (15) and combination of the result with fluid mass balances over a differential length of the ICS and ECS yields the following ordinary differential equation for $\bar{u}_L$:

$$\frac{d^2 \bar{u}_L}{dz^2} - \frac{16 L_p (1 + 1/\gamma)}{R_L^3} \bar{u}_L$$
$$= -\frac{16 L_p \bar{u}_{L0}}{R_L^3 \gamma} - \frac{2 L_p}{\mu R_L} \frac{d\Pi_S[C_S(z, R_F)]}{dC_S} \frac{dC_S(z, R_F)}{dz} \tag{18}$$

where $\bar{u}_{L0}$ is the average axial velocity at the lumen inlet (or exit). The presence of the osmotic pressure derivative term required that the above equation be solved numerically. The obtained distribution of $\bar{u}_L$ was used to evaluate the local ECS velocities, u_S and v_S, which were then inserted into eq. (17), and the latter was also solved numerically. Because of the coupling between the HFBR hydrodynamics and ECS protein distribution, the solutions of eqs. (17) and (18) were iterated at each new time level until convergence was attained.

A subsequent study by Patkar et al. [26] simplified the above two-dimensional KCM for closed-shell HFBRs by omitting the ECS and ICS radial gradients of velocities, pressures, and concentration. This modification was justified by the agreement of the one-dimensional model predictions with the experimental ECS protein distribution data (see also Section IIIB1). The following equation was used to calculate the radially averaged concentrations of ECS proteins:

$$\frac{\partial \bar{C}_S}{\partial t} = D \frac{\partial^2 \bar{C}_S}{\partial z^2} - \frac{\partial}{\partial z}(\bar{u}_S \bar{C}_S) \tag{19}$$

where the radially averaged axial velocity $\bar{u}_S$ is linked through a local fluid mass balance to $\bar{u}_L$, which was obtained as a solution of eq. (18) with $C_S(z, R_F) = \bar{C}_S(z)$. Since the flow distribution was found not to change significantly over a single time increment, the hydrodynamic solution was lagged one step behind the solution of eq. (19). This measure greatly reduced the computational effort by avoiding the iterative procedure at each time level. An even simpler analysis was proposed by Taylor et al. [25] to explain the final polarized ECS protein distribution and steady-state HFBR hydrodynamics. In this model, a critical initial protein concentration C_{S0}^{crit} was considered, above which the steady-state protein distribution extends from $z = 0$ to $z = L$ and the ECS flow is effectively shut down over the whole length of the reactor because of osmotic effects. At initial concentrations below C_{S0}^{crit}, the ECS divides at $z = z_s$ into two zones: a protein-free upstream region ($z < z_s$) supporting Starling flow, and a stagnant downstream region ($z \geq z_s$) where all of the proteins have accumulated. The ECS pressure in the upstream region is approximately constant and equal to

$$P_S^{up} = P_{LL} + \frac{8\mu \bar{u}_{L0}}{R_L^2}\left(L - \frac{z_s}{2}\right) \tag{20}$$

where P_{LL} is the absolute pressure at the lumen outlet ($z = L$). The local osmotic pressure $\Pi_S(z)$ at $z \geq z_s$ was found as a difference between P_S^{up} and the known ICS pressure $P_L(z)$, the latter assumed to vary linearly with z according to the Hagen-Poiseuille law. The concentration distribution $C_S(z)$ was then calculated implicitly from $\Pi_S(z)$ using eq. (16), and the corresponding z_s value was determined by root finding and ensuring that the protein mass balance was satisfied. C_{S0}^{crit} was readily obtained as the average ECS concentration when $z_s = 0$.

A KCM describing ECS protein transport in closed-shell HFBRs under conditions of high cell density was developed by Koska et al. [28]. The hydrodynamic analysis of such systems was similar to that presented by Apelblat et al. [121] for protein-free reactors. By using Darcy's law to describe fluid flow in the ECS and with other assumptions analogous to those made in the cell-free studies outlined above [24–26], the following equations for the lumen and shell pressures were derived:

$$\frac{d^2 P_L}{dz^2} = \frac{16 L_p}{R_L^3} [P_L(z) - P_S(z, R_F) + \Pi_S(z, R_F)] \tag{21}$$

$$\frac{1}{r} \frac{\partial}{\partial r} \left(r \frac{\partial P_S}{\partial r} \right) + \frac{\partial^2 P_S}{\partial z^2} = 0 \tag{22}$$

Equation (21) is subject to known-pressure or known-flow-rate boundary conditions at the lumen inlet and outlet. Equation (22) is subject to no-flux boundary conditions at $z = 0$, $z = L$, and $r = R_K$, and to the following condition expressing the balance of fluid fluxes at $r = R_F$:

$$\frac{\partial P_S}{\partial r} = -\frac{L_p}{k_S} \frac{R_L}{R_F} [P_L(z) - P_S(z, R_F) + \Pi_S(z, R_F)] \tag{23}$$

where k_S is the constant and isotropic ECS Darcy permeability. For a cell-filled ECS, k_S can be treated as a measurable system parameter; approximated by the hydraulic permeability of tissue in vivo [47]; or estimated using the Carman-Kozeny equation [125] for a packed bed of spheres. The time-dependent ECS protein distribution was governed by the equation

$$\frac{\partial C_S}{\partial t} = D K_d \left[\frac{\partial^2 C_S}{\partial z^2} + \frac{1}{r} \frac{\partial}{\partial r} \left(r \frac{\partial C_S}{\partial r} \right) \right] - K_c \left[\frac{u_S}{\varepsilon} \frac{\partial C_S}{\partial z} - \frac{v_S}{\varepsilon} \frac{\partial C_S}{\partial r} \right] \tag{24}$$

where $C_S = C_S(z, r)$ is the interstitial ECS protein concentration (i.e., mass of protein per unit volume of ECS fluid), K_d and K_c are the diffusive and convective hindrance factors, respectively, and ε is the porosity of the cell mass in the ECS. For the sake of comparison with the above two-dimensional formulations, simpler forms of eqs. (22) and (24) were also considered in which radial gradients were neglected, i.e.

$$\frac{d^2 \overline{P}_S}{dz^2} = -\frac{2 R_L L_p}{(R_K^2 - R_F^2) k_S} [P_L(z) - \overline{P}_S(z) + \overline{\Pi}_S(z)] \tag{25}$$

and

$$\frac{\partial \overline{C}_S}{\partial t} = D K_d \frac{\partial^2 \overline{C}_S}{\partial z^2} - K_c \left(\frac{\overline{u}_S}{\varepsilon} \frac{\partial \overline{C}_S}{\partial z} \right) \tag{26}$$

where the overbar denotes radially averaged quantities. The boundary conditions for eqs. (24) and (26) are analogous to those pertaining to eqs. (17) and (19), respectively. Numerical methods, again, were employed to solved the coupled hydrodynamic and protein transport problems.

A recent study by Labecki et al. [29] considered a closed-shell hollow-fiber system containing two proteins: one completely retained in the ECS (BSA; MW = 69 kDa), and one only partially rejected by the membranes (myoglobin; MW = 17 kDa). Fluid balances for the lumen and shell sides of the representative tortuous (i.e., axially expanded under wet conditions) Krogh cylinder yielded the following equations:

$$\frac{d^2 P_L}{d\zeta^2} = \frac{16 L_p}{R_L^3} [P_L(\zeta) - P_S(\zeta) - \Delta\Pi_{eff}(\zeta)] \tag{27}$$

$$\frac{d^2 P_S}{d\zeta^2} = -\frac{16 L_p}{\gamma R_L^3} [P_L(\zeta) - P_S(\zeta) - \Delta\Pi_{eff}(\zeta)] \tag{28}$$

which are similar to eqs. (6) and (7), except that: (1) $\zeta = z L_F / L$ denotes the coordinate along the tortuous fiber; (2) γ was determined experimentally as a ratio of the ECS and ICS hydraulic permeabilities in the axial direction, rather than being evaluated from eq. (8); and (3) the effective osmotic pressure difference was calculated from

$$\Delta\Pi_{eff} = \sigma(\Pi_{L,M} - \Pi_{S,M}) - \Pi_{S,A} \tag{29}$$

where the subscripts M and A refer to myoglobin and albumin, respectively, and σ is the osmotic reflection coefficient for the leaking protein. The membrane hydrodynamics were modeled by assuming Hagen-Poiseuille flow through pores treated as tortuous cylinders having a uniform, effective radius r_p. This approach yielded the following relationship between r_p and other membrane properties:

$$r_p = \sqrt{8 L_p (R_F - R_L) k_m} \tag{30}$$

where k_m is the so-called membrane constant, defined as

$$k_m = \frac{\tau^2}{\varepsilon(R_L)} \tag{31}$$

and treated as a fitting parameter. Here, τ is the pore tortuosity, i.e., the ratio of the effective pore length to membrane thickness, and $\varepsilon(R_L)$ is the membrane surface porosity at the luminal surface of the fiber. The convenience of the above definition of k_m is that τ and $\varepsilon(R_L)$ do not have to be known separately, as they appear in the model formulation only in the lumped form expressed by eq. (31). The local transmembrane myoglobin flux from the ICS to the ECS, evaluated per unit area of the membrane at $r = R_L$, was modeled by considering a quasi-steady mass balance of the protein for a differential membrane element [29], which yielded

$$J_M(\zeta) = K_c v_m(\zeta, R_L)\Phi\left[\overline{C}_{L,M}(\zeta) + \frac{\overline{C}_{L,M}(\zeta) - \overline{C}_{S,M}(\zeta)}{\exp[Pe(\zeta)(R_F/R_L - 1)] - 1}\right] \quad (32)$$

where

$$v_m(\zeta, R_L) = \frac{L_P}{\mu}[P_L(\zeta) - P_S(\zeta) - \Delta\Pi_{eff}(\zeta)] \quad (33)$$

and

$$Pe(\zeta) = \frac{K_c v_m(\zeta, R_L) R_L k_m}{K_d D_M} \quad (34)$$

is the local pore Peclet number. K_c and K_d in eqs. (32) and (34) are the convective and diffusive hindrance factors, respectively, while Φ is the myoglobin partition coefficient. Available theories of flow through membrane pores [126] were used together with eq. (30) to express σ, Φ, K_c, and K_d as functions of k_m and thus to reduce the number of unknown model parameters to one. The time-dependent distribution of myoglobin in the ECS was obtained from:

$$\frac{\partial \overline{C}_{S,M}}{\partial t} = D_M \frac{\partial^2 \overline{C}_{S,M}}{\partial \zeta^2} - \frac{\partial}{\partial \zeta}(\overline{u}_S \overline{C}_{S,M}) + \frac{2R_L}{R_k^2 - R_F^2}J_M \quad (35)$$

while solution of a similar equation for $\overline{C}_{L,M}$ showed that the concentration of myoglobin in the lumen, at ICS flow rates of practical interest, was approximately constant with z and equal to the inlet value. Consequently, transient changes in $\overline{C}_{L,M}$ could be calculated by assuming that the mass of protein transferred across the membrane during each time interval, obtained by integrating $J_M(\zeta)$ over the length of the fiber, was instantaneously mixed into the total lumen volume. The distribution of BSA in the ECS was obtained as a solution of eq. (19). The radially averaged shell velocity that appears in the convective terms of the ECS protein transport equations was evaluated as

$$\overline{u}_S = -\frac{\gamma R_L^4}{8\mu(R_k^2 - R_F^2)}\frac{dP_S}{d\zeta}. \quad (36)$$

This model [29] also included the influence of the fiber-free ECS manifolds on the distribution of proteins in the HFBR. This effect was accounted for, in a simplified way, by assuming an instantaneous equilibration of proteins between the manifold and the concentric portion of the ECS at each axial position within the manifold length; in other words, the same axial concentration profile was assumed to exist in the ECS manifold and in the ECS of the fiber bundle. Although protein adsorption was not explicitly included in the model formalism, the possibility of pore plugging by myoglobin molecules, depending on the local ICS and ECS myoglobin concentrations, was considered.

2. Porous Medium Model (PMM)

A porous medium approach to the modeling of hollow-fiber bioreactors has been proposed by Labecki et al. [27,101]. In this model, the intra- and extracapillary spaces of the fiber bundle are treated as two interpenetrating porous media with uniformly distributed sinks and sources of fluid. Axial pressure gradients in the membranes are neglected, and the fluid and membranes are assumed to be incompressible. Consequently, any fluid that disappears from one subspace (ECS or ICS) must instantly appear in the other at the same position. Unlike in the various Krogh cylinder models, the spatial domain of the PMM corresponds to the dimensions of the whole fiber bundle in the reactor. Thus, the model is able to account for interfiber flows and macroscopic radial and angular concentration gradients which are normally present during open-shell operations, and may also be caused by gravity [13], heterogeneity of fiber distribution [127], or radial pressure variations in the lumen manifolds [128], even during closed-shell operations. The flow properties and protein concentrations in the PMM are averaged over a representative elementary volume (REV), which is small enough to be treated in a differential sense, but must also contain a sufficient number of fibers to assure local homogeneity [129]. The PMM should be distinguished from KCM formulations that treat the ECS of the Krogh cylinder as a porous medium [28,32,121].

In the PMM formulation presented by Labecki et al. [27, 101], two-dimensional cylindrical coordinates were used, gravity effects were neglected, and the proteins were confined to the ECS. Application of the Darcy and steady-state continuity laws to the lumen and shell spaces yielded the following equations governing the hydrodynamics:

$$-k_{S,r}\frac{1}{r}\frac{\partial}{\partial r}\left(r\frac{\partial P_S}{\partial r}\right) - k_{S,z}\frac{\partial^2 P_S}{\partial z^2} = \frac{2nR_LL_pL_F}{R_{HFBR}^2L}(P_L - P_S + \Pi_S) \qquad (37)$$

$$k_{L,z}\frac{\partial^2 P_L}{\partial z^2} = \frac{2nR_LL_pL_F}{R_{HFBR}^2L}(P_L - P_S + \Pi_S) \qquad (38)$$

where R_{HFBR} is the inner radius of the HFBR cartridge; n is the number of fibers; the hydrostatic pressures, P_L and P_S, are functions of both z and r; and the osmotic pressure, Π_S, is evaluated as a function of the local ECS protein concentration. The radial component of the Darcy permeability in the lumen was assumed to be zero, since the fibers are not directly interconnected. The axial permeability components, $k_{L,z}$ and $k_{S,z}$, could be determined experimentally or estimated theoretically using simple models. For instance, in the absence of cells, a one-dimensional Krogh cylinder analysis with appropriate modifications to account for wet fiber expansion, yields

$$k_{L,z} = \frac{nR_L^4 L}{8R_{HFBR}^2 L_F} \tag{39}$$

$$k_{S,z} = \frac{R_{HFBR}^2 L^3}{4nL_F^3}\left(-\ln\varphi - \frac{3}{2} + 2\varphi - \frac{1}{2}\varphi^2 \right) \tag{40}$$

where

$$\varphi = \frac{nR_F^2 L_F}{R_{HFBR}^2 L} \tag{41}$$

is the fraction of the reactor volume occupied by the fibers. The ECS radial permeability, $k_{S,r}$, for a cell-free reactor was estimated from Happel's [130] cylindrical cell model for flow perpendicular to an assemblage of parallel cylinders, i.e.

$$k_{S,r} = \frac{R_{HFBR}^2 L^3}{4nL_F^3}\left(-\ln\varphi + \frac{\varphi^2 - 1}{\varphi^2 + 1} \right) \tag{42}$$

At sufficiently high cell densities, the ECS permeability may become essentially isotropic and should therefore be modeled in a different manner, e.g., by using the Carman-Kozeny equation [125] (see also Section IIIA1).

In the absence of protein sinks and sources, the protein distribution within the ECS was governed by the transient convective-diffusion equation:

$$\varepsilon_S \frac{\partial C_S}{\partial t} = \frac{\partial}{\partial z}\left(D_z \frac{\partial C_S}{\partial z} - u_S C_S \right) + \frac{1}{r}\frac{\partial}{\partial r}\left(rD_r \frac{\partial C_S}{\partial r} - rv_S C_S \right) \tag{43}$$

where ε_S is the effective ECS porosity, which equals $1 - \varphi$ in the absence of cells, and C_S is the interstitial protein concentration in the ECS. D_z and D_r, the effective protein diffusivities in the axial and radial directions, respectively, were modeled as

$$D_z = D\varepsilon_S L^2 / L_F^2 \tag{44}$$

and

$$D_r = 2D\varepsilon_S/(2 - \varepsilon_S) \tag{45}$$

where D is the diffusivity in an unobstructed fluid. When cells are present in the ECS, both ε_S and the expressions for the effective diffusivities (or diffusivity, if isotropic conditions prevail) should include the porosity of the cell mass. The superficial shell velocities u_S and v_S, as well as the lumen velocity u_L, were obtained from the calculated ECS and ICS pressure fields and the respective hydraulic permeabilities by using Darcy's law.

The initial condition for eq. (43) is the known, e.g., uniform or steady-state, concentration distribution at $t = 0$. The general boundary conditions for the pressure eqs. (37) and (38) and the concentration eq. (43) are as follows:

1. At $z = 0$ and $z = L$: known $P_L(0, r)$ or $u_L(0, r)$ (and hence dP_L/dz), which includes the case of a closed ICS port, i.e., $dP_L/dz = 0$; $\partial P_S/\partial z = 0$ and $\partial C_S/\partial z = 0$, i.e., no fluid and protein fluxes through the ECS boundary.

2. At $r = 0$ (cartridge centerline): $\partial P_S/\partial r = 0$ and $\partial C_S/\partial r = 0$, i.e., symmetry of the ECS pressure and concentration profiles.

3. At $r = R_{HFBR}$ and $z_{man} < z < L - z_{man}$ (outer cartridge boundary between the ECS manifolds): $\partial P_S/\partial r = 0$ and $\partial C_S/\partial r = 0$, i.e., no fluid and protein fluxes through the ECS boundary.

4. At $r = R_{HFBR}$ and $0 \leq z \leq z_{man}$ or $L - z_{man} \leq z \leq L$ (outer ECS boundary within the length of the manifold): known $P_S(z, R_{HFBR})$, usually assumed constant over z; or known $v_S(z, R_{HFBR})$ (and hence $\partial P_S/\partial r$), which includes the case of a closed ECS port, i.e., $\partial P_S/\partial r = 0$; at an open ECS inlet port—known $C_S(z, R_{HFBR})$, usually assumed constant over z; at an open ECS outlet port—either known $C_S(z, R_{HFBR})$, usually assumed constant over z, or $\partial C_S/\partial r = 0$, i.e., convective protein flux only; at a closed ECS port—$\partial C_S/\partial r = 0$, i.e., no protein flux.

In practice, it is very difficult to determine the velocity distribution at an inflow or outflow ECS boundary, even if the flow rate (i.e., the average velocity) is known. In that case, rather than assuming an arbitrary flow profile such as a constant velocity, it may be more accurate to specify a constant pressure at the boundary and then adjust its value iteratively until the desired total flow rate is obtained. If such fixed-flow experimental conditions are simulated, then a similar iterative procedure may be necessary to account for the variations in the ECS inlet or outlet pressure due to the changing osmotic effects in the ECS. As in the analogous Krogh cylinder models, the coupled fluid flow and protein transport equations are solved numerically, with the hydrodynamic solution lagged one time step behind the concentration solution.

It can easily be shown that, in the absence of radial gradients and with the axial hydraulic permeabilities calculated from expressions derived using the Krogh cylinder approach, the one-dimensional PMM and KCM become identical. However, there is a notable difference between the two-dimensional KCM and PMM: in the former, the radial dimension is considered only on the microscale of a single fiber, while in the latter it accommodates the whole HFBR cartridge. The third spatial dimension can easily be added to the PMM formulation presented above, and the model domain can be extended to include the adjacent fiber-free regions, although the hydrodynamics in these regions will be governed by the Navier-Stokes equations rather than Darcy's law.

B. Model Predictions of Protein Transport in HFBRs

1. Closed-Shell Systems

The models described above generally predict that downstream polarization of ECS proteins can take place under typical operating conditions in a closed-shell HFBR. In addition, PMM simulations demonstrate that similar convective polarization may occur in the transverse direction if significant radial pressure gradients exist in the entrance or exit lumen manifolds [27,128]. Figure 7 illustrates cell-free and cell-packed examples of the temporal evolution of the axial polariza-

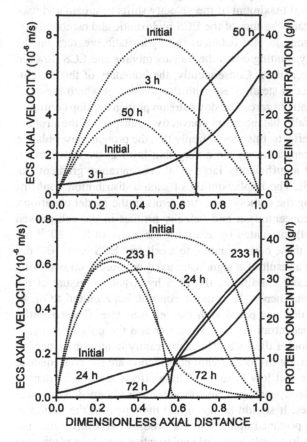

Figure 7 Transient radially averaged axial ECS protein (BSA) distributions (solid lines) and axial velocities (dotted lines) in a closed-shell HFBR. Q_L = 500 mL/min, average ECS protein concentration 10 g/L; upper panel: cell-free ECS, $k_S = 10^{-9}$ m²; lower panel: cell-packed ECS, $k_S = 5 \times 10^{-14}$ m². (Adapted from Ref. 28.)

tion process, starting from a uniform initial distribution of a protein which is completely retained in the ECS. The radially averaged concentration and ECS axial velocity profiles plotted in Figure 7 were generated using a KCM by Koska et al. [28], with the parameters K_d, K_c, and ε assumed to be unity under the conditions of the experiments conducted to support the model predictions. The one-dimensional [eqs. (21), (25), and (26)] and two-dimensional [eqs. (21), (22), and (24)] KCM formulations yielded results that were graphically indistinguishable. After a sufficiently long period of closed-shell operation, division of the ECS into protein-free and protein-rich regions becomes clearly noticeable. The boundary between these two regions shifts downstream with time until steady state is reached. At the same time, parallel changes in the distribution of axial ECS velocity occur: the local maximum of the velocity shifts upstream and usually decreases. At steady state, the sum of the ECS hydrostatic and osmotic pressures in the protein-rich zone locally balances the hydrostatic pressure on the lumen side, thus effectively shutting down the transmembrane and ECS flows in the downstream part of the reactor. Consequently, the boundary of the protein-concentrated zone coincides at steady state with the position at which the axial ECS velocity drops essentially to zero. The downstream protein buildup continues until the diffusive flux locally balances the convective transport of the protein. In the absence of osmotic effects, this would imply that the protein must polarize to an extremely high concentration within a very thin layer at the downstream end of the ECS, such that a sufficiently large axial concentration gradient can develop in that region [25]. The implausibility of such a distribution confirms the importance of including the osmotic pressure terms in the model equations.

The transient ECS concentration and velocity profiles in the cell-packed case (Figure 7, lower panel), simulated by assuming a k_S value of 5×10^{-14} m^2, are qualitatively similar to those corresponding to a cell-free ECS ($k_S = 10^{-9}$ m^2; Figure 7, upper panel). As a result of a significant axial pressure drop occurring in the ECS under cell-packed conditions, the ECS hydrostatic pressure at the downstream end is lower and hence the osmotic contribution required to offset the lumen pressure is smaller, compared with the cell-free case. Consequently, the steady-state protein concentrations are also lower and the polarized region extends over a larger portion of the ECS. The significantly longer times needed to produce a comparable extent of downstream polarization are the consequence of a diminished Starling flow at low k_S values. These effects are also evident in Figure 8, which plots the steady-state protein concentrations obtained at different ECS hydraulic permabilities. It should be noted that the length of the period of protein polarization has important implications for product harvesting from the ECS. If the k_S value is very low, only very little polarization may take place over the 1- or 2-day periods between harvests, thus lowering the harvested product concentration [28]. In addition, because the ECS hydraulic resistance is extremely high, most of the harvested fluid may actually be drawn from the luminal space in the vicinity of the ECS outlet port, thus leaving much of the product protein

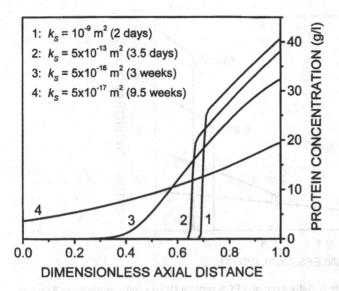

Figure 8 Steady-state radially averaged ECS protein (BSA) concentrations in a closed-shell HFBR at different ECS hydraulic permeabilities. Q_L = 500 mL/min, average ECS protein concentration 10 g/L; times needed to reach steady state are given in brackets. (Adapted from Ref. 28.)

inside the reactor. One must nonetheless remember that the predictions discussed above were made assuming a uniform initial protein distribution in the ECS, no significant protein production during the polarization process, and a homogeneous distribution of cell mass in the ECS, all of which are only approximations.

Axial redistribution of proteins in the ECS can also be slowed down by a decrease in the membrane hydraulic permeability in the absence of cells. Taylor et al. [25] found that the time needed to reach steady state was, roughly, inversely proportional to L_p. At Q_L = 570 mL/min, a steady-state distribution of protein present at 10 g/L in the ECS of a cartridge for which $L_p = 1.25 \times 10^{-13}$ m was attained in about 2 h; hence, at the same ICS flow and in the same reactor geometry but at $L_p = 6 \times 10^{-15}$ m, as used by Koska et al. [28], steady state would be reached in approximately 42 h. Reduction of L_p by several orders of magnitude was found to have little influence on the steady-state protein distribution, except to diminish slightly the large axial concentration gradient that exists at the boundary between the protein-rich and protein-free zones [25].

As expected, the steady-state protein distribution in a closed-shell HFBR also depends on the ICS flow rate and on the total protein content in the ECS. The effects of these two factors are shown in Figures 9 and 10, respectively, which were adapted from the study by Taylor et al. [25]. For their reactor geometry, the radially averaged lumen entrance velocities of 0.5, 2, 5, and 10 cm/s,

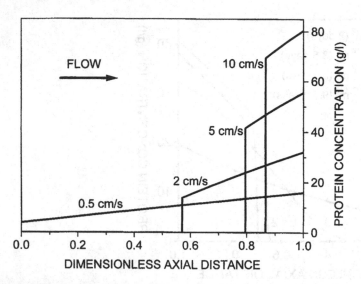

Figure 9   Steady-state radially averaged ECS protein (BSA) concentrations in a closed-shell, cell-free HFBR at different ICS inlet velocities. Average ECS protein concentration 10 g/L. (From Ref. 25.)

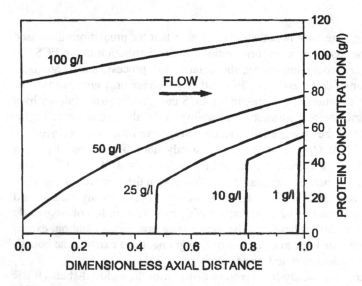

Figure 10   Steady-state radially averaged ECS protein (BSA) concentrations in a closed-shell, cell-free HFBR at different average ECS protein concentrations. ICS inlet velocity 5 cm/s (Q_L = 570 mL/min). (From Ref. 25.)

used to generate the curves in Figure 9, correspond to ICS flow rates of 57, 228, 570, and 1140 mL/min, and to lumen Reynolds numbers of 1.6, 6, 16, and 32, respectively. At lower lumen velocities, the role of convection in the establishment of the steady-state concentration profiles in the ECS is diminished, and hence the protein distributions are more uniform (Figure 9). This is due to the fact that smaller osmotic pressure gradients are needed to compensate for the decreased ICS axial pressure drop. The opposite effect is observed at higher ICS flow rates: the concentration profiles within the protein build-up zone are steeper. Similarly, at higher protein contents in the ECS (Figure 10), more homogeneous steady-state concentration distributions are observed, because the osmotic pressure gradient in the downstream region cannot exceed the maximum value determined by the flow conditions in the lumen. It should be noted that the simple steady-state model outlined in Section IIIA.1 [25] generated concentration profiles that look virtually the same as those depicted in Figure 10. The critical initial concentration, above which the protein extended over the whole length of the reactor and the axial ECS velocities were everywhere zero, was found to be 45.1 g/L for the conditions of Figure 10. At average protein concentrations below this value, axial velocity profiles similar to those shown in the upper panel of Figure 7 existed in the protein-free part of the ECS. Generally, in the various cases discussed above, the time needed to reach steady state was shorter if the steady-state distribution was closer to the initial one—i.e., more uniform; on the other hand, it was longer when the ECS axial velocities were smaller.

By comparing the results of their model simulations with those obtained using the two-dimensional formalism of Taylor et al. [25], Patkar et al. [26] concluded that a one-dimensional KCM [eq. (19)] was sufficient to reproduce the downstream-polarized protein concentration profiles for a closed-shell HFBR. Their theoretical predictions were verified experimentally using a system containing BSA and/or transferrin, both of which were entirely retained in the ECS. The axial protein distributions were determined according to the method developed by Piret et al. [13,61]; i.e., the HFBR cartridges were frozen by immersion in liquid nitrogen and cut into disk-shaped segments, which were then thawed and analyzed for proteins. The average concentrations of transferrin used in the experiments conducted by Patkar et al. [26] were more than an order of magnitude lower than the BSA concentrations, as is typical for these two proteins in most protein-containing media for cell culture [38,76]. The steady-state distributions of transferrin for different amounts of BSA in the ECS (Figure 11) were found to bear a close resemblance to the distributions of BSA itself (see also Figure 10; note that different reactor geometries were used in [25] and [26]). This can be easily explained by the facts that osmotic pressure had the same effect on the distribution of either protein in the ECS; and that transferrin, because of its relatively low concentration, contributed only negligibly to the osmotic effects. A practical implication of this observation is that more uniform distributions of

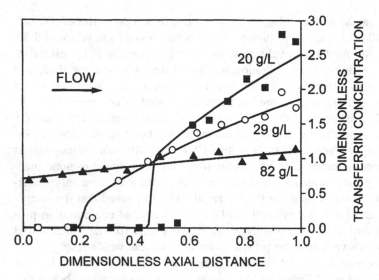

Figure 11 Experimental (symbols) versus theoretical (1-D KCM, solid lines) steady-state distributions of ECS transferrin at the average BSA concentrations of 20 g/L, 29 g/L, and 82 g/L. Q_L = 600 mL/min, average transferrin concentration 0.02 g/L; C_{S0}^{crit} = 36 g/L. (Adapted from Ref. 26.)

growth factors can be obtained by loading the ECS with a sufficiently large quantity of a relatively inexpensive nonleaking protein like BSA. As Figure 12 demonstrates, the periodic switching of the ICS flow direction can be another effective means of improving the uniformity of protein distribution in the ECS. This also provides an example of the successful application of the one-dimensional KCM to the prediction of experimental transient BSA and transferrin concentration profiles measured in a closed-shell cartridge operated in a flow-cycling mode (Figure 12).

Experiments and model simulations were conducted by Labecki et al. [29] to investigate the dynamics of protein (myoglobin) leakage through hollow-fiber membranes. The effects of the initial myoglobin placement (ECS or ICS), presence of BSA in the ECS, the lumen flow rate, the ECS manifolds, and other factors on myoglobin transport and the distribution of both proteins in the ECS were examined. Experimental data points and the corresponding best-fit model curves plotting the changes of ICS myoglobin concentration with time are shown in Figure 13. When the myoglobin was initially added to the ECS (Figure 13, lower panel), markedly faster leakage of myoglobin occurred at the lower ICS flow rate (10 mL/min) and in the presence of BSA. Both of these factors significantly diminished the extent of downstream polarization of myoglobin in the ECS. Consequently, the amount of protein present at any time in the downstream

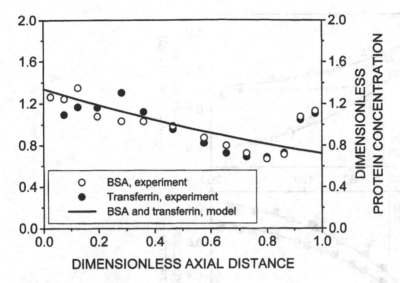

Figure 12 Experimental versus theoretical (1-D KCM) distributions of transferrin and BSA in the ECS of a cartridge operated in a lumen flow-cycling mode. Flow-switching time 60 min, Q_L = 600 mL/min, average transferrin concentration 0.018 g/L, average BSA concentration 4.9 g/L. (Adapted from Ref. 26.)

ECS manifold was reduced (Figure 14), and more myoglobin was transferred by diffusion to the lumen side. By a similar mechanism, BSA reduced the transmembrane transport of myoglobin when the latter was initially placed in the ICS (Figure 13, upper panel). The importance of including the ECS manifolds in the model is obvious from inspection of the two best-fit curves simulating experiment No. 2 (Figure 13, upper panel): the dotted curve, corresponding to the model version which neglects the manifolds, agrees very poorly with the experimental data points. The influence of BSA on the distribution of myoglobin in the ECS was somewhat more complicated than its effect on the nonleaking transferrin, discussed earlier. If myoglobin was initially added to the shell side, the protein would first polarize downstream, pass gradually across the membranes, and then accumulate in the lumen side. At this point, its further redistribution would be similar to that in the case when it was initially added to the ICS; namely, the protein continuously transfers from the ICS to the ECS, and then is convectively polarized to the boundary of the BSA-rich zone at the downstream end of the ECS (there is no convective flow within the latter region). As a consequence, a local maximum of myoglobin concentration is formed in the ECS, as can be seen in Figure 15. Although the distribution of myoglobin, in this case, is not very uniform, entrapment of this protein in the fiber-free manifold regions has been considerably reduced (cf. Figure 14). A more homogeneous distribution could

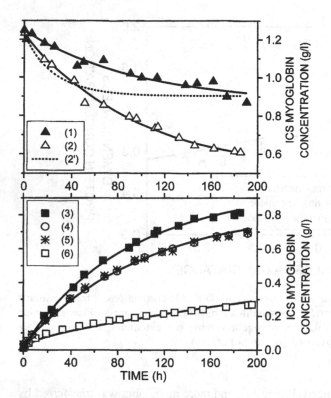

Figure 13 Experimental data (symbols) and best-fit KCM predictions (lines) of transient ICS myoglobin concentrations under various experimental conditions in a closed-shell HFBR. (1) Q_L = 300 mL/min; $C_{L0,M}$ = 1.25 g/L; $C_{S0,M}$ = 0, $C_{S0,A}$ = 15.8 g/L; (2) Q_L = 300 mL/min; $C_{L0,M}$ = 1.23 g/L; $C_{S0,M}$ = 0; $C_{S0,A}$ = 0. (3) Q_L = 300 mL/min; $C_{L0,A}$ = 0; $C_{S0,M}$ = 2.79 g/L; $C_{S0,A}$ = 16.5 g/L. (4) Q_L = 10 mL/min; $C_{L0,M}$ = 0.03 g/L; $C_{S0,M}$ = 2.72 g/L; $C_{S0,A}$ = 0. (5) Q_L = 10 mL/min; $C_{L0,M}$ = 0.03 g/L; $C_{S0,M}$ = 2.75 g/L; $C_{S0,A}$ = 16.7 g/L. (6) Q_L = 300 mL/min; $C_{L0,M}$ = 0.01 g/L; $C_{S0,M}$ = 2.62 g/L; $C_{S0,A}$ = 0. Dotted curve (2′): best-fit of (2) neglecting ECS manifolds; $C_{L0,M}$, initial myoglobin concentration in the lumen; $C_{S0,M}$, initial myoglobin concentration in the ECS; $C_{S0,A}$, initial BSA concentration in the ECS. (Adapted from Ref. 29.)

be engendered by using higher concentrations of BSA. Improved fits of some of the experimental data were obtained by assuming a partial decrease in the membrane porosity with time due to pore plugging by the myoglobin molecules, according to a Langmuir-type adsorption mechanism [29]. This is reasonable, considering that the effective myoglobin radius (1.75 nm) was very close to the average pore radius (about 2.1 nm) of the membranes investigated.

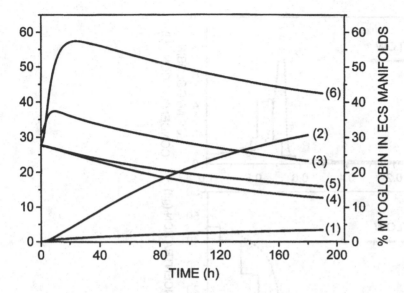

Figure 14 Best-fit KCM predictions of the total myoglobin fraction present in the ECS fiber-free manifold regions as a function of time; the following data indicate, respectively, the ICS flow rate, initial myoglobin placement, and the presence or absence of BSA: (1) Q_L = 300 mL/min, ICS, BSA; (2) Q_L = 300 mL/min, ICS, no BSA; (3) Q_L = 300 mL/min, ECS, BSA; (4) Q_L = 10 mL/min, ECS, no BSA; (5) Q_L = 10 mL/min, ECS, BSA; (6) Q_L = 300 mL/min, ECS, no BSA; initial protein concentrations are the same as for the respective cases in Figure 13.

2. Open-Shell Systems

In the vertically operated closed-shell cartridges used for all of the studies in Section IIIB1, radial gradients of pressure and concentration were usually small, and hence the protein distribution in the HFBR could be described relatively well using the Krogh cylinder approach. However, more significant radial gradients are expected to exist during open-shell operations, for which the porous medium model (PMM) (Section IIIA2), whose spatial domain includes the whole fiber bundle, is expected to produce better predictions. An example of the PMM application to open-shell HFBRs is the study of dead-end inoculation (cf. Figure 3b) carried out by Labecki et al. [27]. Vertical cartridges were again used to minimize gravitational influences and preserve, to the greatest extent possible, the axial symmetry of the system. After each experiment, the cartridge was frozen in liquid nitrogen and sectioned into axial segments, each of which was additionally cut into two radial and four angular pieces to determine the variation of protein concentration in all three spatial dimensions. The upper panel in Figure 16 plots

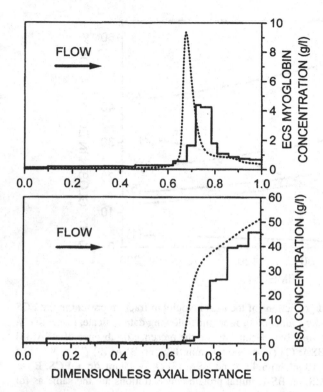

Figure 15 Experimental (solid step curves) and KCM-predicted (dotted lines) ECS concentration profiles for myoglobin (upper panel) and BSA (lower panel) after 190.5 hours of experiment (1): $Q_L = 300$ mL/min; $C_{L0,M} = 1.25$ g/L; $C_{S0,M} = 0$; $C_{S0,A} = 15.8$ g/L. The results of model simulations do not account for experimental losses of the proteins (3% for myoglobin and 21% for BSA). (From Ref. 29.)

the radially and angularly averaged experimental and PMM-predicted protein distributions for a typical inoculation experiment. The lower panel shows the angularly-averaged concentration profiles for the inside bore and the outside annulus of the fiber bundle; the difference between these two profiles is a measure of the radial concentration variation. As can be seen from Figure 16, the PMM was able to capture all the essential features of the ECS protein distribution: the position of the protein front, the relative magnitudes of the inside-bore and outside-annulus concentrations at different axial positions, and the existence of local concentration maxima in the first upstream segment and near the protein front. These maxima were caused by protein filtering at the surface of the membranes due to decreases in the radial and axial velocity components, respectively, as the ECS fluid drained continuously to the lumen side. It should be noted that KCM simulations of the same experiment (results not shown) predicted the maximum

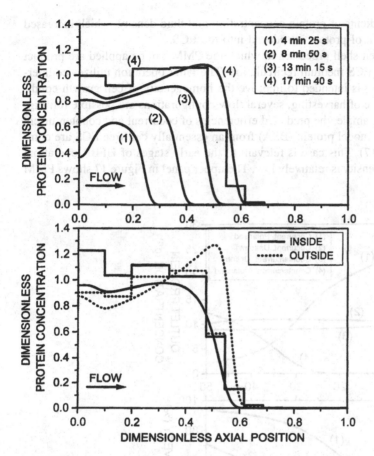

Figure 16 Measured (step lines) and predicted (2-D PMM, smooth curves) ECS protein (azoalbumin) concentrations in a dead-end inoculation experiment, nondimensionalized to the inlet concentration. Flow rate 3 mL/min; inlet concentration 12.15 g/L. Upper panel: radially and angularly averaged concentrations, PMM-predicted transients, and the experimental profile at $t = 17$ min 40 s. Lower panel: angularly averaged concentrations in the inside bore (solid lines) and outside annulus (dotted lines) of the fiber bundle at $t = 17$ min 40 s. (Adapted from Ref. 27.)

near the protein front to be about 25% higher than the experimental one. Furthermore, since the KCM cannot account for macroscopic radial flows in HFBRs, it could not predict the existence of the upstream maximum or the observed radial concentration variation. On the other hand, the agreement of the PMM-predicted and measured concentrations was poorer for those experiments in which the inoculum volume was several times the ECS volume. This was partly because, in these cases, the protein front had reached the downstream ECS manifold (not included in the model), and partly because of natural convection effects induced by vertical gradients of protein concentration and fluid density, which increased with the amount of protein introduced into the ECS.

Other open-shell processes to which the PMM can be applied are product harvesting and ECS medium recirculation. The latter operation utilizes counter-current flow and is intended to improve the homogeneity of cell growth conditions. In the case of harvesting, several flow configurations are possible (cf. Figure 5). As an example, the predicted efficiencies of cocurrent and countercurrent harvesting of a model protein (BSA) from an essentially cell-free ECS are compared (Figure 17). This case is relevant to the early stages of HFBR operation, when the cell density is relatively low. The upper panel in Figure 17 shows PMM

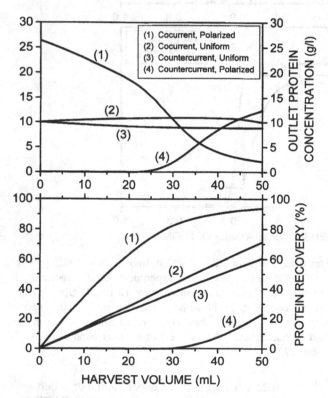

Figure 17 PMM predictions of outlet protein (BSA) concentration (upper panel) and fraction of protein removed from the ECS (lower panel) in cocurrent and countercurrent harvesting from a cell-free ECS. The outlet concentration is averaged over the length of the ECS manifold. The initial protein distribution was either uniform or steady-state downstream-polarized, with the average concentration 10 g/L; Q_L = 200 mL/min; ECS flow 1 mL/min; ECS volume 74.1 mL.

predictions of the ECS outlet concentration profiles as functions of the harvest volume for the average ECS protein concentration of 10 g/L and two extreme cases of the initial protein distribution: uniform and steady-state downstream-polarized. The corresponding product recovery curves are plotted in the lower panel of Figure 17. In the model simulations, gravity effects were neglected, a steady ECS flow of 1 mL/min was assumed, and the transient changes of ECS pressure drop due to osmotic variations were neglected. Although some of the conditions assumed in these simulations were idealized, certain interesting trends can be noticed from Figure 17. Firstly, cocurrent harvesting appears to be more efficient than countercurrent harvesting regardless of the initial protein distribution. Secondly, if the product protein is downstream-polarized prior to harvesting in the cocurrent configuration, then its concentration in the harvest will be maximized. Thirdly, by choosing cocurrent flow and taking advantage of the downstream polarization phenomenon, harvest volumes of only 30% of the ECS volume (74 mL, excluding manifolds) might be sufficient to recover 75% of the protein at a mixing-cup concentration that is twice its average concentration in the ECS. A continuous countercurrent flow configuration, on the other hand, can be useful in reducing the nonuniformity of growth factor distribution which results from downstream polarization. This is illustrated in Figure 18, which

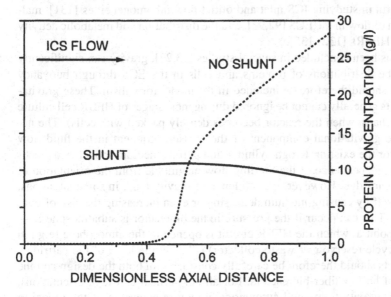

Figure 18 PMM predictions of steady-state ECS protein (BSA) distribution in the presence (solid line) or absence (dotted line) of countercurrent ECS recirculating flow through an external shunt. $Q_L = 200$ mL/min; ECS flow 0.5 mL/min; average ECS protein concentration 10 g/L.

compares the PMM predictions of the steady-state radially averaged BSA concentrations for a closed-shell reactor and for an open-shell system in which the ECS fluid is recirculated at a low flow rate countercurrently to the main lumen flow.

IV. ALTERNATIVES AND PROSPECTS

In spite of many successful predictions of fluid and solute behaviors in hollow-fiber cartridges, the models outlined in the previous section make numerous idealized assumptions about the conditions in the bioreactor, such as uniform fiber packing, absence of cells or homogeneous distribution of cell mass, axial symmetry of the cartridge, retention of proteins in the ECS (except in [29] and [73]), or negligible effects of the ECS and ICS manifolds (except for the simple treatment of the ECS manifold effect in [29]). Further refinement of the existing models, particularly of the PMM, is necessary in order to account for the various nonidealistic phenomena encountered in real HFBR systems. Employment of suitable experimental methods, particularly noninvasive techniques such as nuclear magnetic resonance (NMR), would be of great assistance in the investigation of these phenomena and in the verification of the models. NMR has already proved useful in studying ICS inlet and outlet flow inhomogeneities [131], maldistribution of flow in the ECS [92,93], and the distribution and metabolic activity of cells in HFBRs [132–135].

As has been concluded in several studies [13,27], gravity can significantly influence the distributions of proteins and cells in the ECS through buoyancy effects and should therefore be included in the model formalism. These gravitational effects generally cannot be ignored during most stages of HFBR cell culture (except, perhaps, when the reactor becomes densely packed with cells). The neglect of the gravitational component of the pressure gradient in the fluid flow equations for the existing Krogh cylinder and porous medium models was justified by the independence of the Starling flow magnitude from the orientation of the HFBR cartridge. However, the influence of gravity may, in some situations, manifest itself by causing medium degassing or even increasing the risk of contamination. This can occur if the pressure in the bioreactor is subatmospheric— i.e., if the point at which the HFBR circuit is opened to the atmosphere (e.g., in the ICS recycle reservoir or waste collector) is below the level of the cartridge. These effects should therefore be carefully considered in both the design and the modeling of hollow-fiber bioreactors. For all but vertical cartridge orientations, inclusion of gravity terms will automatically require extension of the model to three spatial dimensions. Such an extension is only possible in the case of the porous medium model, since the Krogh cylinder approximation is not able to deal with macroscopic radial or angular gradients in the reactor.

Considering the important role played by the ECS manifolds in the redistribution of cells and proteins, additional flow and solute transport equations should be applied to these fiber-free regions, with appropriate boundary conditions specified at the interface between the manifold and the porous region of the fiber bundle. The effect of the lumen manifolds, which determine the cross-sectional distribution of pressure at the inlet and outlet of the fiber bundle, could also be taken into account. In the case of leaking solutes, a source/sink term, like eq. (32), should be included in the PMM mass balance equations, such as eq. (43). For asymmetric membranes, extensions of the model could account for the accumulation of proteins in the pores of the spongy matrix, whose volume can comprise a significant fraction of the reactor. Electrostatic interactions between the solute molecules and the pores or surface of the membrane should be included in the model if they significantly affect the leakage process or lead to changes in membrane properties due to fouling. Ultimately, any HFBR modeling endeavor would not be complete without accounting for the presence of cells and including the kinetics of their growth. This task poses a considerable theoretical challenge, especially since the cells grow from a fairly dilute suspension during the startup phase of HFBR operation to a high-density, heterogeneous, relatively immobile mass during the stationary phase. Nonuniformity of cell distribution and the potential existence of flow channels within the cell-packed ECS can be of crucial importance from the point of view of effectively replacing growth factors and harvesting product proteins. Unfortunately, this effect is difficult to model, considering the variations in fiber and cell heterogeneity that can be expected to exist among different cartridges and different culture experiments. A complete model of a cell culture HFBR should also include the kinetics of product generation, as well as the transport and consumption or production of nutrients and metabolites.

In the pursuit of higher bioreactor productivities and optimal conditions for cell growth, a number of operational modifications to the conventional HFBR system (Table 2) as well as alternative HFBR designs (Table 3) have been proposed. Of course, in spite of their ingenuity and potential for improved performance, each of them has its limitations. For instance, many of the proposed systems do not seem to have solved the problem of cell sedimentation resulting in growth heterogeneity. This difficulty could be addressed by fixing the cells within a collagen or agar gel matrix in the ECS [66], but such a matrix would exclude much of the ECS volume otherwise available to the cells and could also diminish harvesting efficiency because of its significant resistance to convective flow [28]. Several alternative designs rely on the use of microporous membranes, thereby compromising some of the most attractive features of ultrafiltration hollow-fiber systems, viz., *in situ* product concentration and the confinement of expensive medium components to the relatively small volume of the ECS. It should be noted, however, that it is very difficult to assess the commercial feasibility of these alternative systems without a detailed economic analysis including, for ex-

Table 2 Examples of Alternative Hollow-Fiber System Configurations and
Operating Modes

System	References	Major advantages	Major disadvantages
Mass culturing technique (MCT) by Bio-Response	81,136	Uniform cell distribution maintained by rotating the growth chamber; cell drains used to remove nonviable cells; long-term productivity maintained	Depletion of growth factors leaking into the product removal loop; dilution of product as a result of using microporous membranes
ECS expansion chamber (Acusyst) by Cellex Biosciences	55,94,96	Improved homogeneity of cell and growth factor distributions; removal of dead cells and cell debris from the ECS	Cells subject to more shear than in closed-shell HFBRs; potential flushing of viable cells out of the ECS; product dilution in the ECS circuit volume; increased chance of membrane fouling due to high transmembrane fluxes; complex servomechanism
ICS flow cycling	13,26	Improved homogeneity of cell and growth factor distributions	Automated valve operation required for flow reversal and for product concentration prior to harvesting
Countercurrent ECS flow circuit		Improved homogeneity of cell and growth factor distributions; simple system	Potential flushing out of ECS viable cells if ECS flow rate too high; external ECS recycle volume should be small to avoid dilution of growth factors and products

Table 3 Examples of Alternative Hollow-Fiber System Designs

System	References	Major advantages	Major disadvantages
Flat fiber-bed reactor	72	Convection-enhanced nutrient delivery at low cell densities; uniform surface growth of anchorage-dependent cells; oxygen delivered separately from other nutrients	Potential membrane fouling; limited possibility of product concentration; potential flushing out of ECS viable cells and nutrient limitation of cells on the downstream side at cell-packed conditions
Radial-flow reactor with a central feed distributor	41,114,137	Convection-enhanced nutrient delivery at low cell densities; oxygen delivered separately from other nutrients	Potential membrane fouling; potential nutrient limitation of outermost cells at cell-packed conditions (may not receive enough nutrients from the central distributor)
Intercalated spiral alternate dead-ended reactor	20,115	Axially uniform growth and high cell viability achieved; relatively easy to model, as axial flows are negligible	Membrane fouling observed (protein gel layer formation), therefore serum not recommended for this design; dilution of growth factors and product in the recycle volume as a result of using microporous membranes
Tecnomouse by Integra Biosciences	138	Improved uniformity of oxygen delivery; oxygen delivered separately from other nutrients	Little product concentration by downstream polarization as a result of low flow rates
Tricentric reactor by Setec	113	Reduced downstream polarization and nutrient gradients as a result of countercurrent flow	Dilution of growth factors and product in the recycle volume as a result of using microporous membranes

Table 3 Continued

System	References	Major advantages	Major disadvantages
Cells cultured in the ICS	140,142	Significant reduction of nutrient and metabolite gradients; uniform cell growth; fibers can be sampled	Dilution of growth factors and products in the recycle volume as a result of using microporous membranes; inability to remove dead cells from inside the fibers

ample, the costs of fabrication and downstream processing. The comparative evaluation of competing systems is further complicated by the need to use the same cell line and to maintain identical or similar culture conditions in different bioreactors.

The simplest and most expedient improvements can be made to HFBR systems which utilize conventional, dialysis cartridge-based bioreactor designs. Examples of such modifications include (Table 2): the MCT (Mass Culturing Technique) system developed by Bio-Response [81,136]; Cellex Biosciences' Acusyst expansion chamber system (see also Figure 5c) [55,94,96]; ICS flow cycling, shown both theoretically and experimentally to produce more homogeneous distributions of proteins and cells in the ECS [13,26]; and countercurrent recirculation of ECS medium in an external shunt, which has been modeled (see Figure 18) but still needs optimization. In the MCT system, continual medium replenishment and antibody removal through microporous membranes take place in a cell exclusion loop, while a separate dialysis circuit with ultrafiltration membranes is used to exchange low-MW nutrients and metabolites with the ECS [81,136]. The cell growth chamber is periodically drained to remove some of the nonviable cells. The MCT system was reported to perform quite impressively by remaining productive for over 5 months, with an average IgG output of about 0.5 g/day [81]. Cellex's Acusyst has also performed well, by providing a stable production of IgG by murine hybridomas at about 0.4 g/day over a period of 60 days [94]. The high efficiency of the expansion chamber system is likely due to enhanced mixing in the ECS as well as effective removal of dead cells, cell debris and any high-MW metabolites from the shell side of the bioreactor. Potential problems associated with open-shell systems include cell damage by shear stresses and the possible removal of viable cells from the ECS, both of which necessitate a careful selection of operating conditions. The dilution of the product in the external recycle volume may also increase the cost of downstream processing.

The systems listed in Table 3 generally require the fabrication of new bioreactor designs and many still have limited commercial availability. Some of them, such as the flat-bed system [72], the radial-flow reactor [41,114,137], and the alternate dead-ended HFBR [20], utilize cross flow to minimize nutrient and metabolite gradients and thus can provide the cells with a more homogeneous growth environment. The radial-flow reactor of Tharakan and Chau [41,114,137] employs a central feed distributor to deliver liquid nutrients to the ECS, while the spent medium is removed with the waste gas stream through the lumina of the hollow fibers. Apart from a high probability of membrane fouling, the major concern about this system is a potential maldistribution of convective flow under cell-packed conditions, leading, in particular, to nutrient limitations in the outer regions of the fiber bundle. In the alternate dead-ended system proposed by Brotherton and Chau [20], two intercalated spiral microporous fiber layers are used to simulate the arterial and venous capillary networks. In this sense, the design is a modification of the earlier dual-circuit concept described by Wei and Russ [70] and by Gullino and Knazek [71]. Although uniform pressure and substrate concentration profiles were shown to exist in this type of bioreactor, serious membrane fouling problems were also encountered when a protein-containing medium was used. In several designs, e.g., flat-bed [72], radial-flow [41], Tecnomouse [138], and Tricentric [113], oxygen—the nutrient that is most likely to become limiting—is delivered to the cell chamber in a gas phase separately from other nutrients, which enables a better control of its supply. The Tecnomouse system by Integra Biosciences utilizes exchangeable, flat cassettes, in which the hollow-fiber bundle is encompassed by a silicone membrane that facilitates uniform oxygenation of the culture [138]. Setec's Tricentric bioreactor uses countercurrent flow, with cells grown in the annular space between two concentric fibers (a similar concentric-fiber reactor was also studied by Cima et al. [139]). Using continuous harvesting, this system was able to produce 550 mg MAb over a period of 3 months [113]. Other alternative designs include reactors in which cells are cultured on the lumen side of sealed hollow fibers, while the medium passes in cross flow through the ECS. An example of this type of system is the bioartificial liver (BAL) developed by Nyberg et al. [140]. Here, the hepatocytes are entrapped inside the fibers using a collagen gel. It should be noted, however, that the main function of the BAL is detoxification rather than biosynthesis. In traditional HFBRs used for production purposes, practical problems such as the inability to remove dead cells from inside the fibers or dilution of the product in the recycle medium have to be taken into consideration.

Ideally, a commercial-scale bioreactor system should include an integrated process control unit. Apart from the need to maintain the pH, temperature, and osmolarity (also other parameters in specific systems, like the pressure in Cellex's Acusyst) at desired levels, the use of efficient control algorithms should help optimise bioreactor performance and reduce operating costs. In a recent study by Dowd et al. [141], adaptive computer software routines were developed to estimate and predict the glucose uptake rate in a series of hollow-fiber hybridoma cell cultures. Based on these predictions, the medium feed rate was appropriately adjusted during the course of the culture, leading to closer control of glucose levels, increased antibody yields, and decreased labor and analytical costs. Implementation of an online glucose sensor together with such control routines could reduce even further operator involvement in the monitoring and optimization of HFBR cell culture.

Some investigators have concluded from their studies that scale-up of axial-flow HFBRs is severely limited and that cross-flow configurations should be more amenable to large-scale application [10,32]. A broader consideration of this issue was given by Chresand et al. [66], who pointed out that hollow-fiber systems are generally limited for scale-up in the direction of medium flow. Consequently, scale-up of cross-flow reactors can be accomplished by increasing fiber length, while axial-flow modules should be scaled up in the radial direction. A practical example of successful scale-up of a conventional (i.e., unidirectional, closed-shell, axial-flow) HFBR is described in a study by Jöbses et al. [97]. They demonstrated that antibody output and cell metabolism (evaluated from the consumption of glucose and oxygen) of a large-scale system containing 10 1-L modules were increased, respectively, 4 to 6 times and 5.5 to 6 times, compared with a pilot-scale system containing ten 200-mL modules. The antibody productivity of the large-scale system was, depending on the cell line, between 1.25 and 3.3 g/day/L of the ECS volume, which amounts to kilogram quantities per year. Although product harvesting under cell-packed conditions may be more difficult for modules having larger diameters, scale-up in the radial direction should be a merely technical task [97].

The coming decades can be expected to witness the growth of HFBR research as well as an increase in the commercial use of hollow-fiber bioreactors both for immobilized enzymes and for cell culture. The expanding HFBR field will include applications such as tissue engineering and artificial organs, which take advantage of the natural, tissuelike geometry of hollow-fiber units. An increasing number of biotechnology companies will likely use HFBRs for the production of 10 to 100-g quantities of MAbs and, perhaps, other important therapeutic and diagnostic biochemicals. Better understanding and control of hollow-fiber culture as a result of extensive fundamental and optimization studies should improve the rate of approval of HFBR processes by the Food and Drug Administration, thus increasing their commercial competitiveness.

REFERENCES

1. Rony PR. Multiphase catalysis. II. Hollow-fiber catalysts. Biotechnol Bioeng 13: 431–447, 1971.

2. Knazek RA, Gullino PM, Kohler PO, Dedrick RL. Cell culture on artificial capillaries: an approach to tissue growth in vitro. Science 178:65–67, 1972.

3. Webster IA, Shuler ML. Mathematical models for hollow-fiber enzyme reactors. Biotechnol Bioeng 20:1541–1556, 1978.

4. Piret JM, Cooney CL. Immobilized mammalian cell cultivation in hollow fiber bioreactors. Biotech Adv 8:763–783, 1990.

5. Butler M. Animal Cell Technology: Principles and Products. New York: Taylor and Francis, 1987.

6. Belfort G. Membranes and bioreactors: a technical challenge in biotechnology. Biotechnol Bioeng 33:1047–1066, 1989.

7. Inloes DS, Smith WJ, Taylor DP, Cohen SN, Michaels AS, Robertson CR. Hollow-fiber membrane bioreactors using immobilized *E. coli* for protein synthesis. Biotechnol Bioeng 25:2653–2681, 1983.

8. Birch JR, Thompson PW, Lambert K, Boraston R. The large scale cultivation of hybridoma cells producing monoclonal antibodies. In: J Feder, WR Tolbert, eds. Large-Scale Mammalian Cell Culture. Orlando: Academic Press, 1985:1–16.

9. Jackson LR, Trudel LJ, Fox JG, Lipman NS. Evaluation of hollow fiber bioreactors as an alternative to murine ascites production for small scale monoclonal antibody production. J Immunol Methods 189:217–231, 1996.

10. Tharakan JP, Gallagher S, Chau PC. Hollow fiber bioreactors in mammalian cell culture. Adv Biotechnol Process 7:153–184, 1988.

11. Starling EH. On the absorption of fluids from the connective tissue spaces. J Physiol 19:312–326, 1896.

12. Waterland LR, Robertson CR, Michaels AS. Enzymatic catalysis using asymmetric hollow fiber membranes. Chem Eng Commun 2:37–47, 1975.

13. Piret JM, Cooney CL. Mammalian cell and protein distributions in ultrafiltration hollow-fiber bioreactors. Biotechnol Bioeng 36:902–910, 1990.

14. Waterland LR, Michaels AS, Robertson CR. A theoretical model for enzymatic catalysis using asymmetric hollow-fiber membranes. AIChE J 20:50–59, 1974.

15. Kim S-S, Cooney DO. An improved theoretical model for hollow-fiber enzyme reactors. Chem Eng Sci 31: 289–294, 1976.

16. Piret JM, Cooney CL. Model of oxygen transport limitations in hollow fiber bioreactors. Biotechnol Bioeng 37:80–92, 1991.

17. Davis ME, Watson LT. Mathematical modeling of annular reactors. Chem Eng J 33:133–142, 1986.

18. Jayaraman VK. Solution of hollow fiber bioreactor design equations for zero-order limit of Michaelis-Menten kinetics. Chem Eng J 51:B63–B66, 1993.

19. Heifetz AH, Braatz JA, Wolfe RA, Barry RM, Miller DA, Solomon BA. Monoclonal antibody production in hollow fiber bioreactors using serum-free medium. BioTechniques 7(2):192–199, 1989.

20. Brotherton JD, Chau PC. Protein-free human-human hybridoma cultures in an intercalated-spiral alternate-dead-ended hollow fiber bioreactor. Biotechnol Bioeng 47: 384–400, 1995.

21. Kleinstreuer C, Agarwal SS. Analysis and simulation of hollow-fiber bioreactor dynamics. Biotechnol Bioeng 28:1233–1240, 1986.

22. Schonberg JA, Belfort G. Enhanced nutrient transport in hollow fiber perfusion bioreactors: A theoretical analysis. Biotechnol Prog 3:80–89, 1987.

23. Salmon PM, Libicki SB, Robertson CR. A theoretical investigation of convective transport in the hollow-fiber reactor. Chem Eng Commun 66:221–248; 1986.

24. Kelsey LJ, Pillarella MR, Zydney AL. Theoretical analysis of convective flow profiles in a hollow-fiber membrane bioreactor. Chem Eng Sci 45:3211–3220, 1990.

25. Taylor DG, Piret JM, Bowen BD. Protein polarization in isotropic membrane hollow-fiber bioreactors. AIChE J 40:321–333, 1994.

26. Patkar AY, Koska J, Taylor DG, Bowen BD, Piret JM. Protein transport in ultrafiltration hollow fiber bioreactors. AIChE J 41:415–425, 1995.

27. Labecki M, Bowen BD, Piret JM. Two-dimensional analysis of protein transport in the extracapillary space of hollow-fiber bioreactors. Chem Eng Sci 51:4197–4213, 1996.

28. Koska J, Bowen BD, Piret JM. Protein transport in packed-bed ultrafiltration hollow-fiber bioreactors. Chem Eng Sci 52:2251–2263, 1997.

29. Labecki M, Weber I, Dudal Y, Koska J, Piret JM, Bowen BD. Hindered transmembrane protein transport in hollow-fiber devices. J Membr Sci 146:197–216, 1998.

30. Kumar RA, Modak JM. Transient analysis of mammalian cell growth in hollow fiber bioreactor. Chem Eng Sci 52: 1845–1860, 1997.

31. Heath C, Belfort G. Immobilization of suspended mammalian cells: analysis of hollow fiber and microcapsule bioreactors. Adv Biochem Eng Biotechnol 34:2–31, 1987.

32. Brotherton JD, Chau PC. Modeling of axial-flow hollow fiber cell culture bioreactors. Biotechnol Prog 12:575–590, 1996.

33. Marshall CL, Boraston R, Brown ME. The effect of osmolarity on hybridoma cell growth and antibody production in serum-free media. In: RE Spier, Griffiths JB, Meignier B, eds. Production of Biologicals from Animal Cells in Culture. Oxford: Butterworth-Heinemann, 1993:259–261.

34. Merten O-W. Culture of hybridomas—a survey. In: AOA Miller, ed. Advanced Research on Animal Cell Technology. Dordrecht: Kluwer Academic Publishers, 1989:367–398.

35. Miller WM, Blanch HW, Wilke CR. A kinetic analysis of hybridoma growth and metabolism in batch and continuous suspension culture: effect of nutrient, concentration, dilution rate, and pH. Biotechnol Bioeng 32:947–965, 1988.

36. Cartwright T. Animal Cells as Bioreactors. New York: Cambridge University Press, 1994.

37. Butler M, Dawson M. Cell Culture. Labfax. Oxford: BIOS Scientific Publishers, 1992.

38. Freshney RI. Culture of Animal Cells: A Manual of Basic Technique. New York: Wiley-Liss, 1994.

39. Griffiths B. Perfusion systems for cell cultivation. In: AS Lubiniecki, ed. Large-Scale Mammalian Cell Culture Technology. New York: Marcel Dekker, 1990:217–250.

40. Griffiths B. Scaling-up of animal cell cultures. In: RI Freshney, ed. Animal Cell Culture: A Practical Approach. Oxford: IRL Press, 1986:33–69.

41. Tharakan JP, Chau PC. A radial flow hollow fiber bioreactor for the large-scale culture of mammalian cells. Biotechnol Bioeng 28:329–342, 1986.

42. Köhler G, Milstein C. Continuous cultures of fused cells secreting antibody of predefined specifity. Nature 256:495–497, 1975.

43. Kitano K, Iwamoto K, Shintani Y, Sasada R. Improvement in cell lines for effective production of human monoclonal antibodies by human-human hybridomas. In: H Murakami, ed. Trends in Animal Cell Culture Technology. Tokyo: Kodansha, 1989:285–290.

44. McCullough KC, Spier RE. Monoclonal Antibodies in Biology and Biotechnology: Theoretical and Practical Aspects. Cambridge: Cambridge University Press, 1990.

45. Barnes D. Attachment factors in cell culture. In: JP Mather, ed. Mammalian Cell Culture: The Use of Serum-Free Hormone-Supplemented Media. New York: Plenum Press, 1984:195–237.

46. Junqueira LC, Carneiro J, Long JA. Basic Histology. 5th ed. Norwalk: Appleton-Century-Crofts, 1986.

47. Swabb EA, Wei J, Gullino PM. Diffusion and convection in normal and neoplastic tissues. Cancer Res 34:2814–2822, 1974.

48. Levick JR. Flow through interstitium and other fibrous matrices. Quart J Experim Physiol 72:409–438, 1987.

49. Ryll T, Lucki-Lange M, Jäger V, Wagner R. Production of recombinant human interleukin-2 with BHK cells in a hollow fiber and a stirred tank reactor with protein-free medium. J Biotechnol 14:377–392, 1990.

50. Knazek RA. Solid tissue masses formed in vitro from cells cultured on artificial capillaries. Fed Proc 33:1978–1981, 1974.

51. Marx U, Roggenbuck D, Wilding M, Tanzmann H, Jahn S. Pitfalls of bioprocessing a human monoclonal multireactive IgM antibody. In: RE Spier, Griffiths JB, Berthold W, eds. Animal Cell Technology. Products of Today, Prospects for Tomorrow. Oxford: Butterworth-Heinemann, 1994:278–280.

52. Evans TL, Miller RA. Large-scale production of murine monoclonal antibodies using hollow fiber bioreactors. BioTechniques 6:762–767, 1988.

53. Sardonini CA, DiBiasio D. Growth of animal cells around hollow fibers: multifiber studies. AIChE J 39:1415–1419, 1993.

54. Pinton H, Lourenco da Silva A, Goergen JL, Marc A, Engasser JM, Rabaud JN, Pierry G. Control of the maximal cell density in a membrane perfusion reactor. In: Spier RE, Griffiths JB, Berthold W, eds. Animal Cell Technology. Products of Today, Prospects for Tomorrow. Oxford: Butterworth-Heinemann, 1994:470–475.

55. Andersen BG, Gruenberg ML. Optimization techniques for the production of monoclonal antibodies utilizing hollow-fiber technology. In: SS Seaver, ed. Commercial Production of Monoclonal Antibodies: A Guide for Scale-up. New York: Marcel Dekker, 1987:175–195.

56. Kidwell W, Knazek R, Wu Y. Effect of fiber pore size on performance of cells in hollow fiber bioreactors. In: H Murakami, ed. Trends in Animal Cell Culture Technology. Tokyo: Kodansha, 1989:29–33.

57. Fleischaker RJ Jr, Sinskey AJ. Oxygen demand and supply in cell culture. Eur J Appl Microbiol Biotechnol 12: 193–197, 1981.

58. Reuveny S, Velez D, Riske F, Macmillan JD, Miller l. Production of monoclonal antibodies in culture. Dev Biol Standard 60:185–197, 1985.

59. Ryan GB, Simpson MT, Jones WT, Nicol MJ, Reynolds PHS. Effect of dissolved oxygen on monoclonal antibody production from hybridoma cultured in haemodialysers. In: RE Spier, Griffiths JB, Berthold W, eds. Animal Cell Technology. Products of Today, Prospects for Tomorrow. Oxford: Butterworth-Heinemann, 1994: 437–456.

60. Kilburn DG, Webb FC. The cultivation of animal cells at controlled dissolved oxygen partial pressure. Biotechnol Bioeng 10:801–814, 1968.

61. Piret JM, Devens DA, Cooney CL. Nutrient and metabolite gradients in mammalian cell hollow fiber bioreactors. Can J Chem Eng 69:421–428, 1991.

62. Mather JP, Tsao M. Expression of cloned proteins in mammalian cells: regulation of cell-associated parameters. In: Lubiniecki AS, ed. Large-Scale Mammalian Cell Culture Technology. New York: Marcel Dekker, 1990:161–177.

63. Adamson SR, Behie LA, Gaucher GM, Lesser BH. Metabolism of hybridoma cells in suspension culture: Evaluation of three commercially available media. In: SS Seaver, ed. Commercial Production of Monoclonal Antibodies: A Guide for Scale-up. New York: Marcel Dekker, 1987:17–34.

64. Davis JM, Hanak JAJ, Lewis GM, Chung R, Faulkner J. Long term serum-free hollow-fiber culture of cell lines producing monoclonal antibodies: metabolic aspects. In: RE Spier, Griffiths JB, Meignier B, eds. Production of Biologicals from Animal Cells in Culture. Oxford: Butterworth-Heinemann, 1991:130–133.

65. Thomas JN. Mammalian cell physiology. In: Lubiniecki AS, ed. Large-Scale Mammalian Cell Culture Technology. New York: Marcel Dekker, 1990:93–145.

66. Chresand TJ, Gillies RJ, Dale BE. Optimum fiber spacing in a hollow fiber bioreactor. Biotechnol Bioeng 32:983–992, 1988.

67. Glacken MW, Adema E, Sinskey AJ. Mathematical description of hybridoma culture kinetics. I. Initial metabolic rates. Biotechnol Bioeng 32:491–506, 1988.

68. Schneider Y-J, Lavoix A. Monoclonal antibody production in semi-continuous serum- and protein-free culture. Effect of glutamine concentration and culture conditions on cell growth and antibody secretion. J Immunol Methods 129:251–268, 1990.

69. Handa-Corrigan A, Nikolay S, Spier RE. Biochemical control of monoclonal antibody secretion in hollow fiber bioreactors. In: Spier RE, Griffiths JB, MacDonald C, eds. Animal Cell Technology: Developments, Processes and Products. Oxford: Butterworth-Heinemann, 1992:489–493.

70. Wei J, Russ MB. Convection and diffusion in tissues and tissue cultures. J Theor Biol 66:775–787, 1977.

71. Gullino PM, Knazek RA. Tissue culture on artificial capillaries. Methols Enzymol 58:178–184, 1979.

72. Ku K, Kuo MJ, Delente J, Wildi BS, Feder J. Development of a hollow-fiber system for large-scale culture of mammalian cells. Biotechnol Bioeng 23:79–95, 1981.

73. Pillarella MR, Zydney AL. Theoretical analysis of the effect of convective flow on solute transport and insulin release in a hollow fiber bioartificial pancreas. J Biomech Eng (Trans ASME) 112:220–228, 1990.

74. Jäger V. Serum-free media suitable for upstream and downstream processing. In: Spier RE, Griffiths JB, Meignier B, eds. Production of Biologicals from Animal Cells in Culture. Oxford: Butterworth-Heinemann, 1991:155–164.

75. Brown BL. Reducing costs upfront: two methods for adapting hybridoma cells to an inexpensive, chemically defined serum-free medium. In: Seaver SS, ed. Commercial Production of Monoclonal Antibodies: A Guide for Scale-up. New York: Marcel Dekker, 1987:35–48.

76. Maurer HR. Towards chemically-defined, serum-free media for mammalian cell culture. In: Freshney RI, ed. Animal Cell Culture: A Practical Approach. Oxford: IRL Press, 1986: 13–31.

77. Sanfeliu A, Damgaard B, Cairó JJ, Casas C, Sola' C, Gódia F. Low-level FCS adaptation and media develoment for the culture of two hybridoma cell lines producing IgG and IgM monoclonal antibodies. In: Spier RE, Griffiths JB, Berthold W, eds. Animal Cell Technology. Products of Today, Prospects for Tomorrow. Oxford: Butterworth-Heinemann, 1994:102–104.

78. Wyatt DE. Adaptation of mammalian cells to protein-free growth. In: Spier RE, Griffiths JB, Berthold W, eds. Animal Cell Technology. Products of Today, Prospects for Tomorrow. Oxford: Butterworth-Heinemann, 1994:144–146.

79. Schönherr OT, van Gelder PTJA. Culture of animal cells in hollow-fiber dialysis systems. In: Spier RE, Griffiths JB, eds. Animal Cell Biotechnology, Vol. 3. Oxford: Academic Press, 1988:337–355.

80. Dhainaut F, Pouget L, Richer-Hers MJ, Mignot G. Optimization of human anti-rhesus IgG production using hollow fiber technology. In: Spier RE, Griffiths JB, Meignier B, eds. Production of Biologicals from Animal Cells in Culture. Oxford: Butterworth-Heinemann, 1991:495–497.

81. von Wedel RJ. Mass culture of mouse and human hybridoma cells in hollow-fiber culture. In: Seaver SS, ed. Commercial Production of Monoclonal Antibodies: A Guide for Scale-up. New York: Marcel Dekker, 1987:159–173.

82. Gramer MJ, Poeschl DM. Screening tool for hollow-fiber bioreactor process development. Biotechnol Prog 14:203–209, 1998.

83. Berg TM, Øyaas K, Levine DW. Growth and antibody production of hybridoma cells exposed to hyperosmotic stress. In: Murakami H, ed. Trends in Animal Cell Culture Technology. Tokyo: Kodansha, 1989:93–97.

84. Ryu JS, Lee GM. Effect of hypoosmotic stress on hybidoma cell growth and antibody production. Biotechnol Bioeng 55:565–570, 1997.

85. Oh SKW, Chua FKF, Choo ABH. Intracellular responses of productive hybridomas subjected to high osmotic pressure. Biotechnol Bioeng 46:525–535, 1995.

86. Wurm FM, Gwinn KA, Kingston RE. Inducible overproduction of the mouse c-myc protein in mammalian cells. Proc Natl Acad Sci USA 83:5414–5418, 1986.

87. Kretzmer G, Buch T, Konstantinov K, Naveh D. The temperature effect in mammalian cell culture: an Arrhenius interpretation. In: Merten OW, Perrin P, Griffiths B, eds. New Developments and New Applications in Animal Cell Technology. Dordrecht: Kluwer Academic Publishers, 1998: 363–366.

88. Ozturk SS, Palsson BO. Chemical decomposition of glutamine in cell culture media: effect of media type, pH, and serum concentration. Biotechnol Prog 6:121–128, 1990.

89.   Zeman LJ, Zydney AL. Microfiltration and Ultrafiltration: Principles and Applications. New York: Marcel Dekker, 1996.
90.   Batt BC, Davis RH, Kompala DS. Inclined sedimentation for selective retention of viable hybridomas in a continuous suspension bioreactor. Biotechnol Prog 6: 458–464, 1990.
91.   Anderson JL, Rauh F, Morales A. Particle diffusion as a function of concentration and ionic strength. J Phys Chem 82:608–616, 1978.
92.   Hammer BE, Heath CA, Mirer SD, Belfort G. Quantitative flow measurements in bioreactors by nuclear magnetic resonance imaging. Bio/Technology 8:327–330, 1990.
93.   Heath CA, Belfort G, Hammer BE, Mirer SD, Pimbley JM. Magnetic resonance imaging and modeling of flow in hollow-fiber bioreactors. AIChE J 36:547–558, 1990.
94.   Gramer MJ, Keznoff SM. Optimizing hollow-fiber bioreactors: EC cycling technology for cell culture production. Gen Eng News 18:21, 1998.
95.   Lowrey D, Murphy S, Goffe RA. A comparison of monoclonal antibody productivity in different hollow fiber bioreactors. J Biotechnol 36:35–38, 1994.
96.   Tyo MA, Bulbulian BJ, Menken BZ, Murphy TJ. Large-scale mammalian cell culture utilizing ACUSYST technology. In: Spier RE, Griffiths JB, eds. Animal Cell Biotechnology, Vol. 3. Oxford: Academic Press, 1988:357–371.
97.   Jöbses I, van Zutphen P, Oomens J, van Os A, Schönherr OT. Scaling-up of a hollow fiber reactor for animal cell cultivation. In: Spier RE, Griffiths JB, MacDonald C, eds. Animal Cell Technology: Developments, Processes and Products. Oxford: Butterworth-Heinemann, 1992:517–522.
98.   Altshuler GL, Dziewulski DM, Sowek JA, Belfort G. Continuous hybridoma growth and monoclonal antibody production in hollow fiber reactors-separators. Biotechnol Bioeng 28:646–658, 1986.
99.   Lysaght MJ. Evolution of hemodialysis membranes. In: Bonomini V, Berland Y, eds. Dialysis Membranes: Structure and Predictions. Basel: Karger, 1995:1–10.
100. Filho GR, Bueno WA. Water state of Cuprophan (hemodialysis membrane). J Membr Sci 74:19–27, 1992.
101. Labecki M, Bowen BD, Piret JM. Two-dimensional analysis of fluid flow in hollow-fiber modules. Chem Eng Sci 50:3369–3384, 1995.
102. Radovich JM. Composition of polymer membranes for therapies of end-stage renal disease. In: Bonomini V, Berland Y, eds. Dialysis Membranes: Structure and Predictions. Basel: Karger, 1995:11–24.
103. Blatt WF, Dravid A, Michaels AS, Nelsen L. Solute polarization and cake formation in membrane ultrafiltration: causes, consequences, and control techniques. In: Flinn JE, ed. Membrane Science and Technology. New York: Plenum Press, 1970:47–97.
104. Zydney AL. Bulk mass transfer limitations during high-flux hemodialysis. Artif Organs 17:919–924, 1993.
105. Meireles M, Aimar P, Sanchez V. Effects of protein fouling on the apparent pore size distribution of sieving membranes. J Membr Sci 56:13–28, 1991.
106. Lee SH, Ruckenstein E. Adsorption of proteins onto polymeric surfaces of different hydrophilicities—a case study with bovine serum albumin. J Colloid Interface Sci 125:365–379, 1988.

107. Norde W. Adsorption of proteins from solution at the solid-liquid interface. Adv Colloid Interface Sci 25:267–340, 1986.

108. Matthiasson E. The role of macromolecular adsorption in fouling of ultrafiltration membranes. J Membr Sci 16:23–36, 1983.

109. Chen V, Kim KJ, Fane AG. Effect of membrane morphology and operation on protein deposition in ultrafiltration membranes. Biotechnol Bioeng 47:174–180, 1995.

110. Mochizuki S, Zydney AL. Theoretical analysis of pore size distribution effects on membrane transport. J Membr Sci 82:211–227, 1993.

111. Langsdorf LJ, Zydney AL. Effect of blood contact on the transport properties of hemodialysis membranes: a two-layer membrane model. Blood Purif 12:292–307, 1994.

112. Meireles M, Aimar P, Sanchez V. Albumin denaturation during ultrafiltration: effects of operating conditions and consequences on membrane fouling. Biotechnol Bioeng 38:528–534, 1991.

113. Kessler N, Thomas G, Gerentes L, Delfosse G, Aymard M. Hybridoma growth in a new generation hollow fiber bioreactor: antibody productivity and consistency. Cytotechnology 24:109–119, 1997.

114. Tharakan JP, Chau PC. Modeling and analysis of radial flow mammalian cell culture. Biotechnol Bioeng 29:657–671, 1987.

115. Brotherton JD, Chau PC. Modeling analysis of an intercalated-spiral alternate-dead-ended hollow fiber bioreactor for mammalian cell cultures. Biotechnol Bioeng 35: 375–394, 1990.

116. Krogh A. The number and distribution of capillaries in muscles with calculations of the oxygen pressure head necessary for supplying the tissue. J Physiol 52:409–415, 1919.

117. Chick WL, Like AA, Lauris V. Beta cell culture on synthetic capillaries: an artificial endocrine pancreas. Science 187:847–949, 1975.

118. Wolf CFW, Minick CR, McCoy CH. Morphologic examination of a prototype liver assist device composed of cultured cells and artificial capillaries. Int J Artif Organs 1:45–51, 1978.

119. Jaffrin MY, Reach G, Notelet D. Analysis of ultrafiltration and mass transfer in a bioartificial pancreas. J Biomech Eng (Trans ASME) 110:1–10, 1988.

120. Catapano G. Mass transfer limitations to the performance of membrane bioartificial liver support devices. Int J Artif Organs 19:18–35, 1996.

121. Apelblat A, Katzir-Katchalsky A, Silberberg A. A mathematical analysis of capillary-tissue fluid exchange. Biorheology 11:1–49, 1974.

122. Beavers GS, Joseph DD. Boundary conditions at a naturally permeable wall. J Fluid Mech 30:197–207, 1967.

123. Saffman PG. On the boundary condition at the surface of a porous medium. Studies Appl Math L:93–101, 1971.

124. Vilker VL, Colton CK, Smith KA. The osmotic pressure of concentrated protein solutions: effect of concentration and pH of saline solutions of bovine serum albumin. J Colloid Interface Sci 79:548–565, 1981.

125. Carman PC. Fluid flow through a granular bed. Trans Inst Chem Eng Lond 15: 150–156, 1937.

126. Deen WM. Hindered transport of large molecules in liquid-filled pores. AIChE J 33:1409–1424, 1987.
127. Costello MJ, Fane AG, Hogan PA, Schofield RW. The effect of shell side hydrodynamics on the performance of axial flow hollow fiber modules. J Membr Sci 80: 1–11, 1993.
128. Park JK, Chang HN. Flow distribution in the fiber lumen side of a hollow-fiber module. AIChE J 32:1937–1947, 1986.
129. Bear J. Dynamics of Fluids in Porous Media. New York: Elsevier, 1972.
130. Happel J. Viscous flow relative to arrays of cylinders. AIChE J 5:174–177, 1959.
131. Pangrle BJ, Walsh EG, Moore S, DiBiasio D. Investigation of fluid flow patterns in a hollow fiber module using magnetic resonance velocity imaging. Biotechnol Tech 3:67–72, 1989.
132. Mancusco A, Fernandez EJ, Blanch HW, Clark DS. A nuclear magnetic resonance technique for determining hybridoma cell concentration in hollow fiber bioreactors. Bio/Technology 8:1282–1285, 1990.
133. Briasco CA, Ross DA, Robertson CR. A hollow-fiber reactor design for NMR studies of microbial cells. Biotechnol Bioeng 36:879–886, 1990.
134. Conroy MJ, Hammer BE, Amiot B, Gramer M. Characterization of a large scale commercial hollow fiber bioreactor using multinuclear NMR spectroscopy and imaging. Third Meeting of Society of Magnetic Resonance, Nice, France, Aug 19–25, 1995.
135. Williams SNO, Callies RM, Brindle KM. Mapping of oxygen tension and cell distribution in a hollow-fiber bioreactor using magnetic resonance imaging. Biotechnol Bioeng 56:56–61, 1997.
136. Brown PC, Costello MAC, Oakley R, Lewis JL. Applications of the mass culturing technique (MCT) in the large scale growth of mammalian cells. In: Feder J, Tolbert WR, eds. Large-Scale Mammalian Cell culture. Orlando: Academic Press, 1985: 59–71.
137. Tharakan LP, Chau PC. Operation and pressure distribution of immobilized cell hollow fiber bioreactors. Biotechnol Bioeng 28:1064–1071, 1986.
138. Tzianabos AO, Smith R. Use of hollow-fiber bioreactor for production in problematic cell lines. Gen Eng News 1: 24, 1995.
139. Cima LG, Blanch HW, Wilke CR. A theoretical and experimental evaluation of a novel radial-flow hollow fiber reactor for mammalian cell culture. Bioproc Eng 5: 19–30, 1990.
140. Nyberg SL, Shatford RA, Payne WD, Hu W-S, Cerra FB. Primary culture of rat hepatocytes entrapped in cylindrical collagen gels: an in vitro system with application to the bioartificial liver. Cytotechnology 8:205–216, 1992.
141. Dowd JE, Weber I, Rodriguez B, Piret JM, Kwok KE. Predictive control of hollow fiber bioreactors for the production of monoclonal antibodies. Biotechnol Bioeng 63:484–492, 1999.
142. Adema E, Sinskey AJ. An analysis of intra- versus extracapillary growth in a hollow fiber reactor. Biotechnol Prog 3:74–79, 1987.

2
Vortex Flow Filtration for Cell Separation in Bioreactor Operations

Georg Roth, Clayton E. Smith,* Gary M. Schoofs, Tana J. Montgomery, Jenifer L. Ayala, Thomas J. Monica, Francisco J. Castillo, and Joseph I. Horwitz
Berlex Biosciences, Richmond, California

I. INTRODUCTION

Large-scale mammalian cell culture is frequently performed in stirred tank or airlift reactors in batch mode, and to achieve high cell densities and product concentrations fed-batch processes have been introduced [1,2]. A further modification consists in extended culture with the use of perfusion systems where fresh medium continuously enters the bioreactor and spent medium, sometimes containing the product, is removed in the effluent stream. The key characteristic of a perfusion system is the retention of cell mass in the bioreactor. Compared to traditional batch or fed-batch systems, perfusion systems allow high cell densities for equivalent product yields with smaller bioreactor volumes [3]. In addition, labor and energy costs can be substantially lower due to less frequent reactor configuration and cleaning. Using this approach, high cell densities (i.e., $> 10^7$ cells/mL) can be maintained for long periods of time. This allows for extended collection of extracellular secreted product through the harvest or purge streams and for occasional or continuous removal of cells at high density through a purge stream, which is required for cell production or for collection of an intracellular product.

Additional benefits for large-scale production can be obtained by adapting adherent cell lines to suspension mode, e.g., eliminating porous microcarriers that are a major cost factor and introduce potentially heterogeneous culture conditions. However, suspension-adapted cells present a challenge since they are diffi-

---

* *Current affiliation*: Consultant, Biotechnology Transfer, Bio-Marin, Berkeley, California

cult to retain in a perfusion system. While simple harvest screens are sufficient to create an effective cell retention system for microcarrier cultures, single-cell suspension cultures require more complex devices.

The characteristics of an optimal cell retention device for use in suspension systems can be defined as follows:

1.   Cells must be retained within the bioreactor. Loss of cells through a retention device must be kept at a minimum to maintain cell density and to minimize cell debris and lysate in the harvest.

2.   Viability of the cell culture must be maintained at high levels. Mammalian cells are negatively affected by mechanical shear stress generated by some retention devices.

3.   Minimal fouling of the retention device. Continuously perfused cell cultures can last for several months; the retention device should operate throughout this period without maintenance. If fouling does occur, the device should be easy to clear or exchange aseptically.

4. The retention device should be cleanable and sterilizable. A reusable unit is usually preferable to one that must be replaced after each run.

Many of the cell retention devices that are commonly used have shortcomings in one or several of the properties defined above. Another important factor is the flexibility of the retention device to support various operational conditions and reactor sizes. Scale-up encompasses not only increases in reactor volume but also in the perfusion rate; i.e., the retention device has to accommodate varying and potentially large flow rates. The discussion below of the most commonly used systems will be divided into internal and external cell retention devices, referring to whether the device is located inside or outside the bioreactor. An example of an internal device is the rotating filter developed from the static harvest screen for microcarrier cultures [4].

The rotating or spin filter is an internal device with the membrane attached to the impeller shaft [5–8]. Membranes are usually made out of steel, but other materials such as polyamide [9] and porcelain [10] have also been used. Due to the setup of the filter this system is severely limited in use and its ability to be scaled up. It requires specially designed bioreactors and does not allow for membrane maintenance without termination of the fermentation. Also, the rotational speed is usually directly coupled to the impeller speed, prohibiting the use of a range of rotational rates. Another problem with this kind of retention device is frequent fouling of the membrane. Literature reports indicate fouling within 1 to 2 weeks of operation [5,8]. To circumvent this problem several approaches have been pursued, among them draft tubes to provide better membrane clearance and pore sizes that exceed the cell diameter [7].

External cell retention devices have been designed that rely on sedimentation, centrifugation, hollow-fiber filtration, acoustic cell filtration and dielectrophoresis. A detailed discussion of the various designs can be found in Woodside et. al. [11]. The most simple external cell retention devices are settling tubes that have been used on small-scale reactors [12–15]. Separation efficiency is a function of flow through the settling device; i.e., higher flow rates lead to higher cell washout and this prevents the use of high perfusion rates that require high flow rates. Scale-up is limited because increasing the dimensions of the settling tube leads to high residence times within the device and this reduces cell viability due to extended residence times in a nonaerated environment. An additional problem is the attachment of cells to the device in regions with a quiescent flow environment. Vibrating the settling tube is one approach to prevention of the adherence problem (12).

Intermittent or continuous centrifugation has also been used successfully for cell retention [16–18]. These systems currently provide the greatest rated capacity of any retention device. Separation efficiency of these systems is usually high, but the cells are subjected to potentially high gravitational forces which may cause cell death. Also, periodic oxygen and nutrient starvation can negatively affect productivity [11,19].

Hollow-fiber bioreactors have been used for the production of biologicals with good results [20–23]. Utilizing hollow fibers as cell retention devices while maintaining the cell culture in a stirred tank bioreactor has also been attempted [24–28]. Cell retention efficiency is very high in theses setups, but cell viability can be affected by high shear rates in the hollow-fiber cartridge. Hollow-fiber cartridges also tend to foul during prolonged operation, causing increased transmembrane pressure, often requiring backflushing to continue operation. Membrane fouling can be reduced via high tangential flows achieved by high recirculation rates. However, higher flow rates also cause higher shear forces, reducing cell viability and density. Another problem for a range of operating conditions is oxygen diffusion limitation, leading to cell death. Similar to settling tubes, scale-up beyond certain limits imposed by the maximal throughput is only possible through use of multiple hollow-fiber cartridges. Advantages of hollow-fiber units are the availability of many different configurations and easy and aseptic exchange during a fermentation run.

Acoustic resonance filtration has also been tested, with good retention efficiency and high cell viability [29,30]. Currently these systems are only available for fairly limited scale [11]. We have observed aggregation of the cells after prolonged exposure to the ultrasonic field limiting its use to cell lines with low clumping potential. A new dielectrophoresis based method of cell retention [31] takes advantage of the difference in the electric conductivity between the culture medium and the cell where an electric field effectively retains the cells. Retention efficiency, as observed in the first devices, is high and only viable cells are retained.

From the comparison of available retention devices the properties of an ideal one emerge. It should be external to the bioreactor to allow for easy exchange during a fermentation, provide scalability by a combination of throughput and size and not necessarily by an increase in the numbers of devices, have short mean residence times, and, in the case of a membrane-based system, prevent membrane fouling.

Fouling can be reduced by the use of hydrophilic membranes and by filtration devices minimizing concentration polarization at the membrane surface. The latter can be achieved by generating a secondary flow at the membrane surface with a rotating filter in a narrow annular gap. This process is called vortex flow filtration (VFF), and several devices have been developed and used for cell retention [32–34]. Their filtration performance is superior to cross-flow filtration [35,36] including applications beyond cell retention, such as blood fractionation [37] and protein fractionation [38–41].

This work describes principles, design, and operation of a cell retention device based on VFF. Its use and evaluation in perfusion cultures of mammalian cells are also discussed.

II. PRINCIPLE OF OPERATION

A. VFF Unit

The Benchmark Gx vortex flow filtration system from Membrex (Fairfield, NJ) was studied and characterized. The system is composed of an electronic control unit (ECU) and a rotary separation unit (RSU). The ECU contains operational controls for the system, houses the drive motor, and serves as a mount for the RSU. The primary function of the ECU is to control the rotation speed of the rotor in the RSU. Desired speed is entered on the keypad in the front of the ECU and is displayed on an LED. The LED also shows transmembrane pressure (TMP) and motor torque during operation.

The RSU mounts onto the ECU using a twist-type bayonet connection. As diagrammed in Figure 1, the RSU is composed of a magnetic drive assembly and a separation chamber. The magnetic drive assembly contains a drive shaft, bearings, seals, and a coupling magnet which rotates a polysulfone rotor. The coupling magnet is in turn rotated by a magnetic ring in the ECU. The separation chamber has a stainless-steel housing and contains the rotor which spins within a fixed cylindrical stainless-steel mesh filter. The filter has a 10-μm pore size and a total surface area of 200 cm^2. Ports in the housing and drive assembly allow for feed inlet and retentate outlet, respectively. Another port in the housing provides an outlet for permeate. Hose barbs allow for connection into those ports. The complete RSU can be autoclaved and then connected to the ECU, allowing for prompt exchange in case of filter fouling or a mechanical breakdown during operation. The dimensions of the Gx unit are 12 inches (width) × 7 inches (depth)

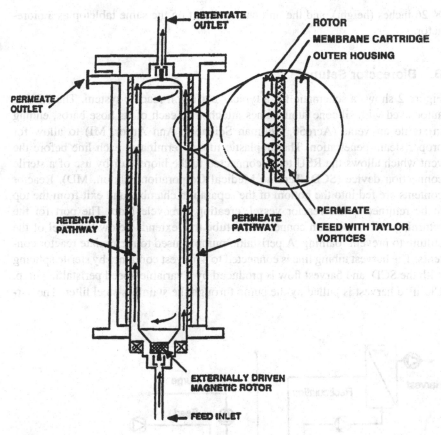

Figure 1 Diagram of a vortex flow filter. Bioreactor content (feed inlet) is pumped through an annular region, with the recycle flow (rententate outlet) exiting from the top of the assembly. The center of the annulus is a rotor that can turn 100 to 2000 rpm. The outer wall of the annulus is a cylindrical filter with an average pore size of 10 μm. The permeate through the filter is removed through an outlet port (permeate outlet), where it can be collected as harvest. Flow rates are controlled by pumps on the feed inlet and permeate outlet lines. The spin rate is set by an external control unit that is coupled to a magnetic drive.

$\times$ 26 inches (height), and the unit can easily fit on the same tabletop as a bioreactor.

B. Bioreactor Setup

Figure 2 shows a schematic for a typical perfusion reactor system. The RSU is
autoclaved with silicone tubing lines attached to each of its hose barbs, ending
in sterile air vents (Acro50, Gellman Sciences, Ann Arbor, MI) to allow for
proper steam penetration. Thermoplastic tubing terminates each line before the
vent which allows the RSU to be connected to the bioreactor by use of a sterile
connection device (SCD; Terumo Medical Corporation, Elkton, MD). Reactor
contents are fed into the bottom of the separation chamber and exit from the top
to be returned to the reactor vessel, creating a recycle loop. The port for the
returning cell suspension connects to a tube that extends below the level of the
culture to prevent foaming. A peristaltic pump is used to recirculate reactor contents. The harvest tubing line is connected to a harvest container by sterile splicing
with the SCD, and harvest flow is produced by a variable-speed peristaltic pump.
Clarified harvest is pulled by the pump through the stainless-steel filter. The har-

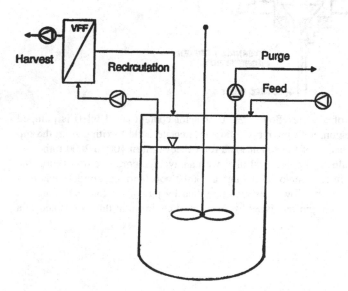

Figure 2   Schematic of a perfusion bioreactor with VFF system. Outlet flows from the
bioreactor consist of harvest flow taken from a recirculation loop with the vortex flow
filter and a purge taken directly from the reactor. Level control regulates the fresh medium
(feed) that enters the bioreactor to keep the reactor volume constant.

vest rate is set to create the desired perfusion rate for the reactor system. When the culture level drops below a set value, a level probe in the bioreactor activates another peristaltic pump to provide fresh medium so that culture volume is maintained near a constant value. A purge pump can be used continuously or intermittently to provide a sink for the built-up cells and debris in the bioreactor and to allow for a steady state.

Transmembrane pressures are monitored by connecting a pressure transducer into the harvest line between the separation chamber and the harvest pump. Fouling of the steel filter will result in an increased pressure drop across the membrane, which is detected by the transducer and is a signal that the RSU needs to be changed. Use of the SCD allows for replacement of a unit in a matter of minutes.

C. Principle of Vortex Flow Filtration

Microfiltration of biological fluids is severely limited by membrane fouling. Concentration polarization, which contributes to fouling, results when cells, cellular debris, or high-molecular-weight soluble molecules (proteins, DNA) are concentrated at the surface of the membrane when they are rejected from transmission. Tangential flow filtration systems were developed to reduce the concentration polarization and hence membrane fouling. In these systems, the feed stream sweeps across the membrane surface, providing a convective flow which limits the concentration boundary layer. However, the flow rate necessary to prevent fouling may result in shear damage to cells traveling within the recirculation loop between filter and bioreactor. Resulting cell lysis may further tax the filtration system, spiraling downward to membrane failure.

The use of Taylor-Couette flow provides an efficient alternative to reduce concentration polarization. The flow profile is primarily generated by rotating the inner surface of an annulus. If a membrane is placed on the stationary outer surface of the annulus, then the stable Taylor vortices generated induce mixing perpendicular to the membrane surface and reduce the extent of polarization. The vortices formed act as whirlpools, carrying particles away from the filter surface which minimizes fouling. The flow rate of the circulating fluid passing through the annulus can be kept low because the required convection comes from the Taylor vortices and not from the feed stream flow.

D. Harvesting Control

There are three independent ways to exert control over reactor harvesting using the VFF system: by changing (1) the speed of the rotor, (2) the harvest rate, and (3) the recirculation rate.

Rotor speed is controlled from the ECU control panel. Typically, rotation speeds of 300 to 1000 rpm have been used. As speed increases, shear rate also increases in the separation chamber according to the relationship [42]:

$$\alpha = \omega R_1 / (R_2 - R_1)$$

where α is the shear rate, ω is the angular velocity, R_2 is the outer radius, and R_1 the inner radius of the annular region. Shear rates of 1000/sec or less have been used with minimal effect on cell viability. For the separation of red blood cells shear rates of up to 20,000/sec have been reported [37]. Therefore, a rotor speed should be selected which provides the adequate vortex action but preserves cell viability.

The harvest rate can be changed by increasing or decreasing the harvest pump speed. Typically, this is increased as cell density increases since more medium will be required with increasing cell mass. Harvest rate is ramped up from zero culture volumes (CV) per day (batch operation) upon inoculation, to several culture volumes per day as the cell density increases. The optimal rate is a strong function of media formulation and cell metabolism and will vary according to these parameters.

The recirculation rate refers to the feed flow rate of the VFF unit. Although the recirculation rate is not as critical for efficient mass transfer through the membrane as in traditional cross-flow filtration systems, the ratio of harvest rate to recirculation rate is of significance. When the feed flow rate equals the harvest flow rate the systems becomes effectively a dead-end filter. For ratios close to 1 a substantial portion of the feed stream exits as harvest, generating a highly concentrated retentate. This will reduce the efficacy of the vortex action cleaning the membrane surface since the mixing will exchange the buildup at the filter with a concentrated bulk rather than a dilute bulk. Too low a ratio can mean that cells are being needlessly subjected to a high shear environment in the recirculation loop. Under the experimental conditions for the studies discussed below, this ratio was less than 0.1 (harvest of 2 CV/day and recirculation of 24 CV/day).

E. Theoretical Approaches to Understanding VFF Behavior

The mechanical behavior of a viscous fluid in a cylindrical annulus with an rotating inner cylinder has been a subject of study since at least 1923 [43]. However, the presence of a porous outer wall (the membrane) significantly increases the complexity. Flow of a Newtonian fluid in the region between a rotating inner cylinder and a stationary outer cylinder consists of three components—an azimuthal flow, the axial flow, and a radial flow component.

The azimuthal component refers to the rotation of fluid around the axis and can be characterized by the Taylor number, Ta:

$$Ta = \frac{\omega \, R_1 \, (R_2 - R_1) \, [2 \, (R_2 - R_1)]^{0.5}}{\nu \qquad\qquad [(R_2 + R_1)]^{0.5}}$$

where ω is the rotational velocity of the inner cylinder, R_1 is the inner radius of the annular region, R_2 is the outer radius of the annular region, and ν is the kinematic viscosity of the fluid. For $41 < Ta < 800$, the system has laminar flow with vortices. For $800 < Ta < 2000$, transitional vortex flow occurs, and for $2000 < Ta < 10,000-15,000$ turbulent flow with vortices occurs.

The flow in the axial direction can be characterized by an axial Reynolds number:

$$Re_a = 2 \, v_z \, (R_2 - R_1)/\nu$$

where again ν is the kinematic viscosity of the fluid, R_1 and R_2 are the inner and outer radii, respectively, and v_z is the velocity in the axial direction. In most cases where $R_1\omega \gg v_z$ the flow field within the annular region is dominated by the rotation of the cylinder. Consequently, the shear force at the wall is mainly dependent on the angular rotation and hence the Taylor number [41]. If the walls of the cylinder are impermeable to the fluid, there can be no net flow in the radial direction and the equations of motion are greatly simplified. Radial flow requires consideration of additional terms in the equations of motion, which makes an analytical solution difficult.

In theory, an understanding of the fluid flow through the VFF coupled with the dynamics of filter fouling would lead to a model that could accurately predict the performance of the system under all conditions. In practice, complete solutions for this flow problem are not yet available. In addition, an understanding of how and why membranes foul is usually achievable only for very well defined solutions and membranes. A fermentation broth presents a wide variety of challenges for a membrane including suspended solids of varying sizes (clumps of cells that may have a radius on the order of 100 μm to lysed cell debris which may be on the order of 1 μm or less); dissolved proteins, nucleic acids, and other macromolecules that have varying hydrophobic/hydrophilic properties; and a large number of dissolved low-molecular-weight compounds that vary in their charges and properties. To date, there is no rigorous solution of the equations of motion and membrane behavior for this system.

Primarily, two approaches have been followed to create some predictive understanding of VFF operation. (These studies often analyze a version of the VFF where the porous membrane was the inner surface. Solutions to the current configuration, however, would follow a similar path.) The first approach has been to approximate solutions for fluid flow through a cylindrical annulus with a rotating and porous inner wall. This will yield a function that can predict the pressure along the surface of the inner annulus at any point. Flux through the filter is modeled by some (usually linear) function of pressure differential, membrane permeability, and available pore area. The permeability and available pore area are time-variant parameters that change with the volume of filtrate processed. Finally, by integrating over the surface of the filter, a time-dependent function is obtained that can predict flow rates and performance. This approach has been utilized by Belfort et al. in an elegant series of papers to characterize a VFF [40,41,44]. Using simplifying assumptions to solve the Navier-Stokes equations, good correlations are demonstrated for theoretical predictions and experimental observations for the filtration by VFF of some well-defined solutions, including cell culture media [41] and particle suspensions of known size [40].

A second approach is to derive an understanding of how mass transfer (characterized by the Sherwood number) is affected by altering hydrodynamic and solution properties, basing the analysis on standard models of membrane fouling including the concentration polarization model, a pore size reduction or blockage model, or both. Empirical experimental data are collected and determination of the most relevant parameters ensues. Holeschovsky and Cooney [35] used this approach coupled with dimensional analysis to generate a power law relationship to predict the Sherwood number under a variety of fluid flow and solute conditions. Balakrishnan and Agarwal took a similar path to describe ultrafiltration of protein solutions using a VFF [38,39].

III. CHARACTERIZATION

To characterize the operation of the VFF cell retention device numerous fermentations were conducted with a variety of operating conditions and cell lines. The characterization encompassed mechanical properties such as sterility, exchangeability, and durability; and dynamic properties including separation efficiency and influence of the system on the biological properties of the cell culture.

For the studies described below, a harvest rate of 2 CV/day was used when the cell densities reached approximately 10^7 cells/mL. Recirculation rate was controlled via the recirculation pump. A sufficiently high rate was chosen to ensure that cells were not exposed to long residence times in the recirculation loop where oxygen deprivation may occur. A rate of 24 CV/day was typical for recirculation of reactor contents.

The evaluation of the mechanical properties of the VFF system revealed no major problems. In only one case over several years of operation did a hairline fracture in the membrane occur, causing minor leakage of cells into the permeate stream. This was easily detected by a change in the transmembrane pressure (see above) and could be corrected by aseptically exchanging the filter unit without interfering with the fermentation. Since all connections are via aseptically welded tubing and the rotor is magnetically coupled, exchanges of the filter unit can be performed any time necessary. Also, the void volume in the filter unit (about 100 mL) is small compared to the bioreactor volume so an exchange causes only minor losses with respect to cell mass. Sterility was maintained throughout all fermentations with typical run times of 30 to 60 days, and several fermentations lasted more than 3 months. None of the fermentations had to be aborted because of a contamination traceable to the cell retention device. Several filter units were rotated between fermentations or kept as backups. Each of them was autoclaved numerous times, and no material failure from fatigue due to repetitive thermal or mechanical stresses has been detected so far.

Separation efficiency of the VFF is mainly a function of the complex hydrodynamic field in the annulus. Therefore, the ratio between membrane pore size and cell diameter is not always critical to the retention performance for a wide range of pore sizes. Even microbial cultures (e.g., *Escherichia coli* and *Saccharomyces cerevisiae*) with much smaller cell diameters than mammalian cell cultures have been successfully maintained in perfusion mode using VFF with appropriate membranes [32]. All cultures discussed here were conducted with only one membrane pore size (10 μm), although a variety of cell lines were used including murine myelomas, Chinese hamster ovary (CHO) and human embryonic kidney (HEK) 293 cells, all with diameters > 10 μm. For all the different cell lines tested the separation efficiency of the VFF was very high ($> 98\%$, and usually $> 99\%$), as measured by examination of the permeate for cells.

Membrane fouling can be detected by changes in the transmembrane pressure (TMP). Buildup of a cell and cell debris layer on top of the membrane or in the pores causes a gradual clogging of the membrane and leads to increasing TMP. Continuous monitoring of TMP is therefore not only useful in detecting fractures in the membrane with an associated breakthrough of cells into the permeate but also an impending loss of permeate flow.

During our cultures few instances of terminal membrane fouling were observed. All of them could be traced to anomalous operating conditions; e.g., after a power outage the rotor did not restart but culture fluid was pumped through the VFF causing a blockage of the membrane. Also, failure to supply fresh medium to the bioreactor while permeate was continuously collected lead to a highly concentrated cell slurry in the recycle loop that caused membrane fouling. In several cases the culture could be rescued by correcting the original problem and exchanging the filter unit.

A cell retention device should not impair the biological properties of the system with which it is used. This includes cell morphology and viability as well as productivity. Visual inspection of cell morphology during all runs indicated no damage to the cells. Measurement of cell viability with Trypan Blue exclusion staining also showed viabilities generally above 90% as can be seen in the graphs for several cultures (Figures 3, 4). Also, no substantial cell lysis was observed as indicated by DNA measurement in the culture fluid (data not shown), thus eliminating the potential loss of cells due to mechanical destruction in the VFF.

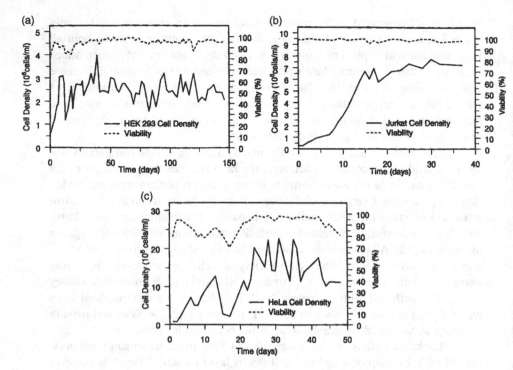

Figure 3   Time profiles of perfusion cultures with a VFF cell retention device. (a) Long-term culture of HEK 293 cells with periodic harvesting. (b) Production of cell mass with Jurkat cells. (c) Culture of HeLa cells with periodic harvesting.

The minimal perturbation of the system on cell viability can be explained by the low shear environment and short mean residence times within the filter unit. The short mean residence times are a result of the small void volume of the filter unit compared to the recirculation volumes. Table 1 shows mean residence times for a wide range of operating conditions, none exceeding 10 min; thus, oxygen limitation did not occur.

The flexibility of the system was first demonstrated by cultures of HEK 293, Jurkat, and HeLa cells where production of cell mass was the objective. Harvested cells were used for a variety of applications, e.g., subsequent infection with virus for generation of gene therapy products (HEK 293; Figure 3a), isolation of cell components for research purposes (Jurkat; Figure 3b), or infection with vaccinia virus for protein expression (HeLa; Figure 3c). For all these applications high cell densities and high viabilites were important. Cells were often removed periodically from the bioreactor and the cell density was allowed to recover for the next harvest. This led to the seesaw characteristic in the cell

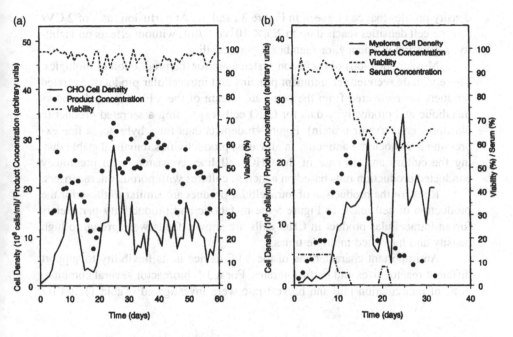

Figure 4 Time profiles of perfusion cultures generating a secreted recombinant biologic. (a) CHO cell culture expressing a tissue plasminogen activator protein. (b) SP2 myeloma cells expressing a monoclonal antibody.

Table 1 Operational Parameters Tested with a 3-L Working Volume Bioreactor

Recirculation flow rate (mL/min)	Recirculation rate (CV/d)	Harvest rate (CV/d)	Mean residence time (s)
16.67	8	2	360
25	12	2	240
25	12	4	240
25	12	6	240
33.33	16	2	180
50	24	2	120
200	96	2	30
300	144	2	20
400	192	2	15
500	240	2	12

Healthy and productive cell cultures were maintained for all operating conditions. The rotor speed for all conditions was 500 rpm.

density profiles that can be seen in Figure 3a and 3c. At perfusion rates of 2 CV/day the cell densities reached up to 2.5×10^7 cells/mL without effects on viability, separation efficiency, or membrane permeability.

Most applications of perfusion systems are for the production of biologics. These include secreted recombinant proteins and intracellular products. Secreted products are recovered from the permeate stream of the VFF. Figure 4a shows metabolic and productivity data for CHO cells expressing a secreted product (a plasminogen activator protein). Figure 4b depicts data for a hybridoma line expressing a monoclonal antibody. In all cases, productivities remained stable during the culture and, in case of the CHO cell line, consistent with previously conducted production runs based on adherent cultures with porous microcarriers.

Runs for the production of intracellular product are similar to those for the production of cell mass. In Figure 5 the metabolic and productivity parameters for an intracellular product in CHO cells are shown. Cells were grown to high density and harvested multiple times.

An important characteristic of the VFF device is its flexibility to support different reactor sizes and perfusion rates. For a 3-L bioreactor several combinations of recirculation rate and harvest rate were investigated (Table 1) and for

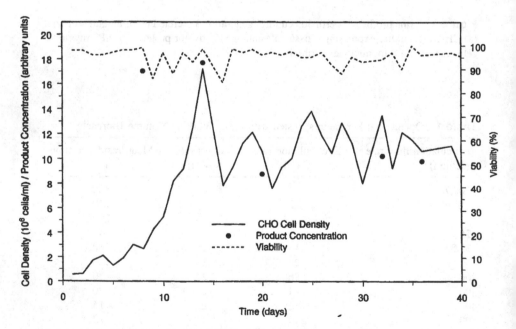

Figure 5   Time profile of a CHO cell perfusion culture generating an intracellular product with periodic harvesting of cell mass.

the presented range no negative influences on the cell culture were observed. A commonly used recirculation rate is 24 CV/d, equivalent to a recirculation flow rate of 50 mL/min for a 3-L working volume bioreactor. Increasing the bioreactor volume to 10 L at the same recirculation rate yields a recirculation flow rate of 167 mL/min, well within the tested range of the same VFF unit.

At the highest tested recirculation flow rate of 500 mL/min a 30-L bioreactor could be supported at 24 CV/d without any modification to the materials used in the recycle loop. If the recirculation rate is reduced to 12 CV/d, even larger bioreactors could be supported by the unit. This demonstrates the wide range of operating conditions supported by only one device. For bioreactors with volumes > 50 L a larger VFF device is commercially available with a filter area of 0.25 m^2 and a typical recirculation flow rate of 3 to 4 L/min.

An unexpected benefit found with the operation at high recirculation rates was the ability to influence the size of cell aggregates in CHO cell cultures by mechanical means. During culture, cells often form aggregates. When the cell aggregates reach a size where cells within cannot efficiently exchange nutrients and metabolic byproducts with the surrounding culture medium, the viability and homogeneity of the cell population suffer and efficiency is reduced. Figure 6

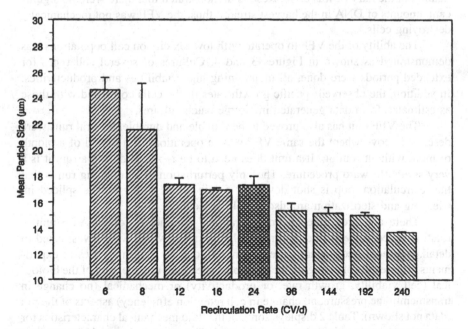

Figure 6 Distribution of particle sizes in a 3-L bioreactor with a CHO cell culture in relationship to the recirculation rate.

illustrates that the mean particle size for a culture broth is reduced as the recirculation rate is increased. Thus, a more homogeneous cell size profile is created. As discussed above, no decrease in productivity, viability, or growth rate was observed at high recirculation rates for any of the cells tested.

IV. DISCUSSION

The vortex flow filtration unit has proven to be an effective and reliable cell retention device in our laboratories. It manifests many of the properties listed earlier that would go into an ideal device: efficient retention, low toxicity, minimal fouling, reusability, sterilizability, low maintenance requirements, and ease of replacement.

As described above, the efficiency of cell retention was shown by two methods, the first being direct examination of the permeate stream coming from the VFF and counting observable cells. The data generated from these experiments suggested that the retention efficiency of the VFF was $\gg$ 99% (data not shown). To address the concern that cell loss may not be visible in the harvest stream because the cells might have been disintegrated by the VFF device, DNA quantification in the harvest was done. Results demonstrated that there were no significant amounts of DNA in the harvest stream; thus, the VFF was not mechanically destroying cells.

The ability of the VFF to operate with low toxicity on cell populations was demonstrated as shown in Figures 3 and 4. Cultures of several cell types for extended periods were done, all maintaining high viabilities and productivities. In addition, the observed specific growth rates of the cells correlated with those as estimated from data generated in simple batch cultures.

The VFF unit has also proved to be reliable and durable. Several runs were described above where the same VFF was in operation for a period of a month or more without fouling. If a unit does need to be replaced during a run, it is a very straightforward procedure. The only perturbation to an ongoing run is that the recirculation loop is shut down for 5 min while a new unit is spliced in. Cleaning and sterilization are also simple procedures.

There are three independent operational parameters for the VFF unit—rotation speed, harvest rate, and recirculation rate—which were investigated in detail. Rotation speeds were varied from 300 to 1000 rpm in a series of experiments. The results of changing the rotational speed did not alter any of the biological (cell viability, growth rate, or productivity) or mechanical (no change in transmembrane pressure, no change in cell retention efficiency) aspects of the run (data not shown). Table 2 displays some of the fluid mechanical characteristics for the system as a function of rotational speed. The upper value of 1000 rpm is a reasonable limit since it yields a shear rate of $\sim$1000/sec, a value which has been shown to produce little damage to cells for short exposure times in relatively

Table 2 Characteristic Values for the Flow Field
Within the Cylindrical Annulus of the VFF

Rotational speed (rpm)	Shear rate (sec⁻¹)	Taylor number
300	316	498
500	527	830
1000	1055	1661

inviscid liquids [42]. The Taylor number calculations demonstrate that the flow field at these values can be described as laminar or transitional flows with Taylor vortices present. Fully developed turbulent flow is not reached. A standard setting of 500 rpm has been used because of the relative insensitivity of the system to the rotational speed over this range. This setting provides well-developed Taylor vortex flow fields with modest shear.

The other two operational parameters that can be independently varied for VFF operation are the rates of recirculation for the cell recycle loop and the rate of harvest removal from the filtration unit. As shown in Table 1, these parameters were investigated independently in bioreactor studies. It was determined that there was a wide window of operating conditions under which the VFF unit performed well. Recirculation rates were varied from 8 to 240 CV/day. Cells did not show any toxic effects with mean residence times up to 6 min. Residence times above this value, however, did lead to loss of cell viability and eventual fouling of the membranes. Thus, mean residence times for cell culture solutions in the VFF unit should not exceed 6 min. This limit does not pose any special problem since the recirculation rate would have to be exceedingly low, or the recycle rate/harvest rate ratio extremely small, before such residence times would be reached in any practical setting. At the other end of the spectrum, the highest recirculation rates that were investigated (240 CV/day in a 3-L reactor) translate to a volumetric flow of 8.33 mL/sec from the bioreactor to the VFF. This would result in an axial velocity of only 3 cm/sec through the VFF unit with a corresponding axial Reynolds number and shear rate (due to axial flow) of order 100. These values are safely at the low end of tolerable flow conditions and could, in theory, be increased by at least a factor of 10 before concern would be warranted regarding the volumetric flux passing through the unit.

Harvest rates ranging from 2 to 6 CV/day were studied in efforts to characterize the performance of the VFF. There are two important concepts that may be very relevant for predicting the scale up of the system from these data. The first is the idea of the recirculation rate/harvest rate ratio. This value is the best indicator of the concentration potential of the VFF unit. For instance, if the R/H ratio is 3, then for every fluid element entering the VFF from the bioreactor,

2/3 of the volume is recycled back to the culture vessel and 1/3 is removed as harvest. If the filter device is 100% efficient in cell retention, then the concentration of cells and other particles that cannot transverse the filter should increase by 50% (3/2 of the entering concentration). Of course, the performance of the filter unit will also depend on the particular cell type, cell concentration, and media. Under the conditions listed in Table 1, an R/H ratio as low as 2 (12/6) was achieved without compromising VFF performance or the culture conditions of the bioreactor. Applying this ratio with some of the higher recirculation rates that have been attempted leads to speculation that the VFF unit could handle significantly greater volumes. This performance is speculative, however, because there is likely to be a limit to the rate of harvest that can transit the filter without causing fouling. This would be the result from the contributions of the radial convective flux in the filter chamber, and be controlled by boundary layer effects near the stationary outer membrane. At some point, the mixing and scrubbing actions of the Taylor vortices could no longer keep up with the radial convection carrying particles back toward the filter. Although increasing the recirculation rate to maintain an R/H ratio that performed well at lower harvest levels would assure that concentrations do not become too high in the bulk for the filter to handle, it would not negate the effects of the higher radial flux. Also, higher recirculation rates would only marginally improve the performance of the unit in terms of mixing and scrubbing since, as has already been discussed, the axial flow is not significant compared to the azimuthal flow and the generation of Taylor vortices to keep the filter clean. For these reasons, additional empirical studies would be required to truly determine what is the upper limit for the harvest capacity of a VFF for a given fermentation.

This chapter has described the theory, design, and operation of a cell retention device based on a vortex flow filter. Studies and discussion have explored the effective range for VFF parameters and some of the biological and mechanical effects of its operation. This device was evaluated for use in the operation of bioreactors in a perfusion mode. For all cell lines tested, the system showed high separation efficiency and no substantial cell damage while maintaining sterility and high cell densities. VFF design has proven efficient and durable for small- and medium-scale bioreactor operations. It offers many advantages in terms of convenience and performance over competing technologies, and should be considered for perfusion fermentation processes of the appropriate scale.

REFERENCES

1. Hu WS, Piret JM. Mammalian cell culture processes. Cur Opin Biotechnol 3:110–114, 1992.

2. Xie L, Wang DIC. Fed-batch cultivation of animal cells using different medium design concepts and feeding strategies. Biotechnol Bioeng 43:1175–1189, 1994.
3. de la Broise D, Noiseux M, Massie B, Lemieux R. Hybridoma perfusion systems: a comparison study. Biotechnol Bioeng 40:25–32, 1992.
4. Himmelfarb P, Thayer PS, Martin HE. Spin filter culture: the propagation of mammalian cells in suspension. Science 2:555–557, 1969.
5. Yabannavar VM, Singh V, Connely NV. Mammalian cell retention in a spinfilter perfusion bioreactor. Biotechnol Bioeng 40:925–933, 1992.
6. Yabannavar VM, Singh V, Connely NV. Scaleup of spinfilter perfusion bioreactor for mammalian cell retention. Biotechnol Bioeng 43:159–164, 1994.
7. Deo YM, Mahadevan MD, Fuchs R. Practical considerations in operation and scaleup of spin-filter based bioreactors for monoclonal antibody production. Biotechnol Prog 12:57–64, 1996.
8. Avgerinos GC, Drapeau D, Socolow JS, Mao J, Hsiao K, Broeze RJ. Spin filter perfusion system for high density cell culture: production of recombinant urinary type plasminogen activator in CHO cells. Bio/Technology 8:54–58, 1990.
9. Esclade LRJ, Carrel S, Peringer P. Influence of the screen material on the fouling of spin filters. Biotechnol Bioeng 38:159–168, 1991.
10. Tolbert WR, Feder J, Kimes RC. Large-scale rotating filter perfusion system for high-density growth of mammalian suspension cultures. In Vitro 17:885–890, 1981.
11. Woodside SM, Bowen BD, Piret JM. Mammalian cell retention devices for stirred perfusion bioreactors. Cytotechnology 28:163–175, 1998.
12. Hansen HA, Damgaard B, Emborg C. Enhanced antibody production associated with altered amino acid metabolism in a hybridoma high-density perfusion culture established by gravity separation. Cytotechnology 11:155–166, 1993.
13. Searles JA, Todd P, Kompala DS. Viable cell recycle with an inclined settler in the perfusion culture of suspended recombinant Chinese hamster ovary cells. Biotechnol Prog 10:198–206, 1994.
14. Perrin P, Madhusudana S, Gontier-Jallet C, Petres S, Tordo N, Merten OW. An experimental rabies vaccine produced with a new BHK-21 suspension cell culture process: use of serum-free medium and perfusion-reactor system. Vaccine 13:1244–1250, 1995.
15. Batt BC, Davis RH, Kompala DS. Inclined sedimentation for selective retention of viable hybridomas in continuous suspension bioreactor. Biotechnol Prog 6:458–464, 1990.
16. Takamatsu H, Hamamoto K, Ishimaru K, Yokoyama S, Tokashiki M. Large-scale perfusion culture process for suspended mammalian cells that uses a centrifuge with multiple settling zones. Appl Microbiol Biotechnol 45:454–457, 1996.
17. Tokashiki M, Arai T, Hamamoto K, Ishimaru K. High density culture of hybridoma cells using a perfusion culture vessel with an external centrifuge. Cytotechnology 3:239–244, 1990.
18. Engler R, Kemp CW. Serum free production of monoclonal antibody in a perfusion system. Norfolk, VA: Waterside Monoclonal Antibody Conference, March 1997.
19. Johnson M, Lanthier S, Massie B, Lefebvre G, Kamen AA. Use of the Centritech Lab centrifuge for perfusion culture of hybridoma cells in protein free medium. Biotechnol Prog 12:855–864, 1996.

20. Castillo FJ, Mullen LJ, Thrift JC, Grant BC. Perfusion cultures of hybridoma cells for monoclonal antibody production. Ann NY Acad of Sci 665:72–80, 1992.

21. Prior C, Bay B, Ebert B, Gore R, Holt, J, Irish T, Jensen F, Leone C, Mitschelen J, Stiglitz M, Tarr C, Trauger RJ, Weber D, Hrinda M. Process development for the manufacture of inactivated HIV-1. BioPharm 8:25–35, 1995.

22. Griffith B. Perfusion systems for cell cultivation. In: Lubiniecki AS, ed. Large-Scale Mammalian Cell Culture Technology. New York: Marcel Dekker, 1990:217–250.

23. Salmon PM, Robertson CR. Membrane reactors. In: Asenjo JA, Merchuk, JC, eds. Bioreactor System Design. New York: Marcel Dekker, 1995:305–338.

24. Cacciuttolo MA, Patchan M, Lamey K, Allikmets E, Tsao E. Large-scale production of a monoclonal IgM in a hybridoma suspension culture. BioPharm 11:20–27, 1998.

25. Merten OW. Concentrating mammalian cells. I. Large-scale animal cell culture. TIB-TECH 5:230–237, 1987.

26. Kiy T, Scheidgen-Kleyboldt G, Tiedtke A. Production of lysosomal enzymes by continous high-cell-density fermentation of the ciliated protozoon *Tetrahymena thermophila* in a perfused bioreactor. Enzyme Microbial Technol 18:268–274, 1996.

27. Brennan AJ, Shevitz J, Macmillan JD. A perfusion system for antibody production by shear-sensitive hybridoma cells in a stirred reactor. Biotechnol Techn 1:169–174, 1987.

28. Zhang S, Handa-Corrigan A, Spier RE. A comparison of oxygenation methods for high-density perfusion cultures of animal cells. Biotechnol Bioeng 41:685–692, 1993.

29. Trampler F, Sonderhoff SA, Pui PWS, Kilburn DG, Piret JM. Acoustic cell filter for high density perfusion culture of hybridoma cells. Bio/Technology 12:281–284, 1994.

30. Pui P, Trampler F, Sonderhoff SA, Groeschl M, Kilburn DG, Piret JM. Batch and semicontinous aggregation and sedimentation of hybridoma cells by acoustic resonance fields. Biotechnol Prog 11:146–152, 1995.

31. Doclosis A, Kalogerakis N, Behie LA, Kaler KVIS. A novel dielectrophoresis-based device for the selective retention of viable cells in cell culture media. Biotechnol & Bioeng 54:239–250, 1997.

32. Roth G, Smith CE, Schoofs GM, Montgomery TJ, Ayala JL, Horwitz JI. Using an external vortex flow filtration device for perfusion cell culture. BioPharm 10:30–35, 1997.

33. Kulozik U. Physiological aspects of continous lactic acid fermentations at high dilution rates. Appl Microbiol Biotechnol 49:506–510, 1998.

34. Kroner KH, Nissinen V. Dynamic filtration of microbial suspensions using an axially rotating filter. J Membr Sci 36:85–100, 1988.

35. Holeschovsky UB, Cooney CL. Quantitative description of ultrafiltration in a rotating filtration device. AIChE J 37:1219–1226, 1991.

36. van Reis R, Leonard LC, Hsu CC, Builder, SE. Industrial scale harvest of proteins from mammalian cell culture by tangential flow filtration. Biotechnol Bioeng 38:413–422, 1991.

37. Zeman LJ, Zydney AL. Microfiltration and Ultrafiltration. New York: Marcel Dekker, 1996.

38. Balakrishnan M, Agarwal GP. Protein fractination in a vortex flow filter. I. Effect

of system hydrodynamics and solution environment on single protein transmission. J Membr Sci 112:47–74, 1996.

39. Balakrishnan M, Agarwal GP. Protein fractionation in a vortex flow filter. II. Separation of simulated mixtures. J Membr Sci 112:75–84, 1996.
40. Belfort G, Pimbley JM, Greiner A, Chung KY. Diagnosis of membrane fouling using a rotating annular filter. 1. Cell culture media. J Membr Sci 77:1–22, 1993.
41. Belfort G, Mikulasek P, Pimbley JM, Chung KY. Diagnosis of membrane fouling using a rotating annular filter. 2. Dilute particle suspensions of known particle size. J Membr Sci 77:23–39, 1993.
42. Schuerch U, Kramer H, Einsele A, Widmer F, Eppenberger HM. Experimental evaluation of laminar shear stress on the behavior of hybridoma mass cell cultures, producing monoclonal antibodies against mitochondiral creatine kinase. J Biotechnol 7:179–184, 1988.
43. Taylor GI. Stability of a viscous liquid contained between two rotating cylinders. Phil Trans R Soc A 223:289, 1923.
44. Dolecek P, Mikulasek P, Beldfort G. The performance of a rotating filter. 1. Theoretical analysis of the flow in an annulus with a rotating inner porous wall. J Membr Sci 99:241, 1995.

3

Crossflow Membrane Filtration of Fermentation Broth

Gregory Russotti and Kent E. Göklen
Merck Research Laboratories, Rahway, New Jersey

I. INTRODUCTION

Crossflow membrane filtration became a popular topic in biotechnology in the late 1960s and early 1970s as methods of fabricating microporous and ultrafiltration membranes were commercialized and new methods for harvesting fermentation broth were sought. Early references in this field from Michaels [40], Blatt et al. [4], Porter and Michaels [51], Henry and Allred [25], and many others sought to apply and characterize this filtration technique, and to develop predictive models of its performance.

By the time Wolf Hanisch wrote the chapter on cell harvesting [24] in the first edition of *Membrane Separations in Biotechnology* [37], several manufacturers were offering a full line of membrane products to support laboratory- and production-scale processing of fermentation broths and other biologically derived products, and these were finding applications in the laboratories of pharmaceutical and biotechnology firms, among other industries. The high level of activity on crossflow filtration-based processes was reflected in the large body of literature which had accumulated in the preceding decade. These efforts were driven by a number of factors. Perhaps most important among them was the need to develop processes appropriate for the recovery of new high-value biologically derived products, the syntheses of which were made possible by an explosive advance in the biological sciences. Also important was the desire to eliminate the personnel exposure to and large-scale handling of filter aid typically associated with vacuum drum filtration of fermentation broth, which was a common harvesting method at that time. Gravatt and Molnar [22] provided one of the first accounts of the development of an industrial-scale crossflow filtration application in that first edition of *Membrane Separations in Biotechnology* [37].

In this chapter we have attempted to provide a broad review of the topic, which we hope will be useful to those in the field and especially valuable to those just entering it. For the most part, changes in the underlying technology have been evolutionary and not revolutionary, such that all that was written by Hanisch in 1986 still applies. In this second edition, we have attempted to provide greater detail on the topic, based on the more extensive literature currently available, along with some insights based on our own experience. After providing a review of the basic concepts of crossflow filtration (Section II), and reviewing typical materials and membrane configurations (Section III), we have attempted to provide a breakdown of the types of processes which are frequently encountered (Section IV) and to provide examples of each type of process (Sections V through VIII). While mentioned in passing, we have left it to other authors of this volume to provide details on new and novel membrane applications which are just now appearing in the literature. Sections IX through XIII provide a practical guide to development, optimization, and trouble-shooting of crossflow processes, with a brief discussion of membrane regeneration in Section XIV. We end in Section XV with a discussion on the scale-up of crossflow processes, including some general strategies and very practical guidelines for system design.

The authors note that while the literature on crossflow filtration is voluminous, examples of scale-up and larger-scale applications remain limited. We are especially thankful to contributors to the literature on this topic and their organizations for allowing the release of such information, which can provide critical guidance to workers in this field. We also wish to thank our own management for having allowed us to release several examples of our efforts to the literature. We are hopeful this will encourage others to release discussions of their efforts to the public domain, and thus accelerate progress in this area.

II. DEFINITIONS AND PRINCIPLES OF CROSSFLOW FILTRATION

A. General Concepts

Crossflow filtration is the most general term to describe the various techniques by which slurries of particles or solutions of macromolecules flow across the surface of a filtration medium with only a fraction of the liquid volume permeating the membrane per pass (Figure 1). Through various mechanisms, depending on the size of the molecule or particle, crossflow reduces the accumulation of materials on the membrane surface, allowing filtration to continue beyond the point which would be possible using traditional dead-end filtration. In this section, the authors attempt to provide a brief review of the major concepts employed

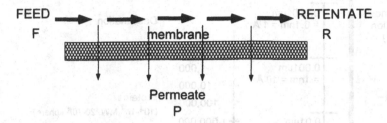

$$F = R + P \qquad F \approx R \gg P$$

Figure 1   Crossflow filtration (CFF)—general schematic.

in developing and analyzing crossflow filtration processes as applied to cell harvest applications, but recognize that a comprehensive review is beyond the scope of this work. This topic is the subject of many excellent texts and review articles, and some of these are provided in the references section at the end of this chapter for those seeking more detail. In this work, flux will always refer to the permeate flowrate per unit area of membrane.

Crossflow filtration (CFF), also known as tangential flow filtration (TFF), can be implemented for a wide variety of purposes by varying the type of membrane employed. Membranes most commonly fall into one of four categories— reverse osmosis (RO), nanofiltration (NF), ultrafiltration (UF), and microfiltration (MF), as illustrated in Figure 2. RO and NF membranes have pores which are so small that only inorganic salts and relatively small organic molecules can pass through along with the solvent; these systems typically operate at elevated pressures, in the 150 to 1000 psi range. UF and MF membranes are more commonly employed in biotechnology applications, and both are used for cell separation and cell harvest as well as for many other purposes. In this section we shall be concerned solely with the concepts on which such processes are based, while the actual materials and configurations employed are presented in the next section. Many details of the operation will depend on the nature of the product and the characteristic of the broth, but in this section we strive to generalize.

While details of an application may vary significantly, there are a few common flow schemes which represent the majority of processes, including most cell harvest applications. The most common representation is of a simple batch concentration system (Figure 3), in which a feed is recirculated between a feed tank and the membrane module, with permeate collected in another vessel. Two common variations of the batch concentration are shown in Figures 4 and 5, the fed-batch concentration and the feed-and-bleed loop system; a combination of these two is also possible. The simple batch system is generally the most efficient,

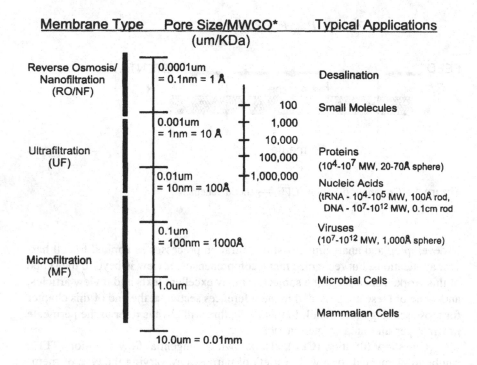

Membrane Type	Pore Size/MWCO* (um/KDa)	Typical Applications
Reverse Osmosis/ Nanofiltration (RO/NF)	0.0001um = 0.1nm = 1 Å	Desalination
	100	Small Molecules
	0.001um = 1nm = 10 Å 1,000 10,000	
Ultrafiltration (UF)	100,000	Proteins (10^4-10^7 MW, 20-70Å sphere)
	0.01um = 10nm = 100Å 1,000,000	Nucleic Acids (tRNA - 10^4-10^5 MW, 100Å rod, DNA - 10^7-10^{12} MW, 0.1cm rod
	0.1um = 100nm = 1000Å	Viruses (10^7-10^{12} MW, 1,000Å sphere)
Microfiltration (MF)	1.0um	Microbial Cells
		Mammalian Cells
	10.0um = 0.01mm	

* correspondence of pore sizes approximated for globular proteins

Typical sizes for different classes of materials from:
Cantor CR, Schimmel PR. Biophysical Chemistry,
Part I, The Conformation of Biological Macromolecules,
p. 4. San Francisco: W.H. Freeman and Company, 1980.

Figure 2 Characteristics of membranes used for CFF.

because, on average, the membrane is exposed to the lowest possible material concentration to achieve a given final concentration, usually resulting in the highest average flux. The fed-batch is common because it provides several practical advantages. A single CFF skid with a small feed tank can serve many different fermenters. After initially filling the feed tank, further transfers from the feed reservoir are made either batchwise or at a relatively slow rate, e.g., to match the permeate rate, so the fermenter outlet design is of little concern. Also, the hold-up volume of the system can be minimized, allowing concentration to proceed to the highest possible level. These benefits must be balanced against the fact that in fed-batch mode, the membrane is exposed to a higher material concentration throughout the process, which will generally result in a lower average

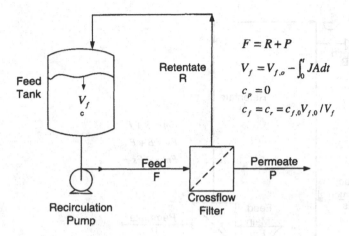

$$F = R + P$$

$$V_f = V_{f,o} - \int_0^t JAdt$$

$$c_p = 0$$

$$c_f = c_r = c_{f,0}V_{f,0}/V_f$$

Figure 3 Batch concentration mode.

flux. The feed-and-bleed mode of operation is employed primarily for very large-scale applications, because it limits the portion of the system which must be designed to handle the high recirculating flow, especially with respect to the sometimes problematic flow out of and back into the feed vessel. Successful operation of the feed-and-bleed loop requires the flows to be carefully balanced to ensure that the fluid in the inner recirculation loop does not become over

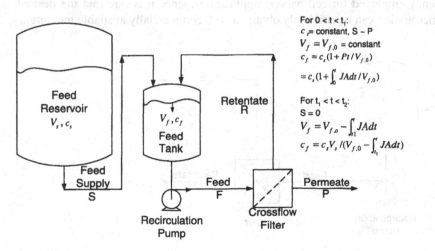

For $0 < t < t_1$:
c = constant, S ~ P
$V_f = V_{f,0}$ = constant
$c_f = c_s(1 + Pt/V_{f,0})$

$\quad = c_s(1 + \int_0^t JAdt/V_{f,0})$

For $t_1 < t < t_2$:
$S = 0$
$V_f = V_{f,o} - \int_{t_1}^t JAdt$
$c_f = c_s V_s /(V_{f,0} - \int_{t_1}^t JAdt)$

Figure 4 Fed-batch concentration mode.

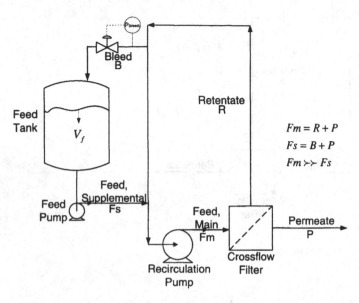

Figure 5   Feed-and-bleed loop configuration.

concentrated and plug the flow channels. If the ratios of the feed, bleed, recirculation, and permeate flows are properly selected, then feed-and-bleed operation approaches the efficiency of a simple batch concentration.

Continuous crossflow filtration, as illustrated in Figure 6, is only infrequently employed for cell harvest applications, since it is rare that the desired concentration can be efficiently obtained with commercially available membrane

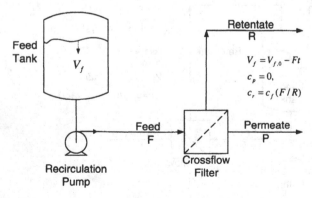

Figure 6   Continuous CFF mode.

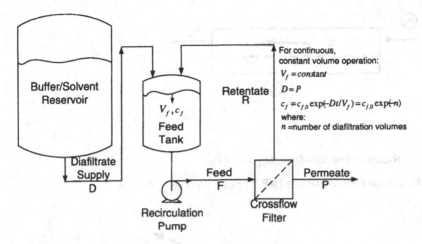

For continuous, constant volume operation:

$V_f = constant$

$D \approx P$

$c_f = c_{f,0} \exp(-Dt/V_f) = c_{f,0} \exp(-n)$

where:

n = number of diafiltration volumes

Figure 7 Diafiltration mode.

configurations. Applications for this type of once-through, or single-pass configuration can be found in other industries, such as water treatment. Special membrane configurations, such as vibratory and rotary crossflow systems, provide an exception to this rule; applications for which the high shear of these devices is not an issue can operate in a true continuous mode. Cheryan [9] provides a detailed discussion of these different operating schemes.

Diafiltration is illustrated in Figure 7; analogous to dialysis, diafiltration allows the concentration of permeable components to be reduced in the retained volume. By adding liquid to the feed tank while continuing to filter, diafiltration can be employed to wash cells of fermentation medium components, to achieve a solvent or buffer switch, or to increase the yield of a soluble product recovered in the permeate. Diafiltration can be performed using either continuous or batchwise addition of the diafiltrate. Continuous addition to the feed tank minimizes the overall diafiltration volume to achieve a given effect, provided the system is not complicated by a concentration-dependent limitation on permeation or flux.

Using these simple concepts, CFF can be used to address several challenges encountered in biotechnology and pharmaceutical production, as illustrated in Sections IV through VIII.

B. Definitions in Crossflow Filtration

Several key parameters must be defined prior to any discussion of crossflow filtration. The cartoons provided in Figure 8 combined with the definitions in

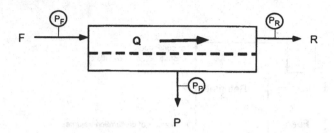

Retentate Pressure Drop, $\Delta P_L = P_F - P_R$

Transmembrane Pressure Drop = TMP = $\Delta P_{TM} = (P_F + P_R)/2 - P_P$

(A)

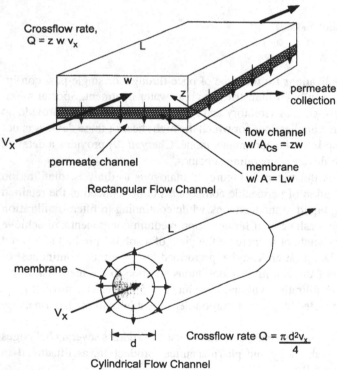

(B)

Figure 8 Definitions in CFF: (a) basic parameters; (b) membrane geometries; (c) characterizing the permeation of soluble components.

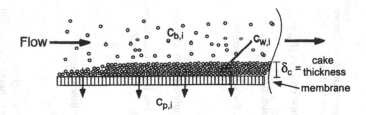

$C_{b,i}$ = concentration of i in bulk fluid

$C_{w,i}$ = concentration of i at membrane wall

$C_{p,i}$ = concentration of i in permeate

Sieving Coefficient, $S_i = \dfrac{C_{p,i}}{C_{w,i}}$

Permeation Coefficient, $p_i = \dfrac{C_{p,i}}{C_{b,i}}$

for $0 \leq C_{p,i} \leq C_{b,i}$

$$0 \leq p_i \leq 1$$

Retention Coefficient, $r_i = 1 - p_i$

or $r_i = 1 - \dfrac{C_{p,i}}{C_{b,i}}$

for $0 \leq C_{p,i} \leq C_{b,i}$

$$1 \leq r_i \leq 0$$

(C)

Figure 8 (continued)

Table 1 provide the most common terms encountered in the literature, although specific symbology may vary among research groups. Special terms which appear in specific mass transfer models are not included.

C. CFF: Mechanisms and Modeling

While abundant models of CFF processes are available, it should be noted at the outset that few of the models are useful in making predictions of filtration behavior of a particular fermentation broth and membrane product *a priori*. Most of these models provide a basis for understanding the underlying process, for organizing experiments and interpreting data, for choosing directions of further research and development. The models account for aspects of the phenomenon which can be described using mass transfer analysis, those elements which contribute to reversible decreases in flux. Most of these analyses do not account for irreversible phenomena, which contribute to what is generally termed fouling (discussed later in this section and in Section IX). Subtleties of these processes, such as the propensity of cells to aggregate under certain growth and/or processing conditions, or the effects of accumulating both cells and proteins at the

Table 1 Common Definitions in Crossflow Filtration for Broth Harvest Applications

Parameter	Symbols	Units	Definition
Flux	J	L/m²-hr or LMH, gal/ft²-hr	Bulk fluid flow rate through the membrane normalized to membrane area. 40 LMH = 0.98 gal/ft²-hr
Permeate "velocity"	v_z	cm/s	Apparent velocity of permeate through the membrane (similar to "superficial" flowrates through packed beds)—equivalent to flux. 1 LMH = 2.78 × 10⁻⁵ cm/s
Pressure (feed, retentate, permeate)	P	bar, kPa psi	Pressure measured at the entrance and exit of the flow channel (P_{in}, P_{out}), and on the back side of the membrane (P_{perm})
Transmembrane pressure	ΔP_{tm} or TMP	bar psi	Pressure drop across the membrane: for average value, $\Delta P_{tm} = [(P_{in} + P_{out})/2]-P_{perm}$
Retentate pressure drop	ΔP_L	bar psi	Pressure drop along the length of the flow channel: for simple geometries, $\Delta P_L = P_{in}-P_{out}$
Crossflow rate	Q	L/hr, gal/min	Bulk fluid flowrate in the membrane flow channels
Crossflow velocity	v_x	m/s	Average velocity of bulk fluid flow through the membrane flow channels: $v_x = Q/A_{cs}$
Flow channel dimension	d z, w, H	mm, cm, in.	Inner diameter for cylinders (hollow fibers, tubular and monolithic membranes); height and width for rectangular channels; radius or half height also used in some works.
Channel length	L	cm, m	Length of filter module flow channel
Membrane surface area	A	m², ft²	Surface area for filtration
Cross-sectional area	A_{cs}	mm², cm²	Cross-sectional area for fluid flow in the membrane flow channels: $A_{cs} = \pi d^2/4$ (cylinders), $= zw$ (rectangular)
Membrane loading	Λ	L/m² gal/ft²	Ratio of feed volume to membrane area: $\Lambda = V_f/A$
Time cycle	τ	hr	Time required to perform a specified CFF process for a given Λ
Membrane pore sized	d_p, MWCO	μm, MWCO	Average equivalent diameter for flow through the membrane for microfiltration membranes; ultrafiltration membranes pore sizes characterized by MWCO

Term	Symbol	Units	Definition
Cell or particle size (diameter)	d_c or a	μm	Average cell diameter for single-cell organisms; filamentous organisms typically characterized by diameter of hyphae
Cell or particle density or suspension concentration	c or φ	cells/mL or fractional	Cells or particles in the feed, expressed either in concentration units (particles per unit volume) or as a volume fraction
Maximum cell/particle density	c_{max} or φ_{max}	cells/mL or fractional	Maximum or close-packed density which can be achieved for a given cell or particle
Wall cell/particle density	c_w or φ_w	cells/mL or fractional	Cell/particle density at the membrane wall (membrane surface); maximum value under pressure independent conditions.
Wall shear rate	γ_w	s^{-1}	dv_x/dz evaluated at the membrane surface; shear stress, $\tau = \eta(dv_x/dz)$ for Newtonian fluids
Suspension viscosity	η_o	g/cm-s, cP	1 g/cm-s = 1 Poise (P)
Permeate viscosity	η_p	g/cm-s, cP	
Cake thickness	δ_c	μm, mm	Thickness of the compressed layer at the membrane wall, usually considered to be the boundary layer thickness in mass transfer modeling
Resistance (total, membrane, concentration polarization layer, fouling layer)	$R, R_{tot}, R_m, R_{cp}, R_f$	m^{-1}	Proportionality factor relating flux and transmembrane pressure: $J = \Delta P_{tm}/\mu R$
Hydraulic permeability	h	LMH/bar, s/g	Ratio of flux to transmembrane pressure; equivalent to resistance w/viscosity lumped $h = J/\Delta P_{tm}$
Membrane hydraulic permeability	h_m	LMH/bar, gal/ft²-hr psi	Change in flux with a change in transmembrane pressure for pure solvent
Sieving coefficient	Si	fractional	Concentration of a soluble component in the permeate relative to its concentration at the retentate surface of the membrane
Permeability	pi	fractional	Fraction of a soluble component which permeates a membrane, relative to the bulk fluid condition
Retentivity	ri	fractional	Fraction of a soluble component which is retained by a membrane, relative to the bulk fluid condition

membrane surface, or the effect of lysis of a small fraction of cells on the filtration of the remaining population, must be examined on a case-by-case basis. As such, there is no substitute for experimentation.

Crossflow filtration, like its parent technology, dead-end filtration, is essentially a pressure-driven process. A pressure gradient through the membrane pores, characterized by the transmembrane pressure differential (ΔP_{tm} or TMP), drives flow of solvent and permeable materials through the pores. Impermeable and semipermeable solutes are convected to the membrane surface (or into the pores) by flow through the pores and, as they accumulate there, present an additional resistance to flow. Osmotic pressure concerns, a key element in the filtration of dissolved species, especially in RO and NF, are typically not a concern in cell harvest applications since particle concentrations are much smaller than the concentration of dissolved species in those other operations. Crossflow of bulk fluid across the membrane surface during filtration is employed to disrupt this accumulation at the membrane surface, minimizing resistance and enhancing flux (Figure 8). Modeling of CFF is substantially more complex than dead-end or stirred-cell filtration, since in any practical system there is a need to account for variations in filtration conditions along the length of the filter's flow path.

Efforts to model filtration behavior during cell harvest applications have generally followed one of two paths—either modifying resistance-in-series concepts originally developed for the dead-end filtration of inorganic suspensions which form well-defined cake layers, or adapting the gel polarization model originally developed for the ultrafiltration of macromolecules (Figure 9). The various gel polarization-based models differ most fundamentally in the mechanism by which the components of the accumulated surface layer are conveyed back toward the region of bulk flow. In addition, some models account for a conveyance of the concentrated layer along the surface of the membrane, rather than assume that the layer is stagnant in the bulk flow direction. Zydney and Colton [80] and Nagata et al. [43] provide a good overview of the evolution of concentration polarization models, originally developed to describe the ultrafiltration of macromolecules, to forms suitable to describe the microfiltration of cells and other particulates.

In concentration polarization theory, a high concentration of retained species accumulates at the membrane surface and provides additional resistance to flow. These modeling efforts focus on the pressure-independent region of operation, where mass transport effects dominate (Figure 10). The steady-state thickness of the boundary layer results from a balance between the convective transport of solute from the bulk to the membrane surface and its back diffusion from the high concentration surface region to the bulk. This problem can be solved by analogy to the problem of convective heat transfer in laminar-flow channels, using the Lévêque solution [49]. Direct application of the ultrafiltration model to microfiltration of particulates using the Brownian motion particle diffusivity

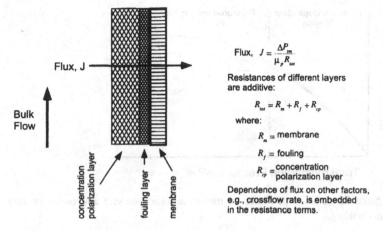

Flux, $J = \dfrac{\Delta P_m}{\mu_p R_{tot}}$

Resistances of different layers are additive:

$$R_{tot} = R_m + R_f + R_{cp}$$

where:

R_m = membrane

R_f = fouling

R_{cp} = concentration polarization layer

Dependence of flux on other factors, e.g., crossflow rate, is embedded in the resistance terms.

Resistance-in-Series Model

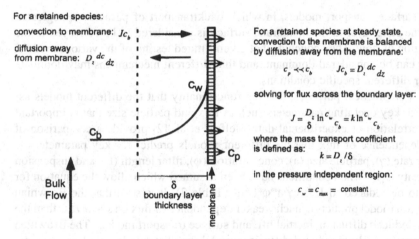

For a retained species:

convection to membrane: Jc_b

diffusion away from membrane: $D_z \dfrac{dc}{dz}$

For a retained species at steady state, convection to the membrane is balanced by diffusion away from the membrane:

$$c_p \lll c_b \qquad Jc_b = D_z \dfrac{dc}{dz}$$

solving for flux across the boundary layer:

$$J = \dfrac{D_z}{\delta} \ln \dfrac{c_w}{c_b} = k \ln \dfrac{c_w}{c_b}$$

where the mass transport coefficient is defined as:

$$k = D_z / \delta$$

in the pressure independent region:

$$c_w = c_{max} = \text{constant}$$

Concentration Polarization Model

Figure 9 Filtration models: (top) resistance-in-series; (bottom) concentration polarization.

results in poor agreement of predicted flux with experimental observations; the model significantly underpredicts the flux. Augmentation of backtransport beyond that due to simple particle diffusion has been attributed to shear-induced diffusion and inertial lift, and several workers have attempted to model these interactions with some success. Belfort et al. [2] also provide a comprehensive review of these efforts specifically as they apply to microfiltration, and also dis-

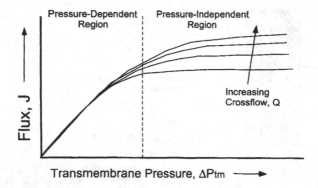

Figure 10 Flux dependence on transmembrane pressure and crossflow rate—typical functionality.

cuss surface transport models, in which backtransport of particles is neglected and the cake layer at the membrane surface is considered to be conveyed in the direction of flow. They conclude that given limited testing of the various models, none can be considered dominant, and that different mechanisms may dominate under different specific conditions.

For practical purposes, it is the functionality that the different models ascribe to key operating parameters such as flux and particle size that is important for correlation of experimental data. Belfort et al. [2] provide a comparison of the dependence of flux that the different models predict for key parameters— shear rate (γ), particle size (a), concentration (φ), filter length (L), and suspension viscosity (η)—when conditions have been assumed which allow the equation for flux to be reduced to $J = c\gamma^n a^m \varphi^p L^q \eta^r$ (Table 2). It is seen that the Brownian diffusion model predicts a much weaker dependence of flux on *shear rate* than the shear-induced diffusion, inertial lift, and surface transport models. The Brownian diffusion and shear-induced diffusion models predict similar weak inverse dependences of flux on *slurry concentration* and *filter length*, whereas the inertial lift and surface transport models predict no effect of these parameters. All the models predict a different dependence of flux on *particle size*, although only the Brownian diffusion model counterintuitively predicts a decrease of flux with increasing cell size. All models except surface transport predict a decrease in flux with increasing *suspension viscosity*, although the dependency varies among the models. Practical experience will suggest to most practitioners that all these parameters will play a role in determining the steady-state flux, but that their pure effect as anticipated in these models may be masked by the many irreversible phenomena which are likely to contribute to observed performance.

Table 2 Parametric Dependence of the Long-Term Flux for Various Transport Mechanisms*

	Brownian Diffusion	Shear-Induced Diffusion	Inertial lift	Surface transport
Shear rate (γ)	increase	increase	increase	increase
	n = 0.33	n = 1	n = 2	n = 1
Particle size (a)	decrease	increase	increase	increase
	m = −0.67	m = 1.33	m = 3	m = 1
Concentration (φ)	decrease	decrease	no effect	no effect
	p = −0.33	p = −0.33	p = 0	p = 0
Filter length (L)	decrease	decrease	no effect	no effect
	q = −0.33	q = −0.33	q = 0	q = 0
Suspension viscosity (η)	decrease	decrease	decrease	no effect
	r = −1	r = −0.33	r = −1	r = 0
Predicted flux (cm/s)†:				
for a = 1 μm	6.3×10^{-5}	2.4×10^{-4}	4.5×10^{-7}	1.1
for a = 50 μm	4.6×10^{-6}	4.4×10^{-2}	5.6×10^{-2}	54.9

* It is assumed that the feed suspensions are dilute and composed of nonadhesive spherical particles which form cake layers that dominate the membrane resistance, so that equations predicting flux for each of the different models can be reduced to $J = c\gamma^n a^m \varphi^p L^q \eta^r$.
† Calculated for $\gamma = 10^3 s^{-1}$, T = 293K, $\eta_o = 0.01$ g/cm-s, $\rho_o = 1$ g/cm³, $\varphi_w = 0.6$, $\varphi_b = 0.01$, L = 10 cm, $R_c a^2 = 253$, cot $\theta = 1$, $k = 1.38 \times 10^{-16}$ erg/mol-K, $H_o = 0.1$ cm.
Source: Ref. 2.

Resistance-in-series models correlate flux to ΔP_{tm}, accounting for the effects of the membrane substrate, concentration polarization and fouling in a group of additive resistance terms: $J = \Delta P_{tm}/\mu_p (R_m + R_{cp} + R_f)$. Additional resistance terms can be added for specific systems. This model is frequently employed to correlate and compare experimental data obtained with a single experimental apparatus while varying the membrane or feed material. Flux, permeate viscosity, and transmembrane pressure are measured directly, allowing calculation of the total resistance, R_{tot}. The intrinsic resistance of the membrane, R_m, is measured with a clean liquid stream (usually pure water) prior to introducing the experimental feed, which allows calculation of the sum of the remaining resistances (or residual resistance, $R_{res} = R_{tot} - R_m$). Correlation of resistance to variations in system performance for specific feeds is then possible (e.g., Junker et al. [27]) and provides a slightly more general result than direct comparison of flux data. Further deconvolution of the residual resistances can be attempted using an experimental protocol which limits the contributions of the various terms at certain points in the procedure (e.g., Cho et al. [10]), although portability of the results is likely to be limited. Substitution of more elaborate expressions for individual resistance terms can significantly extend the usefulness of the resistance model. Noting that the thickness of the concentration polarization layer is expected to be proportional to ΔP_{tm}, Cheryan [9] has $R_{cp} = \Phi(\Delta P_{tm})^n$. Very good correlation of the flux for the filtration of corn starch hydrolysate (which has filtration characteristics not unlike that of a typical fermentation broth) through a 0.2-μm ceramic membrane was obtained by fitting experimental data for a broad range of operating conditions to this model, where Φ is further specified as a function of the crossflow rate [62]. It should be noted that an alternate but essentially equivalent terminology is the use of hydraulic permeability, h, which is simply the inverse of resistance with viscosity lumped into the resistance term:

$$h = 1/R' = J/\Delta P_{tm} \text{ (where } 1/R = \mu J/\Delta P_{tm})$$

Choosing the optimum operating condition is a key issue in the development of a CFF process. Returning to the generally observed flux versus ΔP_{tm} curves (Figure 10, or see Figures 5 and 33 in [50]), the authors typically choose to operate in the transition between pressure-dependent and -independent regimes, and at as high a crossflow rate as is practical for the system under study. From traditional gel polarization theory, one may define the transitional ΔP_{tm} as the pressure at which the species concentration at the membrane surface, c_w, rises to the "gel" concentration for that species, c_{max}. By operating at that transitional ΔP_{tm}, *ceteris paribus*, the thickness of any polarization layer is minimized while obtaining the maximum flux. Increasing ΔP_{tm} above the transitional value increases the thickness of the polarization layer with no improvement in the flux. While the concentration polarization model suggests no penalty for operating further out in the pressure-independent regime, it is suspected that the polarization layer compacts over time and becomes an irreversibly bound fouling layer [43], which would result in a time dependent decrease in filtration performance. Operating at the highest possible crossflow rate (Q) translates into the highest possible shear rate at the membrane surface, which will maximize flux [50]. In practice, increases in Q are limited by the interdependence of Q and ΔP_{tm}, and by other practical factors.

While the preceding discussion has focused on understanding factors which determine filtration flux, the permeation of soluble components through membranes is often as critical to overall process performance as the bulk fluid transport; this is clearly the case in some of the examples presented in later sections. Nilsson [45] provides a good discussion on the mechanisms of protein permeation through UF membranes and the causes of retention for cell-free systems; when cells are added to the mix, the complexity of the problem quickly multiplies. Many of the factors which result in restricted solute permeation are difficult to model or might be considered to result from irreversible fouling phenomena; some of these are discussed in the next section.

It might be expected that electrostatic effects could dominate permeation behavior under some circumstances, and this has been clearly demonstrated in several recent papers from the group of Zydney [7,39,52]. Working with several proteins, Burns and Zydney [7] showed a maximum in transmission could be achieved at their isoelectric point, while a secondary maximum was observed at a pH between the isoelectric points of the protein and the membrane. Menon and Zydney [39] showed that the specific ionic composition, as well as the pH and ionic strength, affected permeation. These investigations showed that the effective size of the proteins could be manipulated by changing pH, ionic strength and ionic species, thus changing their extent of electrostatically-based exclusion from the membrane pores. These results were obtained under pristine experimental conditions using pure proteins in clean solutions. For real-world fermentation broth harvests, it is unlikely that such effects could be predicted (or would be measured) based on the properties of a single broth component, since one would need to consider the presence of cells, proteins, salts, and other broth components all interacting at once; the behavior of any one component would likely be masked by the presence of all the others. As with the efforts to model bulk fluid flux, these investigations of membrane permeation for pure systems provide a good guide to the types of experimental factors which should be investigated during a course of filtration optimization.

Several systems have been developed in an attempt to enhance flux and permeation by employing additional mechanisms of particle transport away from the membrane surface. These include systems which provide extremely high shear rates at the membrane surface and those which induce hydrodynamic instabilities which result in the formation of vortices in the flow path. Sustained high shear rates are provided by rotary and vibratory membrane systems [33]. Systems which generate Taylor vortices are available commercially, and employ flow through the annular space between concentric rotating cylinders to form the desired flow patterns [14,31,47]. More recently, it has been demonstrated that Dean vortices are induced by flow in a curved channel, such as that in a hollow fiber membrane formed into a helix, and that in many cases they result in substantial enhancements in flux [6,12,20]. Additional discussion of these types of systems appears in other chapters of this volume.

D. Irreversible Phenomena/Fouling

The models discussed in Section IIC focused on flux variations due to reversible phenomena. Contrary to most practical applications, most models ignore cell-cell, cell-membrane, and other component interactions as well as simple physical problems which reduce membrane performance. In this work, the term *fouling* will be restricted to flux- or permeation-restricting phenomena which cannot be reversed under common conditions of crossflow filter operation.

The simplest mode of fouling to conceptualize is *pore plugging*, which occurs when a fraction of the recirculating particles are sufficiently small to enter the pore structure of the membrane, but cannot pass through it, and so physically blocks further flow. Feeds which have a broad particle size distribution are most susceptible to fouling by pore plugging of microfiltration membranes. While cells themselves may be too large to cause pore plugging, the disruption of cells through the rigors of processing can lead to the release and/or generation of smaller particles and debris which contribute to pore plugging. In their work, Liew et al. [34] demonstrated that fouling due to pore plugging for *Candida utilis* broth resulted primarily from smaller particles present in the broth medium, and hypothesized that the presence of a cell layer on the membrane surface actually protected the pores from exposure to these finer particles, almost acting like filter aid. Fouling effects were more pronounced for aged broths, presumably due to a greater percentage of lysed cells. Ultrafiltration membranes are often employed for cell harvest applications, despite their higher intrinsic hydraulic resistance, since they are less prone to this type of pore plugging problem.

Flow reduction due to irreversible accumulation of particles on the membrane surface is common. This type of fouling can be due to *cell-cell interactions* or *cell-membrane interaction*. Cell-cell interaction, or cell aggregation, is very specific to the system under study and generalizations are difficult. The hydrophobicity/hydrophilicity of the cell and membrane surfaces are key characteristics which impact these interactions. Hydrophobic membrane surfaces are generally considered more susceptible to fouling through cell and protein adsorption; use of hydrophilic membranes can moderate this effect. Regenerated cellulose membranes are relatively hydrophilic, but are less available commercially in forms appropriate for cell harvesting. It is a reality that the majority of polymeric membranes suitable for cell harvest applications are hydrophobic in nature— primarily polysulfone (PS) or polyvinylidinedifluoride (PVDF). Permanent chemical modification of the membrane surface to make it more hydrophilic (e.g., Millipore's hydrophilic PVDF, Durapore GVWP) can be beneficial, as can be the treatment of the membrane with a wetting agent just prior to processing. Ceramic membranes are an interesting subset of microfilters which have a significantly more hydrophilic character than most polymers, and as such may reduce this problem. But their significantly higher cost per installed unit area limits their utility.

The hydrophobicity of the cells can vary greatly among different strains of a given species. It is also possible to change the cell surface characteristic through variations in the growth medium, especially through changes between defined and complex media [34]. These interactions can be minimized by careful selection of the membrane material, but in many cases it is a matter of picking the "least bad" performer. In some cases it may be found that a soluble broth component is adsorbing to the cell or membrane surface and mediating the aggregation interaction. The effect of these phenomena can be amplified by *cell deformation*, where the interstitial space between cells is reduced or eliminated, restricting flow. One might expect that animal cells would be most susceptible to this type of effect; however, animal cell harvests are typically performed at very low transmembrane pressures in order to avoid cell lysis, such that this is probably not a factor. A similar effect for systems containing a broad size distribution of rigid particles can also be considered; this can come into play during the filtration of mycelial broths and broth homogenates (lysates). It is also possible for soluble components to form a gel layer at the membrane which becomes the controlling resistance to flow.

In addition to considering bulk fluid flow fouling, it is important to consider *permeation fouling*—fouling with respect to the permeation of soluble components through the membrane. This is important since many harvest applications seek the recovery of these species in the permeate rather than the cells themselves, and other applications seek the elimination of these components from the broth concentrate prior to further processing. Despite the use of microfiltration membranes with pore sizes in the micron range, it is not uncommon to encounter incomplete permeation of solutes, usually proteins but smaller components as well. This is most frequently attributed to the formation of a gel layer which becomes the controlling membrane for passage of these soluble components. While the CFF process can be developed to optimize performance of these suboptimal systems, modification of the fermentation process is often the key to an efficient overall process [21].

Proteins can play a dominant role in fouling during cell harvest, affecting both overall flux and solute permeation. Sirkar and Prasad [63] noted the importance of protein aggregation and adsorption to membranes in UF filtration behavior, and suggested that these mechanisms are a large contributor to time dependent variations in system performance. Nilsson [45] attempted to organize published results on protein fouling of UF membranes into a theoretically grounded framework, but the complexity of the problem defied simple conclusions. In their review, Marshall et al. [36] conclude that proteins reduce performance of ultrafiltration membranes through deposition on the membrane surface as a dynamic cake layer, whereas they reduce the performance of microfiltration membranes through pore plugging, despite their small size relative to the pore size. Bowen and Hall [5] suggest that the mechanism of protein fouling of microporous membranes is concentration dependent, with pore plugging dominant under dilute conditions while surface deposition (cake formation) comes into play at higher concentrations. While most such studies have been conducted in the absence of cells, one can expect that these phenomenon will come into play during cell harvest operations in the presence of soluble protein, whether the protein is present by design or due to cell lysis.

E. The Effects of Antifoams

Antifoams added during the fermentation are often believed to be key players in membrane fouling, but relatively little has been written about this topic. One can examine the effect of the antifoam on flux, on soluble component permeation, and on the ability to clear the antifoam itself from the cell concentrate. In approaching this problem, one must be aware of the nature of the antifoam in use; many of the most common are nonionic polymers and have an inverted cloud point, i.e., they are soluble in water at low temperatures, but form a separate phase at higher temperatures. While the cloud point temperature is a function of antifoam concentration, for many common antifoams it is just slightly below ambient. Silicone oils are the other common class of antifoams; most do not exhibit cloud point behavior at temperatures of interest.

It is often observed that the presence of an antifoam during CFF processing will significantly reduce the flux. Among other factors, the extent of this effect will depend on the concentration of the antifoam and the temperature relative to the cloud point. Flux reductions are very system specific but have been reported in the range of 10% to 80% relative to the antifoam-free system for a variety of different surfactants, including P2000 (polypropylene glycol 2000), Ucolab N115 (polypropylene-polyethylene glycol), Tegosipon T52 (silicon oil dispersion), DISFOAM CE-1 (polyoxyethylene alkylether), and X81-10 (fatty alcohol ester) [8,30,74]. Minier et al. [42] did not observe significant membrane fouling in the presence of a variety of different antifoams on several types of membranes, but they employed an antifoam concentration of 100 ppm (~0.01 v/v%), which is tenfold lower than is commonly used in practice. Studying fouling as a function of both antifoam concentration and temperature for a pure solution of a polyglycol antifoam (Mazu DF-60-P) and a polysulfone membrane, McGregor et al. [38] observed severe fouling when the antifoam concentration was increased to 0.1% and the cloud point temperature of 25°C was approached. The extent of fouling was much less at temperatures below the cloud point (15°C and below). For their simple system, McGregor et al. also found that antifoams which did not exhibit cloud point-type behavior, especially those which are silicone oil-based, produced the least fouling effect.

McGregor et al. [38] showed that fouling effects generally increased at higher membrane loading, Λ. This agreed with the results of Yamagiwa et al. [76] for similar experiments in which solutions of a hydrophobic non-ionic surfactant (DISFOAM CE-120R) were filtered through a polysulfone ultrafilter. Yamagiwa et al. showed that antifoam adsorption onto the membrane was proportional to its concentration in the filtered solution for both a polysulfone and a polyolefin membrane. However, a corresponding decrease in flux was only observed for a hydrophobic polysulfone membrane; adsorption of the antifoam onto a hydrophilic polyolefin membrane had little effect on flux, with no correlation to antifoam concentration. This divergent behavior was correlated with the wetting properties of the antifoam-membrane pairs. It was determined by contact angle measurements that DISFOAM readily wet the polysulfone, allowing it to be adsorbed into its pores, whereas it wet the polyolefin membrane poorly, suggesting that surfactant droplets would stick to that membrane surface but would not flow into its pores [76].

If cell concentration is the basic goal of the process, the problem can be minimized by filtering at a temperature below the cloud point (where applicable). If this does not provide enough of an improvement, and changes in membrane material and antifoam are not possible, it is usually necessary to take a brute force approach and design around this problem by increasing the membrane area.

If the process objective is to recover a soluble component such as a protein in the permeate, or to clear the antifoam from the cell concentrate, the problem becomes much more difficult. Kroner et al. [30] demonstrated that the presence of as little as 0.1% of P2000 (polypropylene glycol 2000) increased the retention for several different enzymes from 5% to 50%, and the effect was more pronounced at higher antifoam concentrations. Such higher retentions are problematic because they result in a disproportionate increase in the extent of diafiltration required to achieve a desired product recovery, increasing not only the scale of the CFF system but downstream volumes as well. Operation at higher ΔP_{tm} has been shown to exacerbate the problem [75]. Clearance of the antifoam from a cell concentrate is often desirable, e.g., to ensure that the surfactant does not further complicate the post-lysis recovery of the product. Passage of an antifoam is typically highly temperature dependent, with almost complete retention by polymeric membranes above its cloud point [74,76]. Passage improves below the cloud point, but is still incomplete (rarely better than 50%), requiring many diafiltration volumes to achieve complete clearance. In the presence of antifoams, similar effects have even been observed for lower molecular weight solutes as well [74,75].

Membranes with a more hydrophilic character generally offer better performance in the presence of antifoams. McGregor et al. [38] showed that a regenerated cellulose membrane exhibited almost no flux loss under conditions which severely fouled a polysulfone membrane. Yamagiwa et al. showed superior performance of a hydrophilic polyolefin [75,76] membrane versus that of a hydrophobic polysulfone with respect to flux, antifoam retention and solute retention, when both were exposed to the same hydrophobic non-ionic surfactant. While antifoam retention for both types of membrane was high above its cloud point, retention for the hydrophilic membrane was significantly lower below its cloud point— ~20% versus ~70% for the hydrophobic membrane. Most ceramic membranes are hydrophilic and also seem to offer a benefit in their ability to maintain performance in the presence of antifoams. In one study, flux with a ceramic membrane was compared to that of a polysulfone UF membrane; flux reduction with the ceramic membrane was only 10% in the presence of antifoam, with much less dependence on temperature than for the polysulfone membrane. While a similar difficulty in clearance of the antifoam was observed, its presence had little effect on the clearance of other, lower-molecular-weight solutes [74]. Minier et al. [42] also showed excellent performance of a number of alumina- and zirconium oxide/carbon-based membranes in the presence of a variety of antifoams, but as noted earlier, they used a very low antifoam concentration (100 ppm). The filtration of a bacterial broth containing up to 30% soybean oil dispersed in the broth might be considered a related processing challenge, and in that case good CFF performance was obtained using a ceramic membrane (Millipore Ceraflo 0.1-μm pore size, alumina-based membrane [13]). One might expect that there is just less affinity between the hydrophilic ceramic membrane surface and the primarily hydrophobic antifoam molecule, lessening the effect of these ubiquitous foulants.

III. MEMBRANE CONFIGURATIONS AND MATERIALS

Membrane configurations used in crossflow filtration can be classified into four major categories: plate-and-frame, tubular, hollow-fiber, and spiral-wound. Figure 11 shows schematics of these four types of membrane modules. In all cases

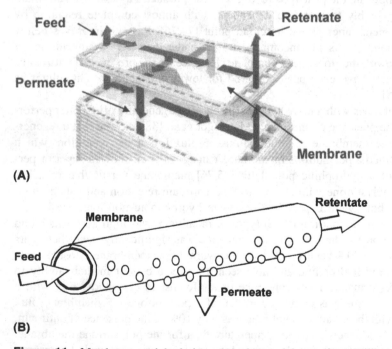

(A)

(B)

Figure 11   Membrane module designs: (A) plate and frame (from Millipore Corporation, Bedford, MA); (B) tubular or hollow-fiber element (from A/G Technology Corporation, Needham, MA); (C) hollow-fiber (from Pall Filtron Corporation, East Hills, NY); (D) spiral wound (from Pall Filtron Corporation, East Hills, NY).

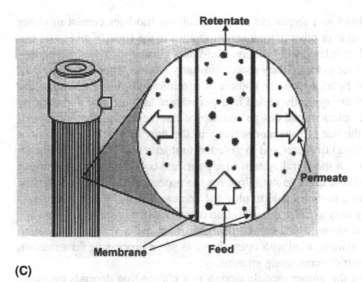

(C)

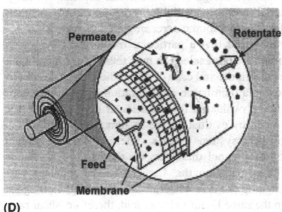

(D)

Figure 11 (continued)

fluid flows tangentially to the membrane surface to reduce concentration polarization and cake formation effects as described in detail in Section II. The plate-and-frame design, also known as the cassette, flat sheet, or stacked sheet design, simply consists of flat membrane sheets which are mounted to plates such that material that passes through the membrane is channeled to a permeate collection port or ports. Fluid can either flow in parallel through all membrane sheets in a stack in the same direction or enter at a sheet on the bottom of the stack and then continue across the next membrane sheet in the stack in the reverse direction such that fluid flows in a serpentine manner. Tubular modules consist of either polymeric or ceramic or other inorganic membranes in the form of tubes that can be supported in a stainless-steel housing. Fluid flows inside the tubes and permeate is collected in the extracapillary space. Similarly, hollow-fiber cartridges consist of polymeric fibers which are housed in a reinforced plastic housing. The ends of the fibers are typically sealed in a permanent fashion with a glue or by heat sealing. The spiral-wound design consists of flat membrane sheets wrapped in a spiral with the use of separator screens in the flow channels to maintain a constant flow channel diameter and to give the unit structural support. Permeate is channeled through the spiral system to a permeate collection core. The mesh screens found in these units do not allow for the processing of fluids containing particulates because particles tend to plug the screens. Hence, spiral wound systems are typically only used for ultrafiltration of particle-free solutions. Likewise, plate-and-frame designs, which can be designed with mesh screens to promote turbulent flow, should be used with open channels when processing fermentation broths or other particle-containing streams.

Selection of the proper module design is a choice that depends on many factors. Some membrane materials are only available in certain module designs. Depending on the findings of initial membrane screening studies (discussed below and in Section IX), one may find that a particular membrane type works best for an application. Membrane material selection may then dictate membrane module selection. Cost, space constraints and ease of use at the final scale should also be considered when choosing a membrane configuration. If possible, more than one module design should be evaluated for a particular application, but it is not uncommon for one to begin by evaluating the system with which one is simply most familiar. Most of these systems offer a variety of channel heights (in the case of stacked sheets) or channel diameters (in the case of tubular or hollow fibers) as well as different path lengths. Larger channel openings will reduce the chance of solids plugging the flow channels but they require a higher volumetric flowrate to result in the same linear velocity and, therefore, shear rate, as smaller openings. Path length will affect pressure drop through the module; longer path lengths will result in a greater transmembrane pressure at a given volumetric flowrate.

Alternatives to the above membrane configurations are rotary modules and modules designed to create flow through curved channels. These systems are similar to tangential flow devices in that fluid flows perpendicularly to the membrane and shear forces are created at the surface of the membrane to prevent accumulation of material. However, in rotary modules the shear is created by rotating parts, either an inner of two concentric cylinders in which fluid flows in an annular gap or by rotating discs where fluid flows between the disks. These flows create vortices, called Taylor vortices in the case of flow between concentric cylinders, which provide high shear at the membrane surface. Flow in curved channels is similar to conventional tangential flow filtration except that the curved flow path creates Dean vortices, which as is the case for rotating systems, provide high shear at the membrane surface. However, unlike rotating systems, they do so without any moving parts. The main advantage of rotating systems and curved channel systems is that transmembrane pressure and shear rate are at least somewhat de-coupled. In conventional tangential flow filtration high shear rate is obtained by increasing crossflow rate which results in a higher transmembrane pressure such that the maximum shear rate which is achievable is limited by the pressure rating of the membrane. In rotating devices one can achieve higher shear rates by increasing rotational speed which does not affect transmembrane pressure. In a properly designed curved channel system, Dean vortices will form resulting in high shear rates. These high shear rates are achieved without excessively high crossflow rates which would result in high transmembrane pressures. These designs are fairly new alternatives to conventional tangential flow systems and will not be discussed further in this work. However, several excellent papers have been written describing the use of rotating systems [14,20,31] and curved channel systems [6,12] for the harvest of fermentation broths as well as for other bioprocessing applications.

Many different membrane materials are commercially available today including cellulose acetate, mixed cellulose esters, polysulfone, polyamide, nylon, polytetrafluoroethylene (PTFE), polypropylene, polycarbonate, polyvinylidinedifluoride (PVDF), and most recently ceramic and other inorganic membranes. When choosing a membrane type, one must consider whether such features as solvent resistance and ability to autoclave or steam-in-place (SIP) are important for one's application. Membrane material selection can be difficult but membrane manufacturers can often recommend one or more particular membrane type to try based on the experience of their customers with various types of fermentation broths and with various types of components. However, it is advisable to examine as many membrane types as possible for a particular application (as discussed in Section IX) since it is difficult to know *a priori* which membrane type will perform best with one's fermentation broth.

IV. CLASSIFICATION OF APPLICATIONS

The optimum filtration process depends very much on the process objective, which in turn depends on the application. It is very useful to differentiate membrane filtration processes of fermentation broth into groups based on two key attributes, the relationship of the product to the cells and the nature of the organism.

In general terms, most products can be thought of as either intracellular or extracellular. For simplicity of discussion in this work, products which are in any way cell-associated will be considered to be intracellular, regardless of the true physical location of the product. Thus, the protein inclusion body in an *E. coli* cell, the recombinant antigen in a yeast cell, the membrane-bound enzyme in a *Pseudomonas* cell and the secondary metabolite which is in anyway distributed in a fungal hyphae (within the cell, dissolved in the cell membrane, or just adsorbed to the outside of the cell due to lack of solubility in the medium) will all be considered to be intracellular or cell-associated products. In contrast, the enzyme which is secreted from a bacterial cell, the soluble secondary metabolite which is secreted into the medium, or the protein growth factor which is secreted from an animal cell in culture are all considered extracellular products.

The central issue in making this distinction is whether the ultimate product will be retained by the membrane along with the cells or if it is desired to obtain the product as a cell-free solution in the membrane permeate. Membrane selection is much simpler for the intracellular case, since a broad range of membrane types are available for the retention of most cells. The key issues for these cases becomes flux, fouling, and the ability to clear smaller components from the retentate; yield of the product is generally a secondary issue. On the contrary, if it is desired to permeate a product while retaining the cells, the selection of the membrane is more difficult and the options are generally more limited. In this case, yield is an issue as the partial permeation or "sieving" of the product can result in the need for extended diafiltration to separate the product from the cells.

Another dichotomy can be made between fermentation broths of single-cell versus multicellular (filamentous) organisms. In this case, single-cell organisms would include all simple bacteria (e.g., *E. coli.* and *Pseudomonas* sp.), yeasts, and most cultures of animal cells grown in suspension. Multicellular cases then include actinomycetes and molds, both of which have a filamentous primary growth structure (although molds can have secondary growth structures such as the formation of pellets). Rather than distinguish culture types based on some intrinsic difference (e.g., procaryotic vs. eucaryotic), we group systems based on this phenotypical criterion because of the very practical observation that single-cell fermentation broths which arise in biotechnology applications are typically grown to relatively low cell densities and have a modest viscosity. In contrast, most commercial cultures of multicellular organisms are grown to high cell densities, which, combined with their filamentous morphology, results in a very high apparent viscosity.

Of importance in making the distinction between single cell and filamentous cultures, it is not the difference in the actual number density or dry weight of the cells but the volume fraction which the cells occupy and the extent to which the broth can be concentrated prior to encountering difficulty in recirculating the cell slurry. Typical values of packed cell volume for commercial cultures of a variety of cell types are presented in Table 3, along with cell density and dry cell weight where appropriate. It is observed that while cell number density varies over a broad range, the volume fraction of solids typically is in the 1 to 10 vol%

Table 3 Fermentation Broth Characteristics of Different Cell Types: Cell Density, Cell Dry Weight, and Volume Fraction Solids

Cell type	Volume fraction cells (wet vol%)*	Cell density (cells/mL)	Cell dry weight (g/L)	Reference
Single-cell organisms				
Animal cells	0.5–2.5	$1–5 \times 10^6$		
E. coli	2–10, 15		30–100	78
Pseudomonas sp.	~5			
Other bacterial broths	5–12			
Saccharomyces cerevisiae	6	$2–7 \times 10^7$	25–60	46, 68
Candida and other yeast	5–15		50	
Filamentous organisms				
Streptomyces and other actin-omycetes (filamentous bacteria)	5–35		30–65	79
Filamentous Fungi	30–55		60–80	

* Volume fraction data from personal communications with M. Chartrain, N. Connors, G. Hunt, C. Seamans, J. Zhang of Merck Research Laboratories, and I-T. Tong of Merck Manufacturing Division.

range, occasionally reaching 15 vol%, for those species which grow as individual cells. With such relatively dilute streams, concentration factors of 5 to 10 or greater can often be achieved before the solids fraction increases to levels sufficient to impede filtration operations. In contrast, the volume fraction of solids is much higher, in the 20 to 45 vol% range and sometimes as high as 55 vol%, for the multicellular, filamentous cultures. Starting with such a high solids content, these cultures can only be concentrated by a factor of two or three at best before the solid content increases to a level which will clog most filtration cartridges. As will be seen in the following sections, this limitation of concentration factor has major consequences for process feasibility. Contributing to this effect is the simpler rheology of the single-cell organisms versus that of the filamentous, since in addition to having a higher apparent viscosity, the filamentous culture will exhibit shear thinning.

V. SINGLE-CELL ORGANISMS/EXTRACELLULAR PRODUCT

One of the primary applications of crossflow filtration is for the recovery of secreted, soluble products, from fermentations of bacteria, yeast, and other single cell cultures (animal, plant and insect). While the characteristics of these cell types differ widely and have broad implications for the recirculation conditions which can be employed, there are key similarities for all these systems which allows this general grouping. As noted in the previous section, the first key similarity is the relatively low volume fraction of solids present, even for well-developed fermentations of these organisms. For the harvest of some animal and insect cell cultures, the cell density is sufficiently low that the harvest can be performed using dead-end depth filtration at small to moderate scales [59].

The second similarity is that these cultures are employed primarily in the production of macromolecular products, most frequently proteins. The major issue in the development of CFF-based recovery steps for these products is the facility with which they permeate the membrane, since large volumes of diafiltrate are required to recover poorly permeating products in high yield, resulting in very dilute product streams. Assuming the appropriate screening experiments have been performed, the MF or UF membrane employed is only rarely the issue in permeation; the secondary membrane formed by accumulation of cells and macromolecular gels on these primary membranes is typically controlling (see Section II). For rapid process development, it is most efficient if there is close integration of the fermentation and CFF harvest development efforts, so that fermentation conditions which produce difficult-to-process broth can be identified and modified as soon as possible.

For these types of applications, it is typical to couple a CFF product recovery step with a second product concentration step, which can also effect a buffer (or solvent) switch. This second concentration step requires an ultrafilter for macromolecular products; in the less frequent case in which such a process is employed to recover a low molecular weight product, the subsequent step can be nanofiltration/RO, chromatography, or even distillation (Figure 12).

An example of such a process is the production of a soluble lipase (MW 29,000) by a strain of *Pseudomonas aeruginosa* [21]. The fermentation broth was to be concentrated 5 to 10 fold to recover as much enzyme in the permeate as possible, and then diafiltered with buffer until the yield target for enzyme recovery had been achieved. Early in the development of this process, both the flux and permeation of the enzyme which could be obtained during concentration of the broth were poor despite screening a wide range of membrane materials and configurations. Diagnostic experiments suggested that neither the cells nor the medium nor the enzyme itself was responsible for poor performance, suggesting the production of some gel-forming byproduct. Reformulation of the fermentation medium resulted in an increase of flux from 8–16 LMH to 24–49 LMH for the concentration experiment, with average enzyme permeation increasing from 10–20% to 50–60%, with no decrease in enzyme production. This improvement allowed development to proceed to the pilot plant, where the process was implemented on the 125- and 1200-L scales (Figure 13).

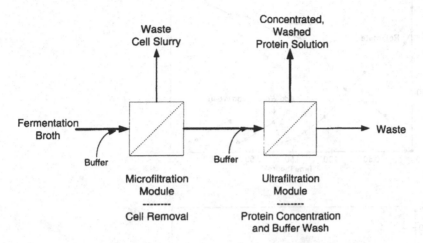

Figure 12   Microfiltration/ultrafiltration schematic for protein recovery.

At the 1200-L scale, a 10-fold concentration followed by three 1:1 batchwise diafiltration washes was achieved in an 18-hour time cycle using 3.7 m^2 of 0.1 μm Millipore Durapore membranes in open-channel Prostak cartridges (hydrophilic PVDF). The system was configured such that the broth recirculated between a 2000-L feed tank and the CFF skid until the volume was reduced to 160 L, at which point a 200-L feed tank was substituted (Figure 14). This reduced the minimum system volume allowing a tenfold concentration before initiating diafiltration. The average flux for the process was 27 LMH, lower than that achieved at smaller scale because of the addition of filter cartridges *in series*. As some flux reduction is noted at the highest cell concentrations, it is expected the lower average flux results from a reduced flux by those final cartridges in the flowpath which see the most concentrated retentate. The enzyme permeation (referenced to the enzyme activity in the feed) remained at 53% throughout the process (in agreement with laboratory results); this seemingly simple behavior most probably results from the coincidental balance of several factors in the process. The enzyme activity yield for the process was 69%, which could have been increased by additional diafiltration washes (flux was stable), but was not due to time cycle constraints. In preparation for ion exchange chromatography, the resulting permeate was concentrated 47- to 130-fold by ultrafiltration using a 30,000-MWCO regenerated cellulose membrane, followed by diafiltration to effect a switch to the column loading buffer [21].

While this type of process might often be employed for the production of a biotechnology product, e.g., a therapeutic protein, the lipase described in Göklen et al. [21] might be considered in the category of a specialty chemical

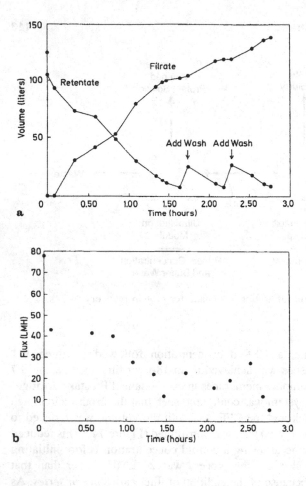

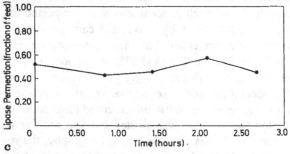

Figure 13 Lipase fermentation broth harvest microfiltration—100-L scale performance. Measurements of (a) retentate and filtrate volume (liters), (b) microfiltration flux (LMH), and (c) enzyme permeation (fraction relative to enzyme activity in the feed), versus time for filtration of 100 L of broth through 1.86 m² of Millipore 0.1-μm Durapore flat-sheet membrane in Prostak cartridges, including two diafiltration wash steps. $\Delta P_{tm} = 0.69$–0.76 bar, crossflow rate = 96.5 Liters/min. (From Ref. 21 with permission from Springer-Verlag.)

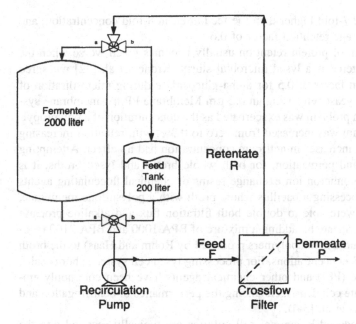

Figure 14 Batch concentration using two feed vessels to allow processing to smaller feed volumes.

since it is used in the reaction of a synthetic drug. Significantly different economic and regulatory constraints exist for the two types of processes. The regulatory burden for the intermediate of a well-characterized synthetic drug is significantly lighter than for a protein drug, but the economics are more stringent. Reuse of the membranes was critical to the feasibility of the lipase process, and in that case it was demonstrated that the membranes could be employed at least eight times with a "standard" cleaning between batches. For highly valued therapeutic proteins and other biotechnology products, reuse of membranes may not be required by the economics, and replacement of membranes with each batch may simplify development. This must obviously be examined on a case-by-case basis.

Difficulties obtaining protein permeations which are high enough to allow feasible recovery by a CFF process have been cited by many workers in this field. Hernandez-Pinzon and Bautista [26] studied the microfiltration of streptokinase (SK) produced by *Streptococcus equisimilis* using 0.3µm and 0.8µm Filtron Serie Omega (PVDF) flat sheet membrane casssettes. For the 0.3µm membrane, they observed an enzyme activity level in the permeate which was roughly half of that of the feed, and a steady activity increase in the concentrated cell slurry, reaching a level 2.7-fold higher than the feed after an 8-fold concentration, and calculated an average retention factor of 0.6.

The problem of protein retention usually becomes much worse when the material being filtered is a lysed microbial slurry. Kroner et al. [32] measured a protein retention factor of 0.5 for alpha-glucosidase during microfiltration of disrupted baker's yeast cells using an 0.2-μm Membrana PP flat membrane system. The retention problem was exacerbated as the concentration of a polypropylene glycol antifoam was increased from zero to 0.3%, with retention increasing to ~75%; further increases in antifoam concentration had no effect. Attempting to improve flux and permeation, for both whole broths and lysed broths, it is possible to add submicron ion exchange resins or chemical flocculating agents to the system. Processing a bacillus whole broth with a polysulfone membrane, Cabral et al. [8] were able to double both filtration flux and alkaline protease recovery in the permeate by adding a mixture of BPA 1000 and BPA 2100 (submicron cationic and anionic polymers produced by Rohm and Haas) to the broth at 0.05% to 0.10% concentrations. For processing lysed cell systems, borax salts, polyethyleneimine (PEI) and other chemical agents have been commonly employed to flocculate cell debris, enhancing the performance of centrifugation and CFF (e.g. Rumpus et al. [54]).

Proteins produced by animal cell cultures are typically secreted into the medium and can be recovered in the permeate of a CFF harvest. Maiorella et al. [35] characterized the filtration characteristics of several animal cell lines (NS-1 murine myeloma, T-88 human/human/murine trioma, D234 h/h/m trioma, 454A murine hybridoma and Sf9 insect cell line) using flat-plate (Millipore Minitan and Prostak systems and Dorr Oliver Ioplate system) and hollow-fiber (Microgon MiniKros and KrosFlo systems) membrane configurations. The presence of 0.1% (w/v) Pluronic polyol F68 in culture media was found critically important to prevent cell damage during processing, especially for serum-free formulations. A region of safe operation (cell viability) and recommended operation (high protein permeation) was defined based on shear rate, (ΔP_{tm} and pore size. For the cultures studied, a rapid falloff in cell viability was observed as shear rate in the membrane device was increased above 3000 s^{-1}. Loss of viability due to ΔP_{tm}-induced deformation (cell damaged from being forced into a pore) was also discussed, and characterized by the critical membrane tension (CMT, a function of ΔP_{tm} and pore size), which is an increasing function of wall shear rate. Above the CMT for a given shear rate, cell damage was expected.

The CMT is a function of the transmembrane pressure and pore size; decreasing pore size allows operation at higher ΔP_{tm} for a given CMT limit. Interestingly, no loss of viability was observed due to pumping a cell slurry through a pipe at Reynolds numbers as high as 71,000; viability loss was observed during recirculation through a rotary lobe pump when tip speed exceeded 250 cm/sec. Maiorella et al. [35] illustrate the benefits of operation of CFF processes for cell culture harvest in a constant permeate rate mode, using a regulating pump on the permeate line to prevent initial rapid fouling of the membrane which can otherwise occur if operating in a constant ΔP_{tm} mode. This mode of operation, with permeate rate regulated to maintain near-constant ΔP_{tm}, is also believed to minimize protein retention by the microporous membrane. This was demonstrated by Maiorella et al. for the filtration of IgM-producing human hybridoma cells with a Microgon 0.2-μm hollow-fiber membrane (hydrophilic cellulose nitrate/cellulose acetate-mixed ester). Operating at a relatively high permeate rate, which resulted in ΔP_{tm} increasing through the process, resulted in a steady reduction of IgM permeation from 100% to ~30%, whereas operating at a lower permeate rate, which resulted in constant ΔP_{tm}, maintained 100% permeation throughout the process.

The recovery of recombinant tissue plasminogen activator (rt-PA, ~65 kDa) from Chinese hamster ovary (CHO) cell culture using AKZO Microdyn 0.2-μm hollow-fiber membranes (polypropylene) has been demonstrated to be a robust commercial process at the 12,500 L scale [71]. The 180 m^2 CFF system operates with a feed recirculation rate of 33,000 L/hr for a wall shear rate of 4000 s^{-1} (interestingly, outside of Maiorella et al.'s "safe" zone). Van Reis et al. [71] also advocate operation of cell culture harvest operations in a controlled permeate rate mode, and have shown this results in stable operation with nearly complete rt-PA permeation. They employed rotary lobe pumps for both broth recirculation and permeate rate control. Operating at a controlled flux of 27 LMH (ΔP_{tm} <5 kPa) and a membrane loading of ~70 L/m^2, they perform a 15-fold concentration and a two-volume constant volume diafiltration in <3 hours, and obtain a >99% yield with only a 6.7% dilution of the product from the original culture titer; ΔP_{tm} only increased above the target value in the final half-hour of processing. An important consideration in their selection of the 0.2-m microporous membrane for this therapeutic product was the absolute retention rating which is available for MF membranes, reducing the load on the subsequent dead-end sterile filtration of the permeate product. They state that UF membranes can have defects which would allow passage of some particles, making the final sterile filtration much more difficult.

Insect cell culture can produce either secreted or cell-associated protein products, usually employing either a recombinant insect cell line or the baculovirus expression system [3]. Working with polysulfone hollow fiber membranes, Trinh and Shiloach [67] characterized the concentration of *Spodoptera frugiperda* (Sf9) cells at a density of 2–2.5 × 10^6 cells/mL. They employed low transmembrane pressures (unrestricted retentate and filtrate outlets), and characterized flux with varying shear rate, pore size and lumen diameter, investigating the ranges 8000 to 16,000 s^{-1}, 0.22 to 0.65 μm, and 0.5 to 0.75 mm, respectively. The optimum combination of parameters (0.45-μm pore size, 0.75-mm lumen ID filter, 14,000 s^{-1}) was scaled up to a 40-L batch size. Working at membrane loadings of ~20 L/m^2, they easily achieved concentration factors of >20-fold in processing times of 1 hour (0.45-μm pore size, 0.75-mm lumen ID, 14,000 s^{-1} shear rate). It is interesting to note that the optimum shear rate is again above the "safe zone" prescribed by Maiorella et al. [35]. Cell viability was reduced only ~3% during the concentration process, suggesting the method is suitable for preharvest cell manipulation as well.

The possibility of selectively recovering one protein in crossflow permeate while rejecting a larger protein along with the biomass in the retentate is highly attractive in concept, but difficult in practice. As noted in Section II.C, it should be possible to manipulate the permeability of a protein through a membrane by varying the membrane structure, pH, ionic strength, and species. More interestingly, Saksena and Zydney [57] have demonstrated large changes in the selective permeation of one protein over another (albumin vs. immunoglobulins (IgG)) through a UF filter in a stirred-cell filter and were even able to favor IgG permeation under certain conditions. It is unknown if this level of permeation specificity could be obtained for filtration of a complex broth; no such cases have appeared to date in the literature. When changing processing conditions to manipulate product permeation, one must be aware of other possible effects, such as changing the tendancy for cell or protein aggregation, or for adsorption to the membrane [63]; cell lysis and product degradation must also be considered.

Recovery of recombinant retrovirus particles from cell culture broth using a combination microfiltration/ultrafiltration scheme has been demonstrated and suggested as a method suitable for larger-scale recovery of these products [1]. The retroviral particles are on average 100 nm in diameter, and so can be permeated through 0.45-μm microfilters and retained using relatively coarse ultrafiltration membranes (100 to 300K molecular-weight cutoff [MWCO]). Loss of infectivity may be due to shear damage to the particle envelope, making control of system hydrodynamics key to high yield.

A common refrain in all discussions of animal and insect cells and viral particle processing is the sensitivity of the cells to shear. This mandates careful selection of filter element channel size and recirculation rates to avoid cell lysis during filtration, complicating downstream purification of the product from the permeate.

VI. SINGLE-CELL ORGANISMS/CELL-ASSOCIATED PRODUCT

Another common biotechnology application of crossflow membrane filtration is the concentration and washing of cultures of single-cell organisms where the product is intracellular or cell associated. Common cases include recombinant yeast cultures producing proteins and antigen particles (the latter particularly common in the production of vaccines), and recombinant *E. coli*, producing proteins in the form of solid inclusion bodies. As noted above, these cultures can typically be concentrated 5- to 10-fold or more, such that this initial filtration can significantly reduce process volumes downstream. The next step is usually cell lysis and recovery of the product from cell debris. Separation of product from cell debris can often be performed by CFF; process considerations relating to this application are similar to that discussed in the previous section.

An example of such a process is the production of an antigen product by fermentation of a recombinant yeast, a strain of *Saccharomyces cerevisiae* grown on a defined medium, as described by Forsyth and Göklen [18]. The initial step prior to cell lysis, antigen extraction and further purification, is concentration and washing of the cells by CFF. In the pilot-scale harvest filtration described, fermentation broth is concentrated from ~540 L to 100 L, then batchwise diafiltered three times with an equal volume of buffer before further processing. As shown in Figure 15, the concentration was performed in a fed-batch mode, with broth fed from the 1000-L fermenter to a 200-L harvest tank which fed the CFF skid. Generally speaking, fermenters make poor CFF feed vessels since their outlet and inlet ports are relatively small and cannot support the high recirculation rates required for filtration; this is especially true toward the end of a concentra-

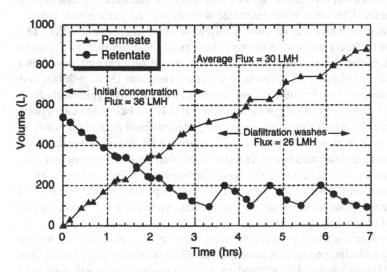

Figure 15 Microfiltration harvest of *Saccharomyces cerevisiae*—500 L scale performance. Measurements of permeate volume (▲) and retentate volume (•) versus time through a fivefold initial concentration, three 1:1 batchwise diafiltration washes and final concentration. Filtered using 2 m² of A/G Technology 0.2-μm polysulfone hollow-fiber membranes with 2-mm lumen diameter. ΔP_{tm} = 0.69–0.76 bar, crossflow rate = 96.5 L/min. (From Ref. 18 with permission.)

tion step when solids content and viscosity are high. In this case, the 200-L harvest vessel provided large diameter lines for transfer to the filtration skid. The use of a dedicated CFF harvest vessel also allows minimization of the system holdup volume.

The filtration employed polysulfone hollow fiber membranes (A/G Technology) with a pore size of 0.1 μm and a lumen diameter of 3 mm. Two cartridges in parallel provided a total filtration area of 4.09 m², for a membrane loading of 132 L/m². The system was operated with a recirculation rate of ~150 LPM using a rotary lobe pump; with an average transmembrane pressure of 12 psi, it produced 3 to 4 LPM of filtrate. The flux was observed to decrease at higher cell density, which is apparent from comparison of the average flux during the concentration phase (36 LMH) to the average flux during the diafiltration phase (26 LMH). The overall average flux, at 30 LMH, was considered to be at the lower limit of acceptability. The broth had a dry cell weight of 21 g/L which was increased to 114 g/kg in the final washed concentrate; wet solids concentration increased from 8% to 44% (v/v). The 3-mm lumen size, while inefficient during the early portion of the operation, allowed the concentration to proceed to the desired endpoint without plugging the fibers.

Forsyth and Göklen [18] compared the retention of cells to the clearance of soluble broth components. Solids as measured by a standard dry cell weight assay are barely detectable in the permeate stream, and the intracellular antigen is below detection. Succinate, an unmetabolized medium component (MW 118), was found to be freely permeable through the membrane, such that its clearance factor matched the theoretically predicted value based on concentration factor and diafiltration volumes. In contrast, a polymer antifoam (MW ~2000) was partially retained during the filtration and its clearance deviated significantly from the calculated value. (Processing was performed at <10°C, below the cloud point of the antifoam.) The data suggest nearly complete antifoam permeation early in the concentration step, while the broth is still relatively dilute and cell cake on the membrane is slight, with retention becoming more of an issue as the concentration proceeds and the cell cake layer thickens. Obviously the importance of these results depends on the sensitivity of the downstream process to the presence of these soluble compounds, since recovery of the antigen in the washed concentrate stream should be nearly quantitative.

In a similar process using a recombinant *Saccharomyces cerevisiae* to produce an intracellular product, Russotti et al. [56] observed high initial flux (~1000 LMH) which rapidly decayed to a pseudo-steady state value of ~20 LMH. The broth, which had a dry cell weight of 6 g/L, was to be concentrated 13-fold in as short a time as possible to avoid product stability issues; membrane loading was ~116 L/m² in these experiments. Their strategy was to exploit the high fluxes observed early in the concentration to the fullest extent possible; this resulted in collection of 35% of the total permeate volume in the first 10 min of the 90-min process. Cake formation was determined to be the main cause of flux decay, such that higher shear was beneficial while higher transmembrane pressure had little effect. The average cell size was 10 μm, although some smaller particles in the 0.5 to 1.0 μm range were observed in the broth. Increasing the membrane pore size from 0.2 to 0.45 to 0.65 μm (Millipore Durapore, hydrophilic PVDF) resulted in increasing flux, confirming that pore plugging was not a primary source of flux decay. In a series of diagnostic runs, it was determined that the presence of soy peptone, a fermentation medium component, and some other soluble byproduct of the fermentation, were primarily responsible for the flux decay, suggesting reformulation of the medium could significantly improve filtration performance.

Kroner et al. [32] discuss the harvesting of recombinant *E. coli* cells using both flat sheet (Millipore Durapore, 0.45 μm) and hollow fiber (Membrana PP, 0.3 μm) membranes. Starting with a broth containing ~5 % (v/v) cells, they observed a rapid initial drop in flux which leveled off to a pseudo-steady value in the 20–28 LMH range, which could be maintained until the cell concentrate exceeded 40% (v/v packed cells). After this, flux dropped sharply, presumably due to a sharp increase in concentrate viscosity. Comparing the concentration behavior of whole broth with that of washed cells, a significant increase in the pseudo-steady flux, from 28 to 48 LMH, was measured for the hollow fiber system, demonstrating the importance of medium components in the development of the gel layer and fouling. Kroner et al. measured this same general trend for flux versus cell concentration for several other types of bacterial broths, including *Bacillus cereus* and *Brevibacterium* species, although the pseudo-steady flux values and the terminal concentration of the cells was very broth specific.

Junker et al. [27] studied the effect of varying *E. coli* strain and fermentation media composition on crossflow filtration performance for a number of fermentation systems employed in the production of recombinant products. Polysulfone hollow fiber filtration modules with a 500K MWCO (Romicon PM500) were employed in the filtration of broths with batch sizes ranging from 15 to 800 L. Membrane loading varied from 7.9 to 87 L/m², generally increasing with scale. They measured the average filtration flux obtained over the course of the concentration for the broths (various combinations of five different *E. coli* strains and nine different medium compositions), and compared them to the flux obtained for the medium alone. They found significant differences in average flux of some strains relative to others, when grown in the same medium. For instance, JM109 filtered better than DH5 when both were grown in M101 medium–1.5-fold better at the 15-L scale and 3.5-fold better at the 180-L and 800-L scales. In this case, the intrinsic differences in filtration performance are amplified at higher membrane loadings employed at the larger scale. Russotti (1996, unpublished data) also observed large divergences in filtration performance for the filtration of different strains of *Saccharomyces cerevisiae* employed to produce a similar recombinant product.

Varying fermentation medium, Junker et al. [27] found that high-strength, complex media total solids (>90 g/L) typically had low cell-free fluxes (≤30

LMH); flux for broths grown on these media was only a weak function of cell density. Cell-free fluxes were higher for more defined media with total solids below 80 g/L, and cell density was a more important factor in overall performance.

These results support the need for close cooperation between the fermentation and isolation development groups to achieve the best overall process performance in the shortest period of time. The ability to select the fermentation system which simultaneously optimizes the fermentation and the isolation processes is an enticing goal, and one which should be attainable for many recombinant products where there is some flexibility in the choice of the organism, strain, and medium.

VII. MULTICELLULAR ORGANISM/EXTRACELLULAR PRODUCT

Many pharmaceutical, agricultural, and food products are produced by fermentation; the majority of them are biosynthesized by either actinomycetes or fungal cultures. An important subset of these products are water soluble and are secreted from the cells, where harvesting of the broth by crossflow filtration is a feasible process option.

To achieve maximum productivity, these fermentation processes are developed to have the highest level of biomass which can be sustained by the oxygen transfer capabilities of a stirred-tank bioreactor. The volume fraction of suspended solids in such systems frequently reaches 25 to 50 vol% at the end of the process. This high volume fraction of solids combined with the filamentous nature of the organisms results in extremely high viscosities, even before any concentration is performed. These systems are usually pseudo-plastic (shear thinning), complicating their characterization, but high apparent viscosities are common even under highly sheared conditions. In practice, these characteristics limit the extent to which the broth can be concentrated, which is an issue because, as was noted in Section V, concentration of a broth to the fullest extent possible prior to initiating diafiltration produces the most concentrated product stream and minimizes diafiltrate volume. Concentration factors of 1.5- to 2.0-fold are typical; in the best of cases a threefold concentration factor can be obtained.

In most cases, the product of such processes is relatively small, with molecular weight <2000. Under these conditions, the product is usually freely permeable through the membrane, even under conditions in which a significant gel layer or cell cake is present. This means that even though limited product can be recovered by the initial concentration of the broth due to its high viscosity/volume fraction of solids, recovery of product during the subsequent diafiltration is relatively efficient. This is in contrast to protein recovery, where partial retention of the macromolecular product by the membrane results in extended diafiltration to achieve an acceptable yield (Section V).

An example of such a process is that employed in the production of cepha-mycin C, an antibiotic biosynthesized by *Nocardia lactamdurans* [22]. The broth is first acidified to solubilize the product, then the batch is concentrated threefold. After initial concentration, the cell slurry is diafiltered with water until 98% of the cephamycin (MW 398) has been recovered in the collected permeate. As is common for many such processes, the product is adsorbed from the permeate and further purified using ion exchange chromatography. Although the process requires only a physical separation (the removal of mycelia and cell debris), screening experiments indicated the use of a UF membrane resisted fouling (Dorr Oliver/Amicon XP-24, vinyl chloride-acrylonitrile copolymer with MWCO of 24,000). This is not uncommon, since cell debris can sometimes cause rapid flux loss with microfilters due to pore plugging. The membranes were provided in a flat-sheet, open-channel cartridge device amenable to pumping of the high viscos-ity broth concentrate. For this 15,000 gallon-scale operation, a feed-and-bleed loop system was employed (see Figure 5), which provided optimum membrane performance while minimizing the amount of high-flow equipment, especially with respect to flow in and out of the feed tank; centrifugal pumps were employed in this process.

Gravatt and Molnar [22] stress that two factors were critical to the commer-cial viability of this process, membrane regeneration and automation. Their sys-tem employed a very large amount of membrane surface (11,520 ft^2); to be eco-nomical, the membrane had to provide consistent performance through many hundreds of batch cycles. The membrane selected was stable to the rigorous cleaning regimen which was employed to regenerate flux between batches. Also, while this harvest procedure was relatively complex, it was amenable to automa-tion, allowing consistent performance and reasonable labor costs.

Another example of this type of application is the isolation of physostig-mine from a *Streptomyces griseofuscus* broth [55]. Physostigmine (MW 275), the precursor of a memory-enhancing drug candidate, is secreted during the fer-mentation. This carbamate is a neurotoxin, so containment was important in addi-tion to the standard criteria for a successful harvest. In preliminary screening experiments, a 300,000-MWCO polysulfone membrane provided better flux than 0.2 μm PVDF or 10,000-MWCO regenerated cellulose membranes (Millipore membranes), again suggesting avoidance of pore plugging by particulates by moving from MF to UF . But even the "best case" 300K membrane only pro-vided an average flux of 15 LMH. Subsequent modification of the fermentation process, by replacing soybean oil in the medium with glucose feeding during cultivation, resulted in an increase of average flux from 15 to 70 LMH, with obvious economic benefits. Moving to pilot scale, the authors were able to main-tain a 70-LMH average flux when using an available ceramic microfiltration sys-tem with 0.2-μm pore size, which allowed for rigorous between-batch regenera-tion. (At that time, the only ceramic UF available to us had a 50,000-MWCO, and this was found to have too high an intrinsic hydraulic resistance, such that observed flux was higher for the microfiltration version of the membrane.) The broth contained 30 vol% cell solids at harvest; based on laboratory experience, a twofold initial concentration was not exceeded to avoid flow channel plugging. Excellent reproducibility was achieved in a five-batch campaign on the 200-L scale with consistent process fluxes; a 90% physostigmine yield was approached by following a twofold concentration by two 1:1 batchwise diafiltrations.

It should be noted that the two large-scale applications described above employed centrifugal pumps. These pumps are available to provide extremely high flowrates—in the 10^2 to 10^3 gpm range—at reasonable costs. These pumps can process broths up to fairly high solids concentrations (50 to 60 vol%), although the shear-thinning nature of the broths may make it difficult to restart recirculation at these high solids levels. When configured in the two-stage feed-and-bleed loop arrangement shown in Figure 5, single-stage centrifugal pumps can easily provide feed pressures of ~100 psi to membrane modules. Most available MF and UF modules have pressure ratings <100 psi, making the higher pressure capability of positive displacement pumps (PD pumps, e.g., rotary lobe pumps) of limited benefit. PD pumps are also usually much more expensive and their scale-up is limited. Downsides to the use of centrifugal pumps are the relatively high shear rate at the tip of the impeller; tip speeds of a centrifugal pump can be expected to be in the range of 20 to 45 m/s versus 0.5 to 5.0 m/s tip speed for the lobe of a typical rotary lobe pump. For these applications, cell damage is not a concern with respect to product recovery, but in some specific systems cell lysis could result in flux reduction below acceptable values.

Haarstrick et al. [23] describes a harvest operation involving a multicellular organism with a secreted product of a different type—the polysaccharide schizophyllan (MW~1.1×10^7), which is produced by the fungus *Schizophyllum commune* (ATCC 38548). In this case, the authors employed a ceramic microfilter with 1.0-µm pores to remove the cells and then flat-sheet hydrophilic PVDF membranes with 0.1-µm pores to concentrate the polymer product. While cell concentrations in this study were significantly lower than those typically encountered in the fungal fermentation of secondary metabolites, only 0.02% to 0.7% (w/v), the presence of the polymer resulted in high viscosity. In the cell separation, retention of the cells was not difficult, but permeation of the polymer was not complete and decreased through the process.

VIII. MULTICELLULAR ORGANISM/CELL-ASSOCIATED PRODUCT

A large sector of industrially important fermentation broths is not well suited to processing by crossflow filtration—the harvest of multicellular microrganisms *with a cell-associated product*. This includes a significant number of fermentations employing strains of *Aspergillus* and *Nocardia* and thousands of less-well-known mold and actinomycetes cultures. In many cases, the metabolic products of these fermentations are not water soluble, and remain associated with the cells until extracted with solvent or by pH change. To maximize productivity of these processes, the fermentations are developed to achieve the highest cell density that can be supported by stirred tank aeration, often achieving 25 to 40 vol% and higher suspended solids at the completion of the process. This high volume fraction of solids combined with the filamentous nature of the organisms results in extremely high viscosities, even before any concentration is performed. These systems are usually pseudo-plastic (shear thinning), complicating their characterization, but high apparent viscosities are common even under highly sheared conditions.

Given these characteristics, the use of crossflow membrane filtration as a harvest technique for a cell-associated product usually makes little practical sense. The limited concentration which can be achieved in these cases has severe repercussions for both equipment size and process efficiency. A separate harvest vessel, if employed, must be almost the size of the fermenter, with the pumping system and membrane area sized accordingly. The volumes of diafiltrates required to adequately wash the cells and then extract the product are very large—typically several times the fermentation volume—and the resulting extract stream is more dilute than the original broth. All these factors make the process a poor choice when compared to its alternatives, such as whole-broth extraction using an immiscible solvent.

In the rare case in which this type of system is grown to a low cell density, crossflow harvest is feasible and the system can be treated as described in Section V. Otherwise, there are some very specific cases in which the benefits of a crossflow filtration harvest outweigh the problems described above. These include cases in which issues of process containment are paramount, and those in which product titer is extremely low (as in the early development of a process).

In the case of a low titer broth, it is possible to implement a CFF-based extraction process which minimizes volumes by coupling the CFF concentration and extraction to an adsorption step. Such a process has been described by Reeder and Göklen [53] for the recovery of a low-titer cell-associated natural product from a fungal broth. In their system, the broth had a relatively modest cell density of 12 vol%, and an initial apparent viscosity of 30 cP (shear rate of 100 s^{-1}) and a very low product titer. Apparent viscosity increased to 100 cP and 300 cP after a 1.8- and 2.9-fold concentration, respectively. Their process is diagrammed in Figure 16. The broth is initially concentrated twofold and then diafiltered with an aqueous wash solution to remove soluble polar impurities. The washed concentrate is then diafiltered with a solvent, in this case methanol, until a solvent concentration sufficient to partially solubilize the product from the cell solids is achieved. As product is detected in the diafiltrate, it is directed over an adsorption

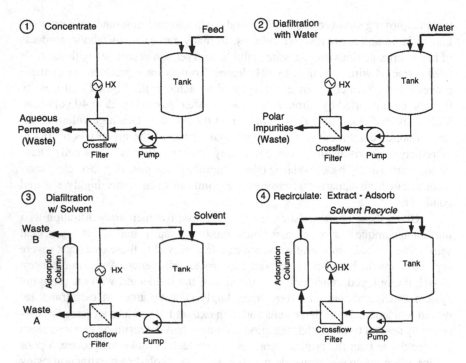

Figure 16   Integrated CFF-based extraction/adsorption process scheme.

column to capture the product on a resin. When the solvent level in the broth reaches the target concentration, the diafiltrate which has passed through the column is recycled to the CFF harvest tank. The filtration-extraction-adsorption is continued until the desired yield is achieved; no additional solvent is required to complete the extraction.

With a properly-sized column (or series of columns), the system can be operated in a closed loop which minimizes the solvent required. The process essentially transfers the product from the fermentation cell solids to the adsorbent solid in the column. With appropriate selection of column elution conditions, the product is recovered in a concentrated and purified solvent stream. This process requires the availability of solvent-resistant membranes; Reeder and Göklen [53] employed a ceramic microfilter (Ceraflo by US Filter) for this purpose. The process achieves the goal of recovering the product in high yield with limited solvent use, but is rather complex and cycle time is long. As such, use of this scheme is limited to those situations noted above—low-titer broth with high-value products and products which require excellent containment.

IX. DEVELOPMENT AND OPTIMIZATION OF A CROSSFLOW FILTRATION PROCESS

When examining the feasibility of utilizing a crossflow filtration process for the harvest of a fermentation broth, one must consider two issues: filtrate flux and product permeation or retention.

Regardless of the application, it is desirable to achieve as high of a flux as possible for several reasons. High flux is beneficial from an economic standpoint since greater flux requires less membrane area and often less pumping capacity. Whether the membrane will be regenerated or not, smaller membrane area requirements will decrease capital costs, for example, of the filtration skid itself, as well as ongoing membrane costs. A smaller pump will also reduce capital costs as well as the cost of utilities associated with the cooling of the system, the requirements for which will increase with increasing pump size. Furthermore, reduction of membrane area and pump size will minimize the total space needed for the crossflow filtration unit operation, which will in turn reduce facility cost. If the process is to be implemented in an existing facility, there may be constraints on the amount of space available, which also illustrates the benefit of achieving the highest flux possible to reduce space requirements. In addition to economics and facility constraints, higher flux can be advantageous if a short cycle time is required, which may be the case if product stability is an issue.

Product permeation or retention is an obvious consideration since the goal of any downstream process unit operation is to obtain as high of a yield as possible. As discussed earlier, product permeation is more of an issue than product retention; i.e., it is more difficult to permeate a product in the cell-free solution than it is to retain a product that is intracellular or is in some other way associated with the cell-containing stream. In either case, one may be interested in permeating or retaining other components of the fermentation as well. Therefore, it is helpful to clearly define goals with respect to permeation or retention of the product as well as these other key components before examining the feasibility of a crossflow filtration process.

Initial experiments should be designed to examine the feasibility of a crossflow filtration process with respect to flux and permeation/retention of the product and other key components of the fermentation. A laboratory scale system, similar to the one shown in Figure 17 can be used for such experiments. Such a system should consist of a feed reservoir that can be agitated, for example, by a magnetic stirrer, a positive displacement pump such as a peristaltic pump, and a membrane device. Any membrane configuration can be chosen for these preliminary experiments though membrane module selection is an important consideration in the development of a crossflow filtration process as discussed in Section II. It is helpful, though, to choose a device that can accommodate many different mem-

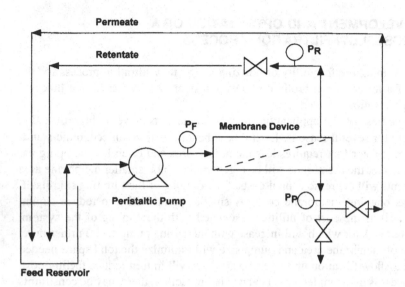

Figure 17 Laboratory-scale crossflow filtration system.

brane types, as discussed below. Membrane manufacturer recommendations on such parameters as starting whole broth volume per membrane area (defined as load), TMP, and crossflow rate are a good starting point. Optimization of these parameters is an important aspect of the development process that should be investigated after first determining whether crossflow filtration is a feasible option, and will be discussed in subsequent sections.

In these initial experiments, whole broth should be concentrated to a degree which will be representative of the final desired concentration factor and flux should be measured over the course of the concentration. Various membrane materials and pore sizes should be examined in these initial screening experiments. Flux may be strongly dependent on membrane material and/or pore size depending of what the main inhibitor of flux is, whether it be cake formation, concentration polarization, physical pore plugging, or surface or pore fouling through chemical adsorption. Hence, this approach can guide one in the ultimate membrane selection and will also allow one to determine the range of fluxes one might encounter for a particular application.

In these same experiments, which are designed to examine flux with a variety of membrane types, samples of the retentate and permeate streams should be assayed for product as well as other compounds of interest. Depending on the characteristics of the product and the other constituents of the fermentation, chemical adsorption to membrane surfaces can occur thereby justifying the need for examining various membrane materials. It is also often necessary to examine a range of membrane pore sizes or MWCOs since it typically cannot be known *a priori* whether a specific compound will permeate or will be retained by a specific membrane type. MWCOs and pore sizes are not absolute; they are usually defined by the manufacturer using some standard solution of compounds of varying molecular weights. These standard compounds may or may not be representative of the compounds present in one's particular application in regard to such characteristics as shape, charge, and hydrophobicity; more often than not, they are not.

It is understood that flux and permeation/retention of product and other components will be further improved later through a series of optimization experiments in which the key operating parameters are varied. However, if flux or permeation/retention is more than a fewfold lower than the value that is necessary to make the process feasible, one must consider more than just optimization of the main parameters. In these instances, it becomes necessary to understand the main cause of poor filtration performance so that either the upstream conditions or the filtration design can be altered appropriately. Such an approach will be discussed in later sections.

X. OPTIMIZATION OF FLUX

Once it is established that crossflow filtration is a feasible option for the harvest of a fermentation broth, flux may be optimized to reduce membrane area requirements and/or cycle time. The main operating parameters, which will affect flux and which must be characterized in such an optimization exercise, are TMP, shear rate or crossflow rate, membrane pore size, membrane load, and process temperature. A laboratory scale system similar to the one shown in Figure 17 may be used for optimizing flux or a pilot-scale system can be used. In either case it is important to use the membrane configuration that will ultimately be used in the final scaled-up process since the effects of TMP and shear rate will be strongly dependent on membrane configuration. An illustration of a pilot-scale system that can be used for optimizing flux is shown in Figure 18. The major differences between a pilot-scale system and a laboratory-scale system are that in the pilot-scale system the feed tank agitation method and the pump type are the same that will be used in the final design and temperature control is attainable by a heat exchanger and/or a jacketed feed vessel.

When examining the effect of the various parameters on flux, it is important that flux be examined over the course of the concentration. However, in order to obtain a preliminary idea of what range of operating parameters will be important to examine, the effect of these on steady state flux can first be investigated. This can be performed by simply recycling the permeate to the feed vessel such

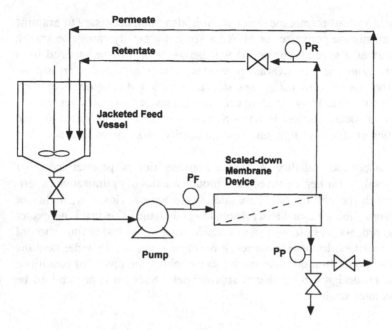

Figure 18   Pilot-scale crossflow filtration system.

that whole broth is not being concentrated at all. Flux should be measured under these conditions of complete recycle at a fixed TMP, shear rate, and temperature over time. Once flux has reached some steady-state value, these parameters can be varied and flux can be measured under various conditions such that a large quantity of data can be obtained in a relatively short amount of time. For example, various TMPs can be examined at a constant shear rate. TMP can be varied by applying back pressure on the retentate side of the membrane at the outlet of the system with the use of a valve such as a diaphragm valve. A typical flux versus TMP relationship is shown in Figure 19, which contains a pressure-dependent regime in which flux increases with increasing TMP and a pressure-independent regime where further increases in pressure have no effect on flux. Generally, one should attempt to operate at the transition point between these two regimes in order to maximize flux and minimize detrimental effects such as compression of the cake on the membrane surface and increased pore plugging. Similarly, various shear rates can be examined at a constant TMP to obtain data such as shown in Figure 20. Increasing shear rate will usually result in an increase in flux due to greater convection forces which will minimize cake formation and concentration polarization effects. Beyond a certain shear rate, flux will level off as seen in Figure 20; one should operate at this critical level of shear rate which results in

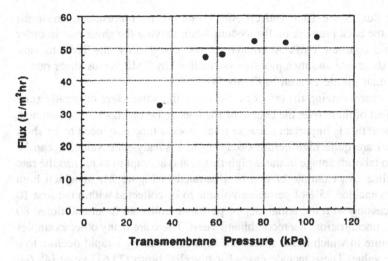

Figure 19 Steady-state flux vs. transmembrane pressure at constant crossflow rate for a ceramic membrane system. (From Ref. 55 with permission from John Wiley & Sons, Ltd.)

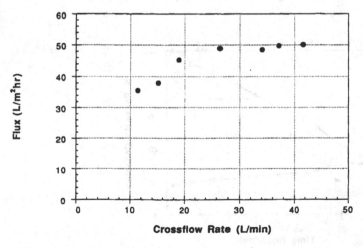

Figure 20 Steady-state flux vs. crossflow rate at constant transmembrane pressure for a ceramic membrane system. (From Ref. 55 with permission from John Wiley & Sons, Ltd.)

the highest flux. Since TMP and crossflow rate are interdependent, one must manipulate the back pressure on the system when varying the shear rate in order to maintain a constant TMP. Alternatively, one could allow the TMP to vary along with shear rate and then plot flux normalized by TMP versus shear rate to obtain a similar profile as seen in Figure 20.

After characterizing the effect of the operating parameters on steady state flux, the effect of these over the course of the concentration should be examined. In some cases, this is important because total process time may need to be short and if fluxes are quite high before decaying to a steady-state value, it can be beneficial to take advantage of these high fluxes and attempt to minimize the rate of flux decline. An example of this is illustrated in Figure 21, in which high initial fluxes enabled 35% of permeate volume to be collected within the first 10 min of processing with the remainder of permeate collected in an additional 80 min for the concentration of a recombinant yeast. There are many other examples in the literature in which initial fluxes are high followed by a rapid decline to a steady state value. These include cases for mycelial broth [23,61] yeast [48,64] and bacteria [19].

In such cases, steady-state data do not always provide a clear choice of the optimum values of these parameters. For example, it is difficult to predict how TMP will affect a flux profile. Some researchers have reported both high initial fluxes as well as high steady-state fluxes with increasing TMPs [56,64]; examples of this effect for the concentration of yeast are shown in Figures 22 and 23. However, in other examples, high TMPs result in higher initial fluxes but a more rapid decline in flux resulting in a lower steady-state flux [48], as illustrated

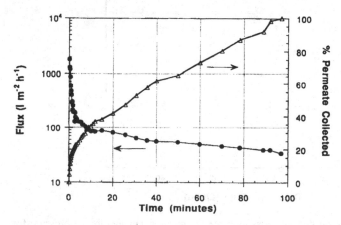

Figure 21 Crossflow filtration performance for the harvest of *Saccharomyces cerevisiae* cells. (From Ref. 56 with permission from Elsevier Science.)

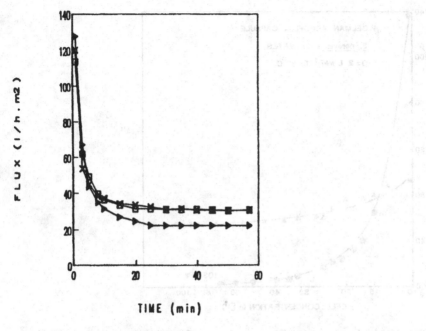

Figure 22 Higher initial and higher steady-state flux with increasing transmembrane pressure: example for yeast cell harvesting from cider. (□) 1000 kPa, (x) 600 kPa, (▶) 200 kPa. (From Ref. 64 with permission from John Wiley & Sons, Ltd.)

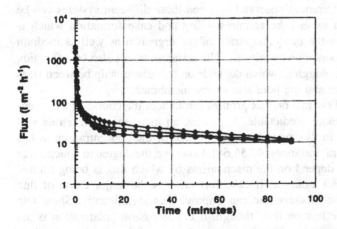

Figure 23 Higher initial and higher steady-state flux with increasing transmembrane pressure: example for harvest of recombinant *Saccharomyces cerevisiae* cells. (●) 55 kPa, (△) 75 kPa, (◊) 95 kPa. (From Ref. 56 with permission from Elsevier Science.)

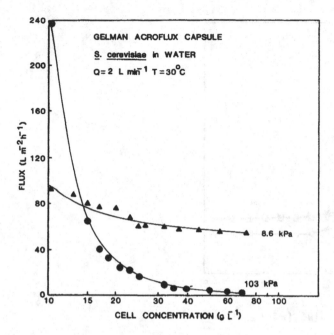

Figure 24 Higher initial but lower steady-state flux with increasing transmembrane pressure: example for filtration of *Saccharomyces cerevisiae* cells. (•) 103 kPa, (▲) 8.6 kPa. (From Ref. 48 with permission from Elsevier Science.)

in Figure 24. The differences observed between these different systems can be attributed to differences in cake compressibility and cake formation which is often dependent upon the particular strain of the organism as well as medium composition and fermentation conditions. The differences can also be a function of the degree of pore plugging, which depends on the relationship between sizes of particulates present and the pore size of the membrane.

The effect of shear rate on flux profiles throughout the course of a concentration is generally more predictable. Typically, an increase in shear rate will result in an increase in flux over the entire course of the concentration, as has been reported in several instances [48,56,64]. However, the degree to which shear rate affects flux will depend on the mechanisms by which flux is being limited in a given system. For example, if cake formation is the major cause of flux reduction, an increase in shear rate can improve flux significantly. Shear rate could have a lesser effect on flux, though, if concentration polarization is the limiting factor and nearly no effect if chemical fouling of the membrane surface or chemical fouling or physical blocking of the membrane pores is inhibiting flux.

Membrane pore size or MWCO is another parameter which can have a strong effect on flux. The choice of this parameter, of course, depends on the goals of the process with respect to permeation and retention of the key components, as discussed earlier. However, the size of the cells and other particulates which may be present in the whole broth must also be considered. If, for example, the membrane pore size is on the same order as the size of the cells, pore plugging can be a problem and higher flux and less flux decline over the course of the filtration can be observed with decreasing pore size. Such a trend was seen by Patel et al. [48] for the concentration of yeast with membranes with pore sizes from 0.2 to 0.007 μm. In some cases, it may even be desirable to choose UF membranes over MF membranes. Gatenholm et al. [19] reported less flux decline and higher steady-state fluxes for the filtration of *E. coli* cells using UF membranes as compared to MF membranes. It should also be mentioned that with UF membranes the controlling resistance is often supplied by the membrane itself, whereas with MF membranes the resistance of the cake is often greater than the resistance of the membrane. Since cake properties can vary over the course of a filtration process, flux using a MF membrane will often vary while flux using a UF membrane will be more consistent. Russotti et al. [55] report such a phenomenon for the concentration of an Actinomycetes broth in which a 50K-MWCO membrane gave a more constant flux than a 0.2-μm ceramic membrane, though in that case fluxes were higher with the MF membrane over the entire course of the concentration. Generally, one should select the largest possible pore size that does not result in significant pore plugging in order to obtain the highest possible fluxes. A practical example of this was in the case of the harvest of a recombinant strain of *Saccharomyces cerevisiae* that had a diameter of approximately 10 μm. It was demonstrated that higher fluxes throughout the course of the concentration were obtained when increasing pore size from 0.2 to 0.45 to 0.65 μm [56]. Even pore sizes as high as 3 μm have been reported to be successful for the filtration of yeast [29].

When developing and optimizing a filtration process, one does not often operate at the volume to membrane area ratio, or load, that will be utilized at full scale. Constraints such as the smallest membrane size available as well as the amount of precious fermentation broth that can be spared early in the development stages of a project often prevent this. At some point in the development process, though, one must consider whether a higher load will affect filtration performance. Regardless of the major inhibitor of flux, whether it be cake formation, concentration polarization, chemical fouling, or physical pore plugging, one can hypothesize that a greater amount of fermentation broth per membrane area translates to a greater amount of the flux inhibitor per membrane area. Therefore, one must examine the effect of load on filtration performance over the course of the concentration to determine if and to what extent flux will be affected. This will allow the specification of the optimum membrane area for a given crossflow filtration application.

The choice of the operating temperature depends upon a number of factors including product stability, whole-broth viscosity, solubility of broth constituents, and antifoam cloud point. Lower temperature is usually beneficial with respect to product stability but the resulting higher viscosity can lead to lower flux. Moreover, some medium components can precipitate at lower temperatures and such precipitation can adversely affect flux. On the other hand, in cases where an antifoam was used in the fermentation, operating at temperatures that exceed the cloud point of the antifoam have been shown to reduce flux [8,30,32]. A general rule of thumb is to operate at the highest possible temperature, that is, below the antifoam cloud point and that does not compromise product stability.

XI. PROBLEM DIAGNOSIS: LOW FLUX

In some instances, flux can be significantly lower than the level which is necessary to make crossflow filtration a feasible option for the harvest of fermentation broth. It is helpful, in these cases, to try to identify the specific cause for low flux. If it can be determined that a particular component of the fermentation is a major detriment to flux, it is often possible to redesign the fermentation to reduce the level of or even eliminate this component. One can take a simplistic approach and view the fermentation broth as consisting of unmetabolized medium ingredients, products of the fermentation, and the cells themselves. There are many examples in the literature, which will be discussed below, which have shown that one or more of these components can be acting individually or in combination with others to inhibit flux.

There are many medium components that can be detrimental to flux. These include hydrophobic compounds, such as lipids, antifoams, and complex medium components as well as compounds that may have a tendency to precipitate out of the fermentation broth. An example of a hydrophobic compound that was shown to inhibit flux is soybean oil, the removal of which from the fermentation medium of a *Streptomyces griseofuscus* culture resulted in a fourfold higher flux [55]. Antifoams, which are commonly used in fermentations, have been shown to result in chemical fouling since their hydrophobic nature causes adsorption to many membrane materials. Reduction in flux due to antifoam presence has been demonstrated in bacteria [30,32] as well as yeast cultures [8]. Complex medium components can also be problematic either by direct interaction with the membrane or by changing the composition of the cake on the membrane formed predominantly by cells. A lesser degree of flux decline was observed during the concentration of a recombinant yeast when soy peptone was eliminated from the fermentation medium [56]. Finally, medium components that precipitate out of solution can reduce flux. Nagata et al. [43] reported that the precipitation of magnesium ammonium phosphate, which occurred during the steam sterilization of the medium, was a major foulant in the filtration of *B. polymyxa* culture.

Medium design can also indirectly affect flux by altering the product profile of the fermentation. In some cases, certain products of the fermentation can be detrimental to filtration performance. For example, Göklen et al. [21] observed that the level of a fermentation byproduct, which was suspected to inhibit flux, could be significantly reduced and flux could, as a result, be significantly improved by redesigning the fermentation medium.

In many applications, it is the cells themselves that are deterring filtration performance, for example, by forming a cake on the membrane surface that offers a great deal of resistance to flux. While cell concentration, in many cases, cannot be reduced—for example, if product concentration is dependent upon cell concentration—the cells can sometimes be altered. Substitution of one cell strain for another can be a simple solution to poor filtration performance as long as this does not adversely affect fermentation performance. Junker et al. [27] have reported differences in flux using different strains of *E. coli*, all of which were suitable for producing recombinant products. The state of the cells at the time of harvest can also affect flux. For example, higher flux was obtained when filtering *A. pullulans* when cells were in the yeastlike form as opposed to when they were filamentous [77]. In that case, more mycelia formed with increasing culture age resulting in lower fluxes for cultures that were harvested at later time points. Cell surface characteristics, which can be affected by the fermentation conditions, can also have a strong impact on flux. Taddei et al. [64] observed that flux was strongly dependent on aeration conditions during the fermentation of yeast and also on the number of nonviable cells present. Both aeration conditions and cell viability were shown to alter the yeast surface, which was believed to cause the differences in flux. Others have also shown that the age of the culture, growth rate, and growth conditions influence the attachment of yeast to membrane surfaces by affecting their surface properties [15,17]. Hence, very often it is possible to change culture conditions in order to affect a change in the state of the cells which can positively influence filtration flux.

An empirical approach which is often successful in identifying the cause of poor flux might be called "divide and conquer." The results from the filtration of whole broth, cells resuspended in buffer, cells resuspended in medium, medium alone, and cell-free supernatant are all compared. The filtration of whole broth can be thought of as a negative control which contains all components of the fermentation—cells, unmetabolized medium components, and products of the fermentation—one or more of which can be the cause of low flux. Filtration of cells can be examined after separating the cells from the other fermentation components, for example, by centrifugation and then resuspending the cells either in buffer or medium. Filtration of cells in buffer allows one to examine the effect of cells alone on filtration performance. Filtration of cells in medium allows one to investigate the interaction that the cells may have with unmetabolized medium components. Filtration of medium alone or of certain medium components will help identify if any unmetabolized medium components are directly fouling the membrane. Lastly, filtration of supernatant from the fermentation whole broth, which can be obtained by centrifugation, will indicate whether either unmetabolized medium components or products of the fermentation are directly fouling the membrane or are interacting with other components to do so.

Such an empirical approach was taken in order to identify the reason for rapid flux decline during the concentration of recombinant yeast by Russotti et al. [56]. In this study, it was shown that a particular unmetabolized complex medium component—soy peptone—was in part responsible for rapid flux decline. Cells resuspended in buffer or cells resuspended in medium without soy peptone resulted in much less flux decline than cells resuspended in medium containing soy peptone. Studies using yeast cultures grown in the presence and absence of soy peptone confirmed this finding since filtration of those without soy peptone had higher fluxes. In a separate example, uninoculated whey permeate which was used for the culture of *Lactobacillus* was shown to give significant flux decline thus showing that medium itself can be the cause for poor filtration performance [65]. Kroner et al. [32] also observed that medium itself can inhibit flux. They reported a 70% increase in flux when bacterial cells were resuspended in buffer as compared to when they were resuspended in medium. This might indicate that the medium was fouling the membrane or that a cell-medium interaction was the cause for lower flux.

An alternate diagnostic approach which aims to identify the mechanistic cause for low flux might be called "bait and switch." After a typical crossflow filtration experiment, one rinses the membrane and filtration system with water a number of times using a shear rate and TMP similar to that used in the filtration process itself. Water flux can then be measured and compared to water flux through a virgin membrane and to process flux. If cake formation and/or concentration polarization is the main reason for low flux, water flux after rinsing with water alone can be much greater than process flux and may even approach the levels of water flux measured through a virgin membrane. This may not be the case, however, if cake formation is irreversible—for example, if the cake is highly compressed. If chemical fouling on the membrane surface or within the pores of the membrane and/or physical pore plugging is the main deterrent of flux, then the water flux after rinsing alone should not be much higher than process flux.

One example of the former situation has been reported for the filtration of *Saccharomyces cerevisiae* [56]. In that case, water fluxes measured at various TMP's after rinsing a membrane with water alone were approximately 85% of the values of water flux for a new membrane. This, along with other data, lead the authors to hypothesize that flux decay was mainly due to the formation of a reversible cake layer. In a separate example, however, using *Saccharomyces cerevisiae*, water flux after a water rinse was only 1% of the value measured for a new membrane suggesting that reversible cake formation was not the major cause for flux decay [64]. In that case, further investigation would be necessary in order to determine what the actual cause might be.

The use of microscopy techniques, such as transmission electron microscopy (TEM), scanning electron microscopy (SEM), or field emission scanning electron microscopy (FESEM), can also identify causes for low flux by allowing one to visualize the membrane surface or membrane pores. Liew et al. [34] utilized SEM to examine the surfaces of both hydrophilic and hydrophobic 0.2-μm membranes after filtration with *Candida utilis*. They found very little adsorption of intact cells to membranes but a great deal of adsorption by cell fragments. FESEM, in some cases, can be better than SEM since it does not require the high beam energies associated with SEM that can damage the sample. Kim et al. [28] used FESEM to examine fouling of ultrafiltration membranes by albumin solutions. Fouling occurred predominantly on the surface of the membrane, either as cake formation or as aggregation of proteins depending upon filtration conditions. No protein was detected inside the membrane pores by FESEM although the authors cautioned that for the pore size of the membrane studied, one or two layers of protein could reduce flux and that such a low level of protein could not be detected by FESEM. Another physical characterization technique that has been used to examine membrane surfaces and membrane pores is atomic force microscopy (AFM). Bowen and Hall [5] used AFM to demonstrate that in-pore deposition of yeast alcohol dehydrogenase occurred under certain conditions while surface deposition occurred under other conditions using 0.1-, 0.2-, or 0.8-μm microfiltration membranes.

XII. OPTIMIZATION OF PERMEATION/RETENTION

As discussed earlier, it is beneficial to set goals with respect to permeation and retention of products and other key components of the fermentation broth early in the development of a crossflow filtration process. Obviously, a reliable assay must be available to quantitate product in the various streams of the crossflow filtration process. In addition, assays to track specific fermentation components of interest should be in place. If one has not identified specific components that must be retained or cleared during the course of the filtration, one may want to obtain general information about what classes of components are being retained or cleared. In this case, one can choose impurities of representative molecular weights, for which there are readily available assays, and quantitate the retention and clearance of these. Forsyth et al. [18] did just that when they examined the retention and permeation of succinate, which was chosen as a representative low molecular weight component, and antifoam, which was chosen as a representative high-molecular-weight component, during the concentration and diafiltration of a recombinant yeast.

When optimizing the permeation and retention of product and other compounds, it is necessary to examine permeation/retention over the entire course of the concentration. Several examples of decreasing product permeation occurring over the course of a concentration have been reported [11,16,23,43]. Changes in the resistance of the cake layer due to, for example, compaction of the cake, or in the resistance of the concentration polarization layer, which can occur with increasing concentration of retained components, are potential reasons why permeation can decrease as a stream is concentrated.

Key operating parameters, such as transmembrane pressure and shear rate, which are known to strongly influence flux, must also be considered in regard to their effect on permeation and retention. Higher transmembrane pressure, for example, can result in a greater degree of permeation of a compound that is only slightly smaller in size than the pore size or the MWCO of the membrane being used. Alternatively, higher transmembrane pressure can increase the compaction of the cake layer on the membrane, which can in turn result in a greater resistance to permeation. Higher shear rate may or may not result in a greater degree of product permeation depending upon the main source of resistance to permeation. If cake formation or the concentration polarization layer is inhibiting product permeation, then it is likely that an increase in shear rate will improve product permeation. However, if product permeation is being prevented due to other factors, an increase in shear rate may have no effect.

XIII. PROBLEM DIAGNOSIS: POOR PERMEATION/ CLEARANCE

In cases where product permeation or permeation of other important compounds is particularly poor, an investigation of the causes of poor permeation can be undertaken. Very often, one or more components of the fermentation broth might be the cause of poor permeation by interacting with the product of interest or by interacting with the membrane to reduce permeation. However, several other possible reasons for product loss should be ruled out before assuming there is a product permeation problem. To ensure that product is not being adsorbed to the membrane material, entrapped in its pores, or is being degraded during the filtration process, one should measure not only product concentration in feed and in the permeate, but also in the concentrated retentate so that a mass balance can be performed. If 100% of the product can be accounted for, it would suggest that product is not being lost for one the above reasons and that there could be a problem with product permeation. If the mass balance cannot be closed, however, one can further examine these possibilities by adding purified product to buffer and again measuring product in the feed, permeate, and retentate streams over the course of a concentration. If the mass balance is still <100%, this would indicate that either product adsorption to the membrane, entrapment in the membrane pores or product degradation is occurring. Product stability data at the temperatures of interest should be obtained to rule out the simple possibility that product is being degraded over the course of the concentration. Furthermore, product stability under relevant flow conditions, preferably using the same pump and flow design as that of the crossflow filtration system but without the membrane, should also be examined to rule out that product is being degraded due to the shear associated with the filtration process.

Once it is proven that the product is stable under the filtration conditions, one can examine whether product is adsorbing to the membrane material. One way to do this is to contact a product-containing stream with the membrane material but to not filter anything through the membrane. This can be accomplished in the crossflow filtration system itself by preventing any flow through the permeate line of the membrane—for example, by closing a valve on the permeate line. It can also be achieved by placing a sample of the membrane material in a solution of a product-containing stream, for example, in a well-mixed container. Product concentration can be measured over time in either of these systems; if product concentration remains constant, then membrane adsorption is not likely to be an issue. If there does appear to be loss due to membrane adsorption, several membrane materials can be tested in a similar fashion. If it is known that product is not being degraded over the course of a concentration and it has been proven that membrane adsorption is not an issue for the membrane of interest, then obtaining a mass balance <100% for a solution of product in buffer would suggest that pore plugging is a problem. To examine this further, one can simply examine membranes of the same material but with increasing MWCOs or pore size to determine if permeation can be increased.

If filtration conditions including process temperature, TMP, shear rate, membrane material, and pore size have been specified such that acceptable product permeation is expected but is not being obtained, then one should examine if a fermentation component is responsible for poor permeation. The same empirical approach, as discussed earlier with respect to determining if a fermentation component is responsible for low flux, can also be utilized to determine the cause for poor permeation. Comparing the permeation of product in whole broth, in buffer with or without cells, in medium with or without cells, and in cell-free supernatant can give insight as to whether the cells themselves, an unmetabolized medium component, or a product of the fermentation is the reason for poor permeation. Göklen et al. [21] employed such a strategy to determine why permeation of an enzyme was poor during the microfiltration of bacterial broth. They first demonstrated that if they spiked product into medium, they obtained 90% product permeation as compared to only 20% for the case of whole broth under similar filtration conditions. This proved that the product itself was not preventing its own permeation, for example, by blinding the membrane, nor were any unmetabolized medium components preventing product permeation. Furthermore, product in a cell-free supernatant stream gave poor permeation thereby proving that cells and other particulates that might be present in whole broth were not the culprit. This led the authors to believe that byproducts of the fermentation were the cause of low enzyme permeation. This hypothesis was supported by the fact that whole broths with less-than-optimal enzyme activity, and, therefore, potentially different spectrums of fermentation byproducts, gave better enzyme permeation. The solution was to redesign the fermentation medium with the goal of reducing or eliminating the byproducts which were inhibiting enzyme permeation. This approach was found to be successful as permeation was improved approximately threefold.

Although fermentation components can directly impair product permeation by interacting with the product or with the membrane, fermentation conditions can also indirectly affect product permeation by altering the physical or chemical state of the product. For example, product aggregation or changes in product charge can have an effect on permeation and can be impacted by fermentation conditions. One must be especially wary of such occurrences if product easily permeates in some cases but not in others when using the same membrane under the same crossflow filtration conditions. Bowen and Hall [5] provide an example where permeation of an enzyme is high only for a narrow range of pH and ionic strength. It was shown using photon correlation spectroscopy that the enzyme existed as discrete molecules instead of aggregates only under these conditions.

Optimization of filtrate flux and of permeation/retention of product and other compounds are the key challenges that one typically faces when developing a crossflow filtration process. In some cases, optimization can be rather straightforward and a good systematic approach is the most efficient way to achieve one's goals. In other instances, there are major obstacles that one encounters which require a great deal of troubleshooting. There are a number of approaches that one could take to perform such optimization and troubleshooting exercises and just a few of these have been described here.

XIV. MEMBRANE REGENERATION

Membrane regeneration is a critical issue when considering the economics of a crossflow filtration process. For many applications, one must be able to regenerate membranes efficiently and reproducibly such that process performance is consistent with respect to flux and permeation. Membrane regeneration efficiency is typically evaluated by comparing water flux after regeneration to water flux through a virgin membrane at the same process conditions—i.e., transmembrane pressure, crossflow rate, and operating temperature [41]. A good practice is to measure water flux at a series of increasing transmembrane pressures, at a constant crossflow rate, for example, at the process crossflow rate, and at a single temperature, for example, at the process operating temperature or at ambient temperature. Figure 25 provides an example of this approach in which 95% of the water flux measured through a virgin membrane was regained after crossflow filtration of a *Streptomyces* broth and subsequent regeneration [55]. More importantly, though, process performance was found to be consistent in this system. Figure 26 shows that similar filtration performance was achieved for multiple batches of this mycelial broth using the same membrane that had been regenerated between processing of batches [55]. It should be emphasized that consistent process performance must be the ultimate criterion when evaluating a membrane regeneration procedure. Although water flux may be a good indication of whether a membrane has been regenerated, this is not always the case. Sims and Cheryan [61] reported that although only 40% to 50% of water flux could be recovered after crossflow filtration of *Aspergillus niger* broth, flux with fermentation broth was not as severely affected.

Membrane regeneration protocols must be evaluated empirically; a good starting point is to follow the membrane manufacturer's instructions. Typical cleaning agents include acids such as nitric acid, phosphoric acid, and hydrochlo-

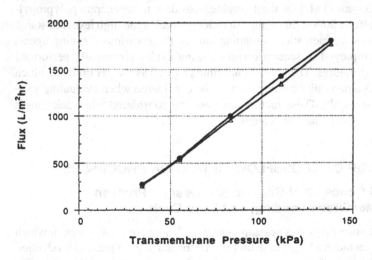

Figure 25 Comparison of water flux for new (•) and regenerated (△) ceramic membrane after filtration of *Streptomyces* broth. (From Ref. 55 with permission from John Wiley & Sons, Ltd.)

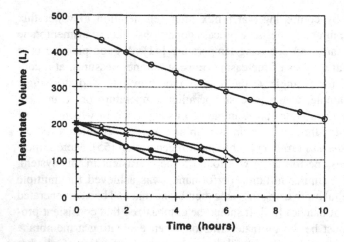

Figure 26 Pilot-scale microfiltration performance for 200 dm³ (•, △, ◊, x) or 450 dm³ (O) batches of *Streptomyces* broth. (From Ref. 55 with permission from John Wiley & Sons, Ltd.)

ric acid, bases such as sodium hydroxide, oxidants such as sodium hypochlorite and enzyme detergents such as tergazyme which contains proteases. Depending on the constituents of the fermentation broth, other enzymes can also be utilized. For example, Santos et al. [58] used amyloglucosidase to regenerate polypropylene and polysulfone hollow-fiber units in order to combat the high levels of starch used in a *Bacillus* fermentation. Combinations of these various cleaning agents are typically employed in sequence and cleaning cycles are usually performed at elevated temperatures. The temperature limits as well as the pH limits to which one can subject a particular membrane must be considered when evaluating various cleaning protocols. These factors must also be considered when selecting a membrane for a particular application.

XV. SCALEUP OF CROSSFLOW FILTRATION PROCESSES

A. General Concepts of Scaleup and Design: Process Scale, Membrane Loading, and Time Cycle

Membrane filtration processes scaleup very well. There are many cases in which it is possible to achieve a linear scaleup from lab bench to pilot plant to production plant, with all key performance parameters maintained, provided this objective is identified at the beginning of the development process. Van Reis et al. [71] provide an excellent example of a well conceived and followed development and scaleup program for the production of recombinant tPA from CHO cell culture by a linear scaleup of a CFF process (see also Section V).

It is more common for broth filtration processes to be performed initially in a laboratory setting, with small fermentation batches concentrated and diafiltered using a small-membrane cartridge and peristaltic pump, with not much thought to future scaleup. At this point in the development cycle, the objective is to make those precious first small lots of product for safety assessment and preclinical efficacy studies; the method used to make the material is not of paramount concern. Often these systems have little or no instrumentation, so the operating conditions of the membrane are poorly characterized. As drug development progresses, the production method quickly assumes more importance. Scalability, productivity, reproducibility and cost become significant concerns as the need to maintain a steady and steadily increasing supply of material to the product development program becomes an issue. An incremental amount of thought and effort, applied during the early development stage, can significantly speed progress through mid and late development.

Scaleup of membrane filtration systems is based on the membrane area required to do the job. But the exact meaning of this statement is elusive. The ratio of fermentation broth feed volume to membrane area, often termed the *membrane loading*, Λ, is a key scaleup parameter which ideally should remain constant as the volume of broth processed increases (some workers define loading as permeate volume to membrane area; in many cases, these values are very similar). The appropriate value of Λ for a given application must be determined experimentally, as discussed in earlier sections. *Time cycle* for the filtration process, τ, is another critical parameter (time is money!); this is especially true if there is a stability issue with the product or the broth. Clearly there is strong relationship between Λ and τ. There is a tendency to scaleup without adding sufficient membrane capacity, as a means of minimizing capital investment, by allowing both Λ and τ to increase. This is perfectly acceptable provided the appropriate experiments are performed to evaluate these new conditions prior to scaleup, to ensure that any of the many pitfalls awaiting the weary researcher are not encountered at the worst possible time—when the large production system has already been purchased and installed. For a linear scaleup, increasing the membrane area to maintain a constant membrane loading with increasing broth volume should provide the same yield, time cycle, and productivity determined during smaller-scale experiments, provided the other key parameters are maintained during operation (e.g. v_x, ΔP_L, ΔP_{tm}). Restricting early-development experiments to the use of membrane devices for which the manufacturer can provide a scaleup path is an important discipline which will speed later-development efforts.

Van Reis et al. [71] have addressed scaleup in a rigorous manner. During their development experiments, they determine a *process capacity*, which is the amount of permeate produced per unit area of membrane (in a controlled permeate rate mode) before the onset of significant protein product retention, i.e., $S <$ 0.9, for their particular system. This process capacity concept determines the maximum practical membrane loading for a given system, while also largely determining the time cycle of the process. In scaling up, they also discuss the need to add a safety factor, defined as the ratio of the process capacity to the actual planned membrane loading, to ensure success upon scale-up; *inter alia*, this accounts for batch-to-batch variations in the filtration characteristics of the fermentation broth. They considered a safety factor of ~3.5 to be appropriate for their particular process application.

In most cases, it is difficult to scaleup without some parameter changing, making a true linear scaleup difficult to achieve. For this discussion, it is assumed the same membrane material from the same manufacturer is available in various-size cartridges. One must be aware of changes in the length and configuration of the cartridge flowpath, L, since such changes affect the local filtration environment. Consider an example in which a fivefold scaleup is desired; this can be achieved by increasing the membrane area by a *parallel expansion* (increasing the number of membrane elements arrayed in parallel fivefold), or by a *series expansion* (increasing the length of the membrane cartridge fivefold), both resulting in an equivalent increase in membrane area which maintains membrane loading constant (Figure 27). The parallel expansion allows for all the key operating parameters to be kept constant, and as such represents the simplest scaleup path. But the total recirculating flow must be increased fivefold, requiring a larger pump, heat exchanger, and associated equipment.

For the series expansion, it is impossible to maintain constant all the key operating parameters that affect permeate flux. For example, transmembrane pressure and crossflow velocity, key inputs to permeate flux, cannot both be maintained constant. If we simplify by assuming that the retentate outlet and permeate outlet pressures are both zero, then ΔP_{tm} is proportional to P_{in} and Q is proportional to P_{in}/L. Increasing L by a factor of 5 while maintaining P_{in} constant will maintain ΔP_{tm}, but Q will fall, probably reducing permeate flux. The effect of such changes on the permeability of partially retained species must also be considered, although quantitative prediction is very difficult. The series expansion is cost-efficient, since it allows membrane area to be increased without increasing the supporting components if the resulting change in filtration performance is acceptable.

A further difficulty with a series expansion is the increased extent of concentration that occurs ''per pass'' through the membrane flowpath. With a short flowpath, the extent of concentration of a fluid element per pass is incremental, and the flow characteristics of the material exiting the membrane cartridge vary only slightly from that in the feed tank. As the length of the flowpath is increased, the extent of concentration can be significant, and especially toward the end of a concentration step, the solids content and viscosity of the exiting material can increase dramatically and be significantly higher than that of the batch in the

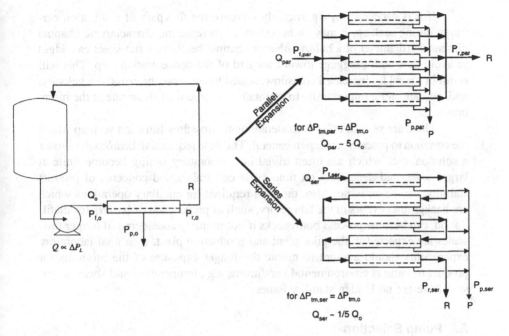

Figure 27 Membrane system scaleup—comparison of parallel and series expansions.

feed vessel. This is especially of concern for broths which filter well and which are meant to be concentrated to the fullest extent possible. In the worst case, this can lead to pluggage of the flowpath, resulting in either deterioration of flux or overpressurization of the membrane cartridge, depending on the type of pump employed. For this reason, a series expansion scaleup should be investigated in an experimental mode before committing a process to this path.

Van Reis et al. [69] have noted that a parallel expansion is readily achieved when hollow-fiber membranes are employed but is more difficult when the use of flat-sheet-type cartridges is desired. Scaled-down hollow-fiber cartridges which provide the same retentate path length are available from most manufacturers which will allow a true linear scaleup. This is not always the case for flat-sheet devices. In addition to the obvious requirements of maintaining constant path length and flow channel spacing, Van Reis et al. have demonstrated the need to scrutinize the effects of entry/exit region design, retentate and permeate manifolding, and the effect of module compression upon assembly (on flow channel spacing), if true linear scaleup is sought. These subtleties can result in differences in the local distribution of ΔP_L, ΔP_{tm}, and other key parameters, changing observed performance.

If it is necessary to significantly increase the flowpath of a filtration cartridge during scaleup, it may be necessary to increase the characteristic channel dimension (diameter of a hollow-fiber or channel height of a flat-sheet cartridge) to avoid flow path pluggage toward the end of the concentration step. This will result in a need for increased crossflow, scaled to conserve the crossflow velocity, and, perhaps more importantly, to maintain the same wall shear rate at the membrane.

There are several other considerations to crossflow filtration scaleup which are common to process scaleup in general. The time required to heat/cool, transfer a solution, etc., which are often trivial in a laboratory setting become finite at larger scale, and depending on time cycle constraints and concerns of product stability, can be an issue. Also, the time required for ancillary operations which are background events in the laboratory, such as preparation of buffers for diafiltration, can become process bottlenecks if not properly considered. If longer time scales are expected in the pilot plant and production plant, then final laboratory experiments should attempt to mimic the longer exposure of the broth and/or product to various environmental conditions, e.g., temperature and shear, to ensure there are no hidden stability issues.

B. Pump Selection

The type of recirculation pump employed in a CFF process can be critical, depending on the nature of the fermentation broth employed. Specifically, positive displacement pumps such as peristaltic and rotary lobe pumps can provide the desired flow with modest shear and low heat input, while centrifugal pumps are much more economical for a given flowrate and there is essentially no limit to their size (Figure 28).

It is typical for most small laboratory CFF systems to employ a peristaltic pump. They are mechanically simple and easy to control, provide good visualization of flow, and expose the broth to only modest levels of shear. However, these pumps are generally not well suited to larger-scale applications. The largest peristaltic pumps available will provide ~160 L/min of flow (and most are much smaller), limiting scaleup. In addition, there is a limited lifetime to the tubing component of the pump, necessitating replacement on a regular basis (assuming that cleaning validation issues do not require its replacement with each new batch), and there is always the possibility of tubing rupture. Under most circumstances, the outlet pressure of such pumps is limited to ~25 psi, due to the pressure rating of the soft tubing employed; this is more than sufficient for many low pressure applications (e.g., animal cell harvest), but may be insufficient for others. Peristaltic pumps do have the benefit that their tubing component, which is the only portion of the pump which contacts the product stream, can be prepared, cleaned, and sterilized (if necessary) with the fermentation system, and can be

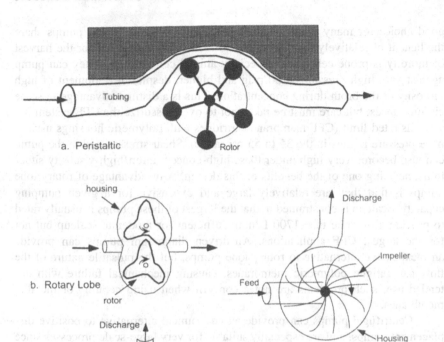

Figure 28 Common pump types used for CFF.

discarded after a single use. This can completely eliminate cleaning concerns which normally arise with other types of pumps, making these the pumps of choice for many small scale biological applications. But given the possibility of tubing rupture and limited scalability, their selection must be carefully considered.

Rotary lobe pumps and other similar moving cavity positive displacement pumps are often employed in the first scaleup from lab to pilot plant, and are a good choice for many moderate-scale production systems. These pumps share the benefit of relatively low shear, and so are usually employed for the harvest of more lysis-prone cell types, especially animal cell cultures. They can pump against very high pressures, 100 psig and higher, despite development of high viscosity of the broth during concentration. This is a distinct advantage in some circumstances, but care must be taken not to overpressurize the CFF system beyond its rated limit (CFF membrane cartridges with polymeric housings usually have pressure ratings in the 35 to 55 psi range). Shear stresses within the pump can also become very high under these high-concentration/high-viscosity situations, negating one of the benefits of this design. A disadvantage of rotary lobe pumps is that they are relatively large and expensive for a given pumping capacity. Scaleup is constrained in that the largest of these pumps is usually rated to provide a flowrate of ~1700 L/min, sufficient for moderate scaleup but not for the largest CFF applications. Air-driven diaphragm pumps can provide an inexpensive alternative to rotary lobe pumps, but the pulsatile nature of the flow can fatigue polymeric membranes, causing mechanical failure with extended use; such problems are not a concern when using ceramic or metallic membranes.

Centrifugal pumps can provide an economical alternative to positive displacement pumps, and are especially suitable for very large-scale processes since they are available at almost any scale desired. While inappropriate for applications which are especially shear sensitive, such as animal cell harvest, these pumps can be employed for the filtration of most bacterial, yeast, and mold broths, especially if the application will tolerate lysis of some fraction of the cells. It has been determined that damage to proteins and cells by the high-shear environment in a centrifugal pump is much less of an issue than originally thought. Especially for globular proteins, the damage that was originally ascribed to pump shear has been found usually to result from the presence of gas-liquid interfaces in the system [44,66,73]. This can result from cavitation in centrifugal pumps, which occurs when the pump and its associated piping are not appropriately sized for an application. Another source of gas-liquid interfaces is from small piping leaks which allow air to be aspirated into the recirculation loop.

Pump shear may be more of an issue with respect to cell lysis than protein damage. While no definitive study of cell breakage from pump recirculation appears in the literature, Shimuzu et al. [60] have demonstrated the negative effect of disrupted yeast particulates (from external bead milling) on flux; this agrees well with the large body of data citing the difficulty of filtering cell debris for protein recovery. Work by Vandanjon et al. [72] have shown that marine microalgae cells can break during repeated passage through pumps (centrifugal and rotary vane), but they also found that passage through throttling valves can have a similar effect. Some manufacturers offer "low-shear" centrifugal pumps, which have larger clearances between the impeller and pump casing, resulting in lower shear at a given pump speed; these pumps are less energy efficient than standard centrifugal pumps, which may not be much of a concern for biotech applications. Given the economic incentive to use centrifugal pumps for large broth harvest applications, the acceptability of their performance will generally have to be determined experimentally.

High-pressure, multistage centrifugal pumps are rarely employed for broth harvest CFF applications, but it is not uncommon to stage two pumps in series to obtain a higher outlet pressure than either pump can provide alone; this is an intrinsic advantage of the feed and bleed loop system (Figure 5). In this manner, it is possible to obtain membrane cartridge inlet pressures of 90 to 100 psi, which can be employed with certain ceramic and metallic membranes. But pressure alone is not always sufficient to achieve the desired flow, and the main problem with centrifugal pumps can be the slip that occurs as broths are concentrated and become more viscous. With such systems, careful sequencing and control of operations can become critical. While it may be possible to use a centrifugal pump to concentrate a broth to a high cell density with high viscosity, it may be difficult to restart flow if pumping is stopped for some reason late in the process cycle, since the pump may lose its prime under these conditions. This can be especially problematic with a shear-thinning mycelial broth.

Frictional heat input is a concern with centrifugal pumps, and can vary depending on how efficiently the pump is being utilized. Pump heat must be countered by providing cooling either via an inline heat exchanger or by jacketing the feed vessel walls. While vessel jacketing can appear more economical, it does pose the problem of a variable heat transfer area, since the fraction of the jacket that contacts the broth will typically be reduced as the broth is concentrated. At the limit, a highly concentrated broth may reach a point where it is contacting only the dish of the feed vessel, so little heat is being removed. The situation is exacerbated by the fact that the more concentrated broth is also more viscous, and therefore will have reduced heat transfer characteristics. For these reasons, most crossflow filtration systems which employ centrifugal pumps employ an in-line heat exchanger to ensure adequate heat removal.

The issue of cleaning was alluded to in the above discussion. In general, standard chemical process industry (CPI) pump designs are suitable for use in CFF systems for production of bulk pharmaceutical products, but special sanitary design units are usually employed when processing biological products—vaccines and therapeutic proteins.

C. Hydraulic Design

The hydrodynamics, or hydraulic design, of a crossflow filtration system must be carefully considered upon scaleup, including the design of the membrane cartridge and the flow path of the CFF system in general. Some of the key scaleup issues with the design of the membrane cartridge have already been discussed in section XVA; this section focuses on the design of the equipment supporting the membrane cartridge operation.

Control of transmembrane pressure is a critical concern, regardless of scale of operation. There are two common schemes for controlling ΔP_{tm} by controlling the pressure in the permeate channels of the membrane. The simplest method employs a throttling valve on the permeate line out of the membrane cartridge; ΔP_{tm} is increased by opening this valve and decreased by throttling it closed, so the pressure on the permeate line approaches that of the retentate side of the membrane. One problem with this method is that momentarily opening the valve too much can result in a high ΔP_{tm}, which can result in membrane fouling for some broths. This can be avoided by careful operational control, either manually or through process automation (i.e., the permeate control valve must be sequenced such that the valve starts in the closed position, and gradually opens until the desired permeate pressure is achieved).

Alternatively, some systems replace the permeate throttling valve with a positive displacement pump which is employed as a flow regulator. This mode of operation is especially popular for animal cell harvest applications, where an undesired high ΔP_{tm} can result in fouling (affecting flux and protein permeation) as well as direct damage to the cells [35,71]. In these systems, the pump speed is set to match the expected flux at the desired ΔP_{tm} (or controlled to maintain a desired ΔP_{tm}), and is usually reduced during the course of the process to match the anticipated dropoff in flux to avoid undesired increases in ΔP_{tm}. If the pump speed is not properly controlled (using a measurement of the permeate pressure) and is set too high, it can actually pull a vacuum on the permeate, negating the desired avoidance of high ΔP_{tm} and possibly exposing product to vapor-liquid interfaces resulting from cavitation on the suction side of the pump. Also, for this type of application, the permeate pump must be of a type that will not permit flow forward through the pump unless the pump is running.

The retentate outlet can be throttled as another means of controlling ΔP_{tm}, particularly to obtain higher ΔP_{tm} at a given flowrate. This can be very useful when the fermentation broth being filtered has a low cell density relative to that for which the filtration system was designed; in such a case, the viscous drag of the broth through the flow channels may be insufficient to develop the desired inlet pressure for a given flow rate. But this method must be approached carefully when using any type of positive-displacement pump (e.g., peristaltic or rotary lobe) since overthrottling the retentate will result in high pressure. This can result in damage to the membrane system, and for peristaltic pumps it can lead to bursting the pump tubing. For this reason, many laboratory systems are not equipped with retentate throttling valves. Larger-scale systems require pressure safety devices after the pump (high pressure cutoff switch or pressure relief valve) to guard against such a possibility.

It should be noted that an alternative basis for process control, which seeks to maintain a constant C_{wall} (the concentration of a retained species at the membrane surface) rather than a constant ΔP_{tm}, has been developed by van Reis et al. [70]. This is based on the predominance of C_{wall} in determination of both flux and permeation in most hydrodynamic models of CFF. Further discussion of this development is beyond the scope of this work.

As the scale of operation increases, control of flow and pressure—transmembrane pressure in particular—can be difficult if there has not been proper attention to hydraulic design of the overall system. In laboratory systems, components are small and can be located close to each other with very short tubing lengths connecting them so pressure drop through the lines is a nonissue. In a pilot or production environment, equipment location can be dictated by non-technical factors such as space constraints, existing equipment availability and other factors that can result in less-than-optimal design. Long line lengths, bends and turns, and passage through valves and other fittings can all impose pressure losses which are difficult to avoid. In addition, significant pressure head can result from different elevations of the components in a large system (for aqueous systems, a good rule of thumb is that a change of two feet in elevation is equivalent to a pressure change of ~1 psi.)

An example of the type of issue which can arise upon scaleup would be a retentate line which must climb 30 feet to return the stream from the outlet of the filtration module to the top of a large feed tank, adding ~15 psi to the retentate outlet pressure (Figure 29a). If the laboratory system operates with a 15-psi feed inlet pressure and zero-gauge pressure on the retentate and permeate outlets, for an average ΔP_{tm} of 7.5 psi, there is no way to exactly match the pressure specifications for the scaleup system. If matching the pressure profiles is a priority, and a more complex system is acceptable, it is possible to add small boot tanks to receive the retentate and permeate flows, maintaining the outlet pressure of these lines from the filter cartridge at zero-gauge pressure; small booster pumps on the boot tanks would then transfer the material on to its final destinations (Figure 29b). Alternatively, a pressure control valve can be added to the filtrate outlet line, and throttled to match the outlet pressure of the retentate (~15 psig), so that when the feed pump is run to achieve the equivalent flowrate, the feed pressure to the system should be ~30 psig and the ΔP_{tm} will be the desired 7.5 psi. This requires good control of the operation and that the membrane cartridges are rated for this higher operating pressure (Figure 29c).

Van Reis et al. [71] address another interesting problem which arises when scaleup requires staging of membrane cartridges in series (they operate in a controlled permeate rate mode). If the permeate lines from the "upstream" and "downstream" cartridges were tied in to a common manifold, there would be a drastic change in local ΔP_{tm} along the flowpath. The upstream cartridges would experience a higher average ΔP_{tm}, while the downstream cartridges would experi-

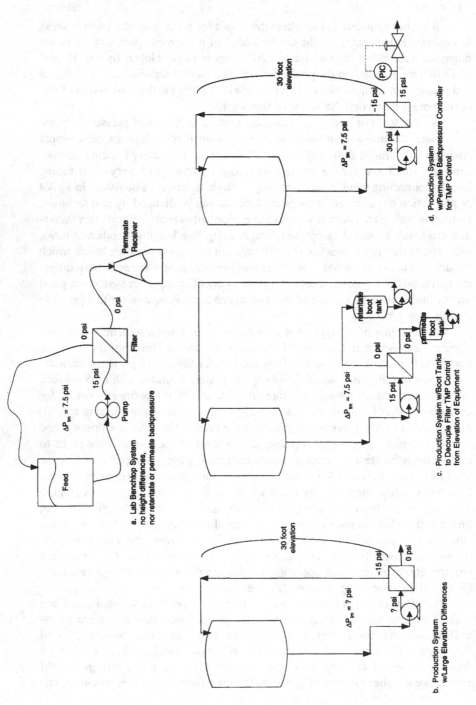

Figure 29 Scaleup example with a change of elevation (lab vs. industrial plant): (a) lab scale system; (b) plant system with large elevation difference problem; (c) plant system with boot tanks and booster pumps; (d) plant system with filtrate back pressure regulation.

ence a lower value, which could result in fouling of the upstream membranes and Starling flow in the downstream. They minimize this problem by controlling the permeate flow from the two staged cartridge housings individually, so the ΔP_{tm} is similar in each (this in addition to optimization of conditions in each housing).

Any number of such situations can be encountered upon scaleup, and in most cases an engineering solution which preserves the desired operating parameters for the filtration can be devised. But the earlier these issues are explored, the less likely that the solution will result in added expense, delay or clumsy workarounds.

A final concern of system design upon scaleup is minimizing system holdup and system "dead zones." System holdup can become an issue when concentrated broth is the product and it is desired to achieve the highest cell concentration possible. The smaller the volume holdup of the membrane cartridge, pump, and associated piping, the smaller will be the final concentrated volume that can be achieved. Ability to efficiently evacuate the system upon completion of processing is essential to achieve high yield (assuming the product is cell-associated) and to minimize subsequent cleaning effort. If it is acceptable to combine a follow flush of the system with the final cell concentrate, this is less critical. But if the dilution this represents is unacceptable, the system must be designed to eliminate liquid "traps" which prevent complete draining to a low point in the recirculation loop. Unfortunately, the goals of having a cleanly draining flowpath, access to components for service, and a compact equipment footprint are often at odds with each other; the CFF process developer must work with the equipment vendor to ensure that the skid design reflects the priorities of the process.

REFERENCES

1. Andreadis ST, Roth CM, LeDoux JM, Morgan JR, Yarmush ML. Large-scale processing of recombinant retroviruses for gene therapy. Biotechnol Prog 15:1–11, 1999.
2. Belfort G, Davis RH, Zydney AL. The behavior of suspensions and macromolecular solutions in crossflow microfiltration. J Membr Sci 96:1–58, 1994.
3. Bernard AR, Kost TA, Overton I, Cavega C, Young J, Bertrand M. Recombinant protein expression in a *Drosophila* cell line: comparison with the baculovirus system. Cytotechnology 15:139–144, 1994.
4. Blatt WF, Dravid A, Michaels AS, Nelsen L. Solute polarization and cake formation in membrane ultrafiltration: causes, consequences, and control techniques. In: JE Flinn, ed. Membrane Science and Technology: Industrial, Biological and Waste Treatment Process. New York: Plenum, 1970:47–97.
5. Bowen WR, Hall NJ. Properties of microfiltration membranes: mechanisms of flux loss in the recovery of an enzyme. Biotechnol Bioeng 46:28–35, 1995.

6. Brewster ME, Chung K-Y, Belfort G. Dean vortices with wall flux in a curved channel membrane system. 1. A new approach to membrane module design. J Membr Sci 81:127–137, 1993.
7. Burns DB, Zydney AL. Effect of solution pH on protein transport through ultrafiltration membranes. Biotechnol Bioeng 64:27–37, 1999.
8. Cabral JMS, Casale B, Cooney CL. Effect of antifoam agents and efficiency of cleaning procedures on the cross-flow filtration of microbial suspensions. Biotechnol Lett 7:749–752, 1985.
9. Cheryan M. Ultrafiltration and Microfiltration Handbook. Lancaster, PA: Technomic Publishing Co., 1998.
10. Cho J, Amy G, Pellegrino J. Membrane filtration of natural organic matter: initial comparison of rejection and flux decline characteristics with ultrafiltration and nanofiltration membranes. Water Res 33:2517– 2526, 1999.
11. Chudacek MW, Fane AG. The dynamics of polarisation in unstirred and stirred ultrafiltration. J Membr Sci 21:145–160, 1984.
12. Chung K-Y, Bates R, Belfort G. Dean vortices with wall flux in a curved channel membrane system. 4. Effect of vortices on permeation fluxes of suspensions in microporous membrane. J Membr Sci 81:139–150, 1993.
13. Conrad PB, Lee SS. Two-phase bioconversion product recovery by microfiltration. I. Steady state studies. Biotechnol Bioeng 57:631–641, 1998.
14. Cooney CL. Holeschovsky U, Agarwal G. Vortex flow filtration for ultrafiltration of protein solutions. In: DL Pyle, ed. Separations for Biotechnology, 2nd ed. New York: Elsevier, 1990.
15. Defrise D, Gekas V. Microfiltration membranes and the problem of microbial adhesion. Process Biochem 23:105–116, 1988.
16. Fane AG, Fell CJD, Waters AG. Ultrafiltration of protein solutions through partially permeable membranes—the effect of adsorption and solution environment. J Membr Sci 16:211–224, 1983.
17. Fletcher M, Pringle JM. The effect of surface free energy and medium surface tension on bacterial attachment to solid surfaces. J Colloid Interface Sci 104:5–14, 1985.
18. Forsyth SK, Göklen KE. Comparison of centrifugation and microfiltration for yeast cell harvests. Paper presented at 213th ACS National Meeting, San Francisco, 1997.
19. Gatenholm P, Paterson S, Fane AG, Fell CJD. Performance of synthetic membranes during cell harvesting of E. coli. Process Biochem 23:79–81, 1988.
20. Gehlert G, Luque S, Belfort G. Comparison of ultra- and microfiltration in the presence and absence of secondary flow with polysaccharides, proteins, and yeast suspensions. Biotechnol Prog 14:931–942, 1998.
21. Göklen KE, Thien M, Ayler S, Smith S, Fisher E, Chartrain M, Salmon P, Wilson J, Andrews A, Buckland BC. Development of crossflow filtration processes for the commercial-scale isolation of a bacterial lipase. Bioprocess Eng 11:49–56, 1994.
22. Gravatt DP, Molnar TE. Recovery of an extracellular antibiotic by ultrafiltration. In: WC McGregor, ed. Membrane Separations in Biotechnology. New York: Marcel Dekker, 1986.
23. Haarstrick A, Rau U, Wagner F. Cross-flow filtration as a method of separating fungal cells and purifying the polysaccharide produced. Bioprocess Eng 6:179–186, 1991.

24. Hanisch A. Cell harvesting. In: WC McGregor, ed. Membrane Separations in Biotechnology. 1st ed. New York: Marcel Dekker, 1986.
25. Henry JD, Allred RC. Concentration of bacterial cells by cross flow filtration. In: Developments in Industrial Microbiology, Vol. 13. Symposium: Concentration of Microbial Cells, 1972.
26. Hernandez-Pinzon I, Bautista J. Microfiltration of Streptococcal fermentation broths: study of the factors affecting the concentration effect. Biotechnol Bioeng 44:270–275, 1994.
27. Junker BH, Timberlake S, Bailey FJ, Reddy J, Prud'homme R, Gbewonyo K. Influence of strain and medium composition of *Escherichia coli* suspensions. Biotechnol Bioeng 44:539–548, 1994.
28. Kim KJ, Fane AG, Fell CJD, Joy DC. Fouling mechanisms of membranes during protein ultrafiltration. J Membr Sci 68:79–91, 1992.
29. Kloosterman J, Kooijman AW, Slater NKH, Tanis E. Filtration of a yeast suspension using an enclosed porous metal filter. Biotechnol Techniques 1:117–122, 1987.
30. Kroner KH, Hummel W, Volkel J, Kula MR. Effects of antifoams on cross-flow filtration of microbial suspensions. In: Drioli E, Nakagaki M, eds. Membranes and Membrane Processes. Stesa, Italy: Plenum, 1986:223–232.
31. Kroner KH, Nissinen V, Ziegler H. Improved dynamic filtration of microbial suspensions. Bio/Technology 5:921–926, 1987.
32. Kroner KH, Schutte H, Hustedt H, Kula MR. Cross-flow filtration in the downstream processing of enzymes. Process Biochem 19:67–74, 1984.
33. Lee SS, Burt A, Russotti G, Buckland BC. Microfiltration of recombinant yeast cells using a rotating disk dynamic filtration system. Biotechnol Bioengin 48:386–400, 1995.
34. Liew MKH, Fane AG, Rogers PL. Hydraulic resistance and fouling of microfilters by *Candida utilis* in fermentation broth. Biotechnol Bioeng 48:108–117, 1995.
35. Maiorella B, Dorin G, Carion A, Harano D. Crossflow-microfiltration of animal cells. Biotechnol Bioeng 37:121–126, 1991.
36. Marshall AD, Monro PA, Trägardh G. The effect of protein fouling in microfiltration and ultrafiltration on permeate flux, protein retention and selectivity: a literature review. Desalination 91:65–108, 1993.
37. McGregor WC. Membrane Separations in Biotechnology. 1st ed. New York: Marcel Dekker, 1986.
38. McGregor WC, Weaver JF, Tansey SP. Antifoam effects on ultrafiltration. Biotechnol Bioeng 31:385–389, 1988.
39. Menon MK, Zydney AL. Effect of ion binding on protein transport through ultrafiltration membranes. Biotechnol Bioeng 63:298–307, 1999.
40. Michaels AS. New separation technique for the CPI. Chem Eng Prog 64:31–43, 1968.
41. Michaels SL. Clean-water permeability as a determinant of cleaning efficacy in tangential-flow filtration systems. BioPharm 7:38–45, 1994.
42. Minier M, Fessier P, Colinart P, Cavezzan J, Liou JK, Renon H. Study of the fouling effect of antifoam compounds on the crossflow filtration of yeast suspensions. Separation Sci Technol 30:731–750, 1995.
43. Nagata N, Herouvis KJ, Dziewulski DM, Belfort G. Cross-flow membrane microfiltration of a bacterial fermentation broth. Biotechnol Bioeng 34:447–466, 1989.

44. Narendranathan TJ, Dunnill P. The effect of shear on globular proteins during ultra-filtration studies of alcohol dehydrogenase. Biotechnol Bioengin 24:2103–2107, 1982.
45. Nilsson JL. Protein fouling of UF membranes: causes and consequences. J Membr Sci 52:121–142, 1990.
46. Park BG, Lee WG, Chang YK, Chung HN. Long-term operation of continuous high cell density culture of *Saccharomyces cerevisiae* with membrane filtration and on-line cell concentration monitoring. Bioprocess Eng 21:97–100, 1999.
47. Parnham CS, Davis RH. Protein recovery from cell debris using rotary and tangential crossflow filtration. Biotechnol Bioeng 47:155–164, 1995.
48. Patel PN, Mehaia MA, Cheryan M. Cross-flow membrane filtration of yeast suspensions. J Biotechnol 5:1–16, 1987.
49. Porter MC. Membrane Filtration. New York: McGraw-Hill, 1979.
50. Porter MC. Concentration polarization with membrane ultrafiltration. Indust Eng Chem Product Res Dev 11:234–248, 1972.
51. Porter MC, Michaels AS. Membrane ultrafiltration. Part 5. A useful adjunct for fer-mentative and enzymatic processing of foods. Chemtech 2:56–61, 1972.
52. Pujar NS, Zydney AL. Electrostatic effects on protein partitioning in size exclusion chromatogtaphy and membrane ultrafiltration. J Chromatogr 796:229–238, 1998.
53. Reeder DH, Göklen KE. An integrated membrane process for product recovery from a dense fungal fermentaiton. Paper presented at AICHE Fall Annual Meeting, Miami, 1998.
54. Rumpus J, Bulmer M, Tirchener-Hooker N, Hoare M. Toward enhanced specificity during the purificaiton of microbial proteins. Biochem Eng 2001. 513–516, 1992.
55. Russotti G, Göklen KE, Wilson JJ. Development of a pilot-scale microfiltration har-vest for the isolation of physosstigmine from *Streptomyces griseofuscus* broth. J Chem Technol Biotechnol 63:37–47, 1995.
56. Russotti G, Osawa AE, Sitrin RD, Buckland BC, Adams WR, Lee SS. Pilot-scale harvest of recombinant yeast employing microfiltration: a case study. J Biotechnol 42:235–246, 1995.
57. Saksena S, Zydney AL. Effect of solution pH and ionic strength on the separation of albumin from immunoglobulins (IgG) by selective filtration. Biotechnol Bioeng 43:960–968, 1994.
58. Santos JAL, Cabral JMS, Cooney CL. Recovery of alkaline proteases by membrane filtration. Bioprocess Eng 7:205–211, 1992.
59. Schorr CL, Singhvi R, O'Hara C, Xie L, Wang DIC. Clarification of animal cell culture process fluids using depth microfiltration. Biopharm Manufacturing 9:35–41, 1996.
60. Shimizu Y, Matsushita K, Wantanabe A. Influence of shear breakage of microbial cells on cross-flow microfiltration flux. J Ferment Bioeng 78:170–174, 1994.
61. Sims KA, Cheryan M. Cross-flow microfiltration of *Aspergillus niger* fermentation broth. In: CD Scott, ed. Eighth Symposium on Biochemistry for Fuels and Chemi-cals. Catlinburg, TN: John Wiley, 1986:495–505.
62. Singh N, Cheryan M. Performance characteristics of a ceramic membrane system for microfiltration of corn starch hydrolysate. Chemical Eng Commun 168:81–95, 1998.

63. Sirkar KK, Prasad R. Protein ultrafiltration—some neglected considerations. In: WC McGregor, ed. Membrane Separations in Biotechnology. 1st ed. New York: Marcel Dekker, 1986.

64. Taddei C, Aimar P, Howell JA, Scott JA. Yeast cell harvesting from cider using microfiltration. J Chem Technol Biotechnol 47:365–376, 1990.

65. Tejayadi S, Cheryan M. Downstream processing of lactic acid-whey permeate fermentation broths by hollow fiber ultrafiltrations. Appl Biochem Biotechnol 19:61–70, 1988.

66. Thomas CR, Nienow AW, Dunnill P. Action of shear on enzymes: studies with alcohol dehydrogenase. Biotechnol Bioeng 21:2263–2278, 1979.

67. Trinh L, Shiloach J. Recovery of insect cells using hollow fiber microfiltration. Biotechnol Bioengin 48:401–405, 1995.

68. Uchiyama K, Morimoto M, Yokayama Y, Shioya S. Cell cycle dependency of rice α-amylase production in a recombinant yeast. Biotechnol Bioeng 54:262–271, 1997.

69. van Reis R, Goodrich EM, Yson CL, Frautschy LN, Dzengeleski S, Lutz, H. Linear scale ultrafiltration. Biotechnol Bioeng 55:737–746, 1997.

70. van Reis R, Goodrich EM, Yson CL, Frautschy LN, Whiteley R, Zydney AL. Constant C_{wall} ultrafiltration process control. J Membr Sci 130:123–140, 1997.

71. van Reis R, Leonard LC, Hsu CC, Builder SE. Industrial scale harvest of proteins from mammalian cell culture by tangential flow filtration. Biotechnol Bioeng 38:413–422, 1991.

72. Vandajon L, Rossignol N, Jaouen P, Robert JM, Quéméneur F. Effects of shear on two microalgae species, contribution of pumps and valves in tangenital flow filtration systems. Biotechnol Bioengin 63:1–9, 1999.

73. Virkar PP, Narendrathan TJ, Hoare M, Dunnill P. Studies of the effects of shear on globular proteins extension to high shear fields and to pumps. Biotechnol Bioeng 23:425–429, 1981.

74. Yamagiwa K, Ikarashi K, Ohkawa A. Effects of antifoam with inverted cloud point on permeation and solute rejection in membrane filtration processes. J Chem Eng Jpn 22:693–695, 1989.

75. Yamagiwa K, Kobayashi H, Ohkawa A, Onodera M. Effect of antifoam fouling on solute rejection by ultrafiltration membrane. J Chem Eng Jpn 26:18–20, 1993.

76. Yamagiwa K, Kobayashi H, Ohkawa A, Onodera M. Membrane fouling in ultrafiltration of hydrophobic nonionic surfactant. J Chem Eng Jpn 26:13–18, 1993.

77. Yamasaki H, Lee M-S, Tanaka T, Nakanishi K. Characteristics of cross-flow filtration of pullalan broth. Appl Microbiol Biotechnol 39:26–30, 1993.

78. Yee L, Blanch HW. Recombinant trypsin production in high cell density fed-batch cultures in Escherichia coli. Biotechnol Bioeng 41:781–790, 1993.

79. Zhang J, Marcin C, Shifflet MA, Salmon P, Brix T, Greasham R, Buckland BC, Chartrain, M. Development of a defined medium fermentation process for physostigmine production by Streptomyces griseofuscus. Appl Microbiol Biotechnol 44:568–575, 1996.

80. Zydney AL, Colton CK. A concentration polarization model for the filtrate flux in cross-flow microfiltration of particulate suspensions. Chem Eng Commun 47:1–21, 1986.

4
Crossflow Microfiltration with Backpulsing

Robert H. Davis
University of Colorado, Boulder, Colorado

I. INTRODUCTION

Crossflow microfiltration is used in biotechnology for applications such as cell recycle and harvesting, separation of recombinant proteins from cell debris, and purification of process streams. The microfiltration membranes used in these processes typically have nominal pore sizes on the order of 0.1 to 1.0 μm. Clean microfiltration membranes have relatively large water fluxes on the order of 0.01 to 0.1 cm/s (1 cm/s = 3.6×10^4 L/m²-h) at typical transmembrane pressures of 1 to 10 psi (1 psi = 6.9 kPa), and they easily pass solutes such as proteins while retaining insolubles such as cells. However, the rejected material may cause severe membrane fouling which reduces the flux by two orders of magnitude and which significantly alters the membrane selectivity [1–9]. This fouling may render the microfiltration process noneconomical when compared with conventional separation methods such as centrifugation and rotary vacuum filtration [10–12]. Several recent reviews provide further discussion of crossflow microfiltration and its applications [11–20].

A common technique for fouling reduction is backflushing [21,22]. Backflushing, sometimes called backwashing, is an in situ method of membrane cleaning by periodically reversing the transmembrane pressure or permeate flow. As shown in Figure 1, the fouling deposit which accumulates on the membrane surface during forward filtration is lifted off and removed during reverse filtration. Approximately two-fold improvements in flux, as well as increases in protein transmission and recovery, have been achieved using periodic backflushing for enzyme separation from yeast cell lysates [5,23]. However, the performance de-

161

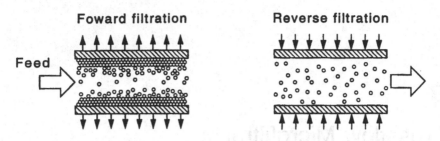

Figure 1 Schematic of forward and reverse filtration. (From Ref. 35.)

graded over time due to irreversible fouling of the membrane by the adhesive cell lysates. For whole cell suspensions and fermentation broths, flux improvements using backflushing have varied from ~ 50% for bacterial suspensions [5,24] to up to 10-fold for yeast suspensions [25,26].

In conventional backflushing operation, the periods of reverse flow through the membrane are typically several seconds in duration and take place at frequencies of once every several minutes or longer. During the relatively long periods of forward filtration between backflushes, significant membrane fouling can occur. Moreover, the foulants are only partially removed by the backflushes or backwashes. An alternative strategy, which has been explored just within the past 10 years, is rapid backpulsing. The concept of backpulsing is similar to that of backflushing, except that the duration of the reverse flow is much shorter, typically less than 1 s, and the frequency of the backpulses is much greater, typically at least once every few seconds. It is thought that such high-frequency backpulses might prevent membrane fouling or at least remove the foulants shortly after they are deposited on the membrane.

High-frequency transmembrane pressure pulsing or backpulsing was first investigated by Victor Rodgers and others for protein ultrafiltration [27–32]. They demonstrated significant improvements in solute flux using backpulsing frequencies of up to 5 s^{-1}. More recent work has focused on rapid backpulsing in microfiltration applications. Gunnar Jonsson and coworkers employed what they refer to as a backshock process to maintain high permeate flux and protein transmission for several days during crossflow microfiltration of beer [33,34]. In this process, backflushing using permeate is performed for a brief duration of 0.1 s once every 5 s. They found that the backshock process is most effective when the feed side was in contact with the porous outer support of the asymmetric hollow fibers employed, rather than in the more typical configuration where the feed is in contact with the inner skin layer of the hollow fibers. As an alternative to using permeate or water as the backpulse fluid, air is sometimes used for dislodging foulants.

Over the past 5 years, my laboratory also has done modeling and experimental work on crossflow microfiltration with backpulsing, primarily focused on biotechnology applications [10,35–37]. In the remainder of this chapter, I describe the primary results of this work. A summary of the modeling work is presented in Section II. Membrane experiments involving whole cells and fermentation broth are described in Section III, and experiments on the separation of protein from bacterial cell debris are described in Section IV. Concluding remarks are provided in SectionV.

II. THEORY OF BACKPULSING

The reverse flow of permeate or other backpulse fluid through the membrane may be controlled either by pressure or by a pump or other positive displacement device. The former case is considered here. A constant transmembrane pressure of magnitude ΔP_f is applied during forward filtration for a duration t_f, followed by a backward transmembrane pressure of magnitude ΔP_b during reverse filtration of duration t_b. The desired transmembrane pressure cycle for backpulsing is then the square-wave pattern shown in Figure 2. The corresponding flux pattern

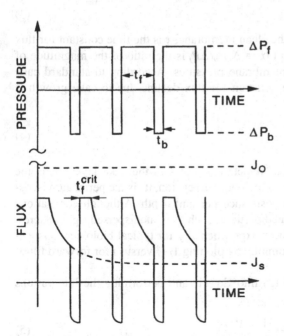

Figure 2 Transmembrane pressure and flux profiles during backpulsing.

is shown in the lower portion of this figure, with each period of positive flux during forward filtration followed by a brief period of negative flux during backward or reverse filtration.

The net permeate flux over one cycle of forward and reverse filtration is defined as the volume of permeate fluid which passes through the membrane during forward filtration, minus the volume of backpulse fluid which passes in the reverse direction through the membrane during backward filtration, per unit time and membrane surface area:

$$\langle J \rangle = \frac{\displaystyle\int_o^{t_f} J_f(t)dt - \int_{t_f}^{t_f+t_b} J_b(t)dt}{t_f + t_b} \tag{1}$$

where $J_f(t)$ is the positive permeate flux during forward filtration and $J_b(t)$ is the magnitude of the negative backpulse flux during reverse filtration. As an initial model, Redkar and Davis [35] assumed that the forward flux declines with time according to cake filtration theory as cells or other insoluble particles are rejected at the surface of the membrane, and that the cake or fouling layer is instantly and completely removed by the backpulse fluid during reverse filtration. Thus,

$$J_f(t) = J_o/(1 + t/\tau)^{1/2} \tag{2}$$

$$J_b(t) = \alpha J_o \tag{3}$$

where J_o is the initial flux for the clean membrane, τ is the time constant for flux decline due to cake growth, and $\alpha = \Delta P_b/\Delta P_f$ is the ratio of the magnitude of the reverse and forward transmembrane pressures. According to standard cake filtration theory [38], the time constant for flux decline due to cake growth is given by

$$\tau = \frac{(c_c - c_b)\,\Delta P_f}{2\hat{R}_c\mu_o c_b J_o^2} \tag{4}$$

where c_c is the concentration of rejected particles in the cake layer, c_b is the concentration of these particles in the bulk suspension, μ_o is the permeate viscosity, and $\hat{R}_c$ is the specific cake resistance per unit depth. Although correlations exist for $\hat{R}_c$ and c_c for well-defined particles such as monodispersed rigid spheres, it is more common to determine τ experimentally for typical biological suspensions. This may be done, for example, by plotting $1/J_f^2$ versus t for forward filtration [39].

Substituting Eqs.(2) and (3) into Eq. (1) and performing the integrations yields [35]

$$\frac{\langle J \rangle}{J_o} = \frac{2\tau\left[(1 + t_f/\tau)^{1/2} - 1\right] - \alpha t_b}{t_f + t_b} \tag{5}$$

For given values of t_b, τ, and α, there is an optimal forward filtration time (or backpulse frequency) which maximizes the net permeate flux [35]. When the frequency is less than the optimum, then the significant membrane fouling during the relatively long periods of forward filtration causes a reduction in the net flux. When the frequency is greater than the optimum, then the significant amount of fluid passing through the membrane in the reverse direction during the frequent backpulses causes a reduction in the net flux.

In experiments with nonadhesive washed yeast suspensions, Redkar and Davis [35] verified the existence of an optimum backpulsing frequency and found good qualitative agreement with the prediction of Eq.(5). However, the measured values of the net permeate flux with backpulsing typically exceed the quantitative predictions of the model. The authors hypothesized that the higher observed fluxes might have been the result of a delay in cake formation during forward filtration and a delay in cake removal during reverse filtration.

The delay in cake formation during forward filtration was investigated through numerical solution of the convection-diffusion equation for concentration polarization and depolarization during repeated cycles of forward and reverse filtration [37]. It represents the time required at the start of each forward filtration period to refilter the clean fluid introduced across the membrane to the retentate side during the previous reverse filtration period, and for the particle or solute concentration at the membrane surface to reach a sufficient concentration that a cake or gel layer begins to form. The theory shows that the delay time due to back diffusion scales as $D(c_c - c_b)/(J_o^2 c_b)$, where D is the particle diffusivity due to Brownian diffusion for submicron particles or shear-induced diffusion for larger particles, and the delay time to refilter the backpulse fluid scales as t_b. For typical parameter values, Redkar et al. [37] estimated that the delay time is approximately 1 s for both microfiltration and ultrafiltration, but it varies greatly with filtration conditions. As depicted in Figure 2, the delay time for cake formation is denoted by t_f^{crit}. Then, again assuming complete membrane cleaning during each backpulse, the forward filtration flux is $J_f = J_o$ for $t < t_f^{crit}$ and then follows cake filtration theory for longer times:

$$J_f(t) = J_o/(1 + (t - t_f^{crit})/\tau)^{1/2}, \quad t > t_f^{crit} \tag{6}$$

The net flux is modified from Eq. (5) to become [37]

$$\frac{\langle J \rangle}{J_o} = (t_f - \alpha t_b)/(t_f + t_b), \quad t_f < t_f^{crit} \tag{7a}$$

$$\frac{\langle J \rangle}{J_o} = \frac{t_f^{crit} + 2\tau\,[(1 + (t_f - t_f^{crit})/\tau)^{1/2} - 1] - \alpha t_b}{t_f + t_b}, \quad t_f \geq t_f^{crit} \tag{7b}$$

As discussed in the next section, good quantitative agreement was obtained between Eq.(7) and backpulsing experiments with washed yeast suspensions [37]. However, in similar experiments with bacterial suspensions, Kuberkar *et al.* [10] found that the measured net fluxes are much lower than predicted by the theory. The authors proposed that the bacterial cells adhere to the membrane, so that only a small portion of the fouling cake is removed during each backpulse. A nonuniform cleaning model based on the mechanism shown in Figure 3 was proposed. During reverse filtration, the backpulse fluid breaks off small pieces of the adhesive cake and then is preferentially channeled through the small holes formed in the cake, resulting in nonuniform cleaning of the membrane. As a first approximation, Kuberkar *et al.* [10] proposed that a fraction β of the membrane is completely cleaned during each backpulse, while the remaining fraction remains

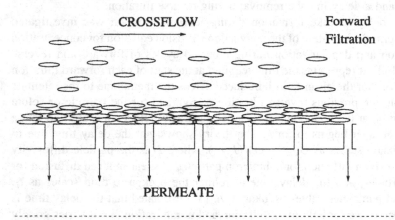

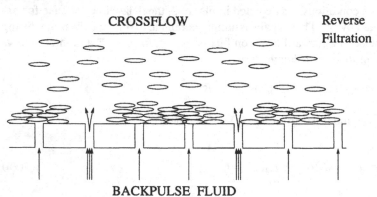

Figure 3   Schematic of nonuniform foulant removed during reverse filtration. (From Ref. 10.)

irreversibly fouled. According to their model, the forward flux, including the delay in cake formation is then $J_f = \beta J_o + (1 - \beta) J_s$, for $t_f < t_f^{crit}$, and

$$J_f(t) = \beta J_o/(1 + (t - t_f^{crit})/\tau)^{1/2} + (1 - \beta)J_s, \ t_f \geq t_f^{crit} \tag{8}$$

where J_s is the long-term or steady flux through the fouled portion of the membrane (see Figure 2).

If it is assumed that the reversible foulants are instantly removed, then the magnitude of the negative flux during backpulsing is simply $J_b = \alpha(\beta J_o + (1 - \beta)J_s)$. Then, substituting these expressions into Eq.(1) yields [10]

$$\langle J \rangle = (\beta J_o + (1 - \beta)J_s)(t_f - \alpha t_b)/(t_f + t_b), \ t_f < t_f^{crit} \tag{9a}$$

$$\langle J \rangle = \frac{\beta J_o(t_f^{crit} + 2\tau [(1 + (t_f - t_f^{crit})/\tau)^{1/2} - 1] - \alpha t_b) + (1 - \beta)J_s(t_f - \alpha t_b)}{t_f + t_b}, \ t_f \geq t_f^{crit} \tag{9b}$$

The parameter β is referred to as the cleaning efficiency and must be determined empirically. Further modifications to the theory have been proposed which relax the assumption that foulants are instantly removed during backpulsing [10,40]. Continued work is underway to better understand foulant removal.

III. SEPARATIONS INVOLVING WHOLE CELLS

A. Materials and Methods

Crossflow microfiltration with backpulsing was performed both with yeast suspensions [35,37] and with bacterial suspensions [10]. For the former, *Sacchromyces cerevisiae* dry yeast (Fleishmann) was rehydrated in deionized water. The cells were washed several times to remove debris and residual proteins. For the latter, *Escherichia coli* bacterial harvests were provided by Amgen, Inc. (Boulder, CO), and used as whole-fermentation broth, diluted fermentation broth, or washed cells resuspended in a buffered (pH = 7) solution containing 8.8g NaCl, 5.6 g EDTA, and 0.55 g NaOH per liter of deionized water.

The experimental apparatus is shown in Figure 4. The fluid in the feed reservoir (1 gal; Alloy Products, Waukesha, WI, model DIV.1 WT304) is pressurized by a high-pressure nitrogen cylinder. The fluid from the feed reservoir is forced by a peristaltic recirculation pump (Millipore Systems, Bedford, MA, model XX 80 000 00) to a flat-plate membrane module (Millipore Systems, Bedford, MA, model Minitan-S). The module has nine parallel channels, each of dimensions 0.4 mm high by 7 mm wide by 50 mm long. The permeate is collected in a permeate reservoir on an electronic balance (Mettler Toledo, Hightstown, NJ, model PG 5002). The retentate is returned to the feed vessel. To prevent

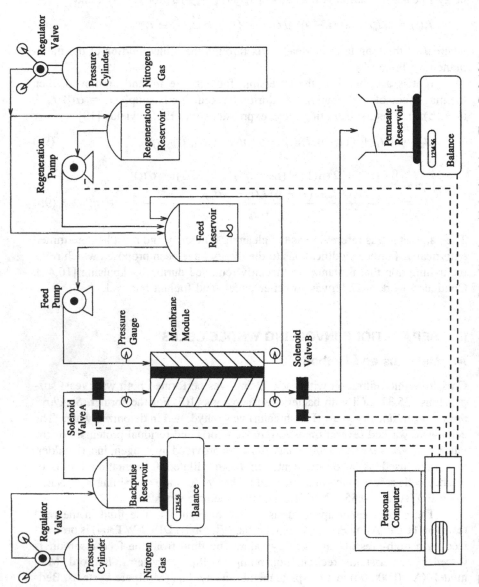

Figure 4 Schematic of experimental apparatus for crossflow filtration with backpulsing. (From Ref. 10.)

changes in the feed concentration of cells, the net loss of liquid from the feed vessel due to permeate collection is replaced from a regeneration reservoir (1 gal; Alloy Products, Waukesha, WI, model DIV.1 WT304) using a peristaltic regeneration pump (Watson Marlow, Wilmington, MA, model 101U). The backpulse fluid is stored in a backpulse reservoir (1 gal; Alloy Products, Waukesha, WI, model DIV.1 WT304) on an electronic balance (Mettler Toledo, Hightstown, NJ, model PM 16-N).

The operation mode of retentate recycle and permeate replacement was chosen to give constant solids concentration and to reflect dialysis conditions. Large-scale operations often use batch concentration (no permeate recycle or replacement), in which the solids concentration increases with time, or a feed-and-bleed mode (retentate recycled to the filter inlet, rather than to the feed reservoir, with a small fraction bled off), in which case a high solids concentration in the retentate is achieved. As shown later, the performance of crossflow filtration, both with and without backpulsing, declines with increasing solids concentration.

The pressures on the feed side, the permeate side, the retentate side, and the backpulse side of the membrane module are measured by pressure gauges (0-30 psi; Ashcrost, Stratford, CT). The pressures in the vessels are controlled by regulator valves (Victor Equipment Company, Denton, TX, model GPS 270A). Backpulsing is controlled by solenoid valves (Burkert Contromatic Corp., Irvine, CA, model 1053-S001-031-00). Neoprene tubing (Masterflex, Vernon Hills, IL, model 6402-15) is used.

The electronic balances, the solenoid valves, and the regeneration pump are connected to a computer (Gateway 2000, North Sioux City, SD, model DX2-66V). Both electronic balances are connected through the serial ports. The solenoid valves and the regeneration pump are connected through a data acquisition board (Data Translation, Marlboro, MA, model DT2801). The components of the setup are interfaced by software programs written in Quickbasic. During forward filtration, solenoid valve B is opened and solenoid valve A is closed. During reverse filtration, solenoid valve A is opened and solenoid valve B is closed.

Cellulose-acetate "acetateplus" membranes manufactured by Micron Separations, Inc. (Westborough, MA) with a nominal pore size of 0.2 μm were used in the membrane module for the experiments with bacteria [10]. No pretreatment was done before the start of an experiment. The membranes are hydrophilic and nearly symmetric. The clean membrane buffer flux declines over the first few minutes to reach a steady value of $J_o = 0.13 \pm 0.02$ cm/s (mean $\pm$ one SD) at 10 psi transmembrane pressure and room temperature (25°C). For the first set of yeast experiments [35], cellulose-acetate "cellgard" membranes, also manufactured by Micron Separations, Inc., with a nominal pore size of 0.2 μm, were used. The clean membrane flux is $J_o = 0.18 \pm 0.02$ cm/s for these membranes at 10 psi transmembrane pressure and room temperature. For the second set of yeast experiments [37], cellulose-acetate membranes manufactured by Sartorious, with a nominal pore size of 0.07 μm, were used in the same module. In this case, the clean membrane flux is $J_o = 0.022$ cm/s at 5 psi transmembrane and room temperature. Unless noted otherwise, a feed flow rate of 6 mL/s was used; it corresponds to an average cross-flow velocity of 24 cm/s and a nominal shear rate at the membrane surface of 3600 s$^{-1.}$

For experiments with bacterial fermentation broth, the feed concentration of proteins and other solubles which pass through the membrane decreased gradually with time due to diafiltration, since they are not present in the buffer used for the backpulsing and regeneration fluids. The effects of this dilution on reducing the observed fouling are not expected to be large, since the volume of permeate replaced in these experiments was only 10% to 20% of the feed volume, due to the low permeate fluxes for diluted and undiluted fermentation broths.

B. Results and Discussion

Figure 5 shows a plot of the average flux per cycle versus total filtration for the first set of washed-yeast experiments with three reverse-filtration pressures of

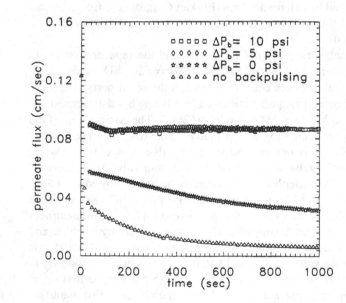

Figure 5 Net flux per cycle versus filtration time for washed yeast with and without backpulsing at $\Delta P_f = 10$ psi, $t_f = 7$ s, $t_b = 2$ s, and $c_b = 10$ g/L dry weight. (From Ref. 35.)

$\Delta P_f = 0$, 5, and 10 psi and a fixed forward transmembrane pressure of $\Delta P_f = 10$ psi. The forward filtration time was $t_f = 7$ s, and the backpulse time was $t_b = 2$ s. For $\Delta P_b = 5$ and 10 psi, the average flux per cycle remained nearly constant at 0.086 cm/s over the entire experiment, which is approximately 50% of the clean-membrane flux. It indicates that the permeability of the membrane was recovered after each cycle and that a long-term reduction of the permeability due to irreversible cake buildup did not occur. For forward filtration without backpulsing, the flux declined rapidly from $J_o = 0.175$ cm/s to $J = 0.006$ cm/s in 1000 s, with $\tau = 0.67$ s [34]. In addition, a nearly steady flux of $J_s = 0.0028$ cm/s was achieved after 5,000 s. Thus, a 30-fold increase in the long-term net flux was achieved by rapid backpulsing with $\Delta P_b = 5$ and 10 psi. For $\Delta P_b = 0$ (in which case the permeate was stopped for a duration of $t_b = 2$ s after each period of forward filtration of duration $t_f = 7$ s, but no reverse flow was applied while the permeate was stopped), however, the average flux per cycle decreased continuously to 0.031 cm/s at $t = 1000$ s. Apparently, reverse flow from the permeate side to the retentate side is required for the yeast cake to be completely removed so that the flux is regained after each cycle. Nevertheless, a considerable flux enhancement over the case of forward filtration without backpulsing is evident. Periodic permeate stoppage, which we refer to as crossflushing, has previously been shown to increase the net flux by up to severalfold [41]. Possible reasons for the flux recovery during crossflushing are diffusion of cells from the cake surface, erosion of the cake, decompression and subsequent washing away of the cake, and a shock wave (due to the rapid valve closing) knocking off part of the cake layer. These mechanisms are also active during backpulsing with $\Delta P_b \neq 0$, but they are as effective as reverse flow of permeate in cleaning the membrane. I also note that diffusion and erosion are not effective in removing a compressed yeast cake, as Redkar and Davis [39] did not observe a significant flux increase upon increasing the feed flow rate during forward filtration.

From Figure 5, it is apparent that the reverse transmembrane pressures of $\Delta P_b = 5$ and 10 psi gave approximately the same flux per cycle, even though more fluid was lost during reverse filtration for $\Delta P_b = 10$ psi than for $\Delta P_b = 5$ psi. Apparently, the higher backpulse pressure removed the cake more completely. This is confirmed by the measurements of the permeate collected on the microbalance from forward filtration only; the average flux during forward filtration is 0.132 and 0.156 cm/s for $\Delta P_b = 5$ and 10 psi, respectively, even through $\Delta P_f = 10$ psi in each case.

Figures 6 and 7 give the results for the second set of washed-yeast experiments [37], with fixed backpulse durations of $t_b = 0.2$ s and 0.1 s, respectively, and various durations of forward filtration, t_f, between backpulses. The forward and reverse transmembrane pressures were both 5 psi. In all cases, the symbols are the global average net flux over the durations of three repeated experiments,

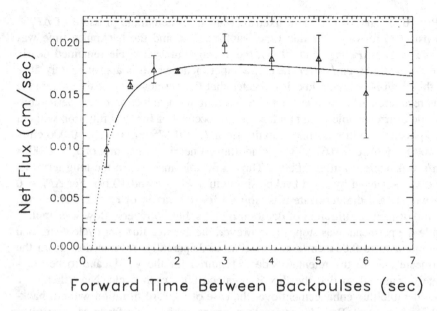

Figure 6 Average net flux versus forward filtration time for washed yeast with $\Delta P_b = \Delta P_f = 5$ psi, $t_b = 0.2$ s, and $c_b = 2.6$ g/L dry weight. The dashed-dotted line is the clean membrane flux, J_o, and the dashed line is the long-term flux without backpulsing, J_s. (From Ref. 37.)

and the error bars represent ± 1 SD. The net flux values were computed by taking the fluid gained in the permeate reservoir, subtracting the fluid lost from the backpulse reservoir, and dividing by the membrane area and the duration of the experiment. Small declines in the net flux of about 10% were observed over the duration (1000 s) of a typical backpulse experiment, apparently due to membrane compaction and incomplete removal of the fouling deposit. The mass measurements from the two electronic balances showed that some fluid bypassed directly from the backpulse reservoir to the permeate reservoir; this could possibly be minimized by incorporating a small delay in opening the valve (B) to the permeate reservoir after closing the valve (A) from the backpulse reservoir at the end of each backpulse.

The dotted lines in Figures 6 and 7 are the predicted net fluxes from Eq.(7a) for $t_f \leq t_f^{\text{crit}}$, and the solid lines are the predicted net fluxes from Eq. (7b) for $t_f > t_f^{\text{crit}}$, where $t_f^{\text{crit}} = 0.29$ and 0.67 s for $t_b = 0.1$ and 0.2 s, respectively [37]. A time constant of $\tau = 4.1$ s was determined from an experiment with forward filtration only. In all cases, good agreement between theory and experiment is observed, except at long forward-filtration times with $t_b = 0.1$ s. For each back-

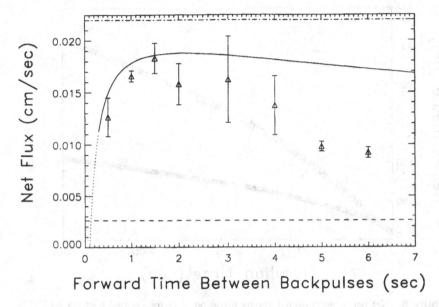

Figure 7 Average net flux versus forward filtration time for washed yeast with $\Delta P_b = \Delta P_f = 5$ psi, $t_b = 0.1$ s, and $c_b = 2.6$ g/L dry weight. The dashed-dotted line is the clean membrane flux, J_o, and the dashed line is the long-term flux without backpulsing, J_s. (From Ref. 37.)

pulse time, there is an optimum forward filtration time which maximizes the flux. Shorter durations of forward filtration do not allow for sufficient collection of permeate relative to that lost during reverse filtration, whereas longer durations of forward filtration lead to flux decline due to cake buildup. For the longest periods of forward filtration, the experimental data fall below the predictions of the theory; this is likely the result of the yeast cakes not being completely removed during backpulsing, especially for the shortest duration of $t_b = 0.1$ s. Most significantly, the maximum flux with backpulsing is $\sim 85\%$ of the clean-membrane flux of $J_o = 0.022$ cm/s, and about sevenfold higher than the long-term flux without backpulsing of $J_s = 0.0026$ cm/s.

Cross-flow microfiltration experiments with bacteria [10] were first done for a washed *E. coli* suspension with dry cell weight of 1.2 g/L. Figure 8 shows the raw data as net permeate (defined as the fluid mass added to the permeate reservoir minus that removed from the backpulse reservoir) collected versus time for experiments done with and without backpulsing. The transmembrane pressure during forward filtration was 10 psi for both experiments, and the transmembrane pressure during reverse filtration for backpulsing was also 10 psi. The backpuls-

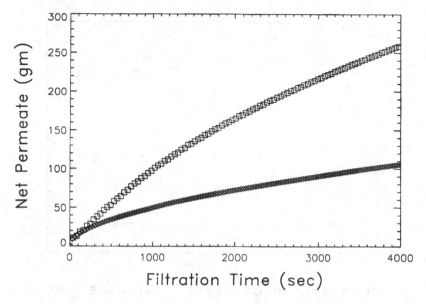

Figure 8 Net permeate collected versus filtration time for washed bacteria with ($\square$) and without ($\Diamond$) backpulsing at $\Delta P_b = \Delta P_f = 10$ psi, $t_b = 1.0$ s, $t_f = 40$ s, and $c_b = 1.2$ g/L dry weight. (From Ref. 10.)

ing experiment was done with a forward filtration time of $t_f = 40$ s and a reverse filtration time of $t_b = 1$ s. In both cases, the rate of permeate collection starts out high but gradually declines over time due to fouling. After about 3000 s, an apparent steady state is reached, with a higher net flux observed for the experiment with backpulsing. The flux values are determined from the slopes of these curves, multiplied by the fluid density and divided by the membrane surface area.

The time constant for cake growth was calculated from the flux decay observed in the normal crossflow experiment without backpulsing. From two repeats, the time constant for cake growth was found to be $\tau = 0.0391 \pm 0.0007$ s. Note that the time constant for the bacteria suspension is much smaller than those for the yeast experiments. The difference occurs because the bacterial cake has a very high specific resistance and so causes very rapid flux decline. The long-term flux in the bacterial experiments without backpulsing was found to be $J_s = 0.00032 \pm 0.00001$ cm/s. This long-term flux is > 400-fold less than the clean membrane flux of $J_o = 0.13 \pm 0.02$ cm/s. The very small values of τ and J_s are indicative of rapid and severe membrane fouling by bacteria. Much of this fouling is irreversible, as the buffer flux after 10 min of backwashing recovered only to 0.0034 ± 0.0003 cm/s. Although this recovered flux is 10-fold higher than the long-term flux, it is still 40-fold lower than the clean membrane flux.

Figure 9 shows a plot of the net flux per cycle versus forward filtration time for backpulsing experiments done using a washed *E. coli* suspension with a dry cell weight of 1.2 g/L and a reverse filtration time of $t_b = 1.0$ s, with $\Delta P_b = \Delta P_f = 10$ psi. As expected, the net flux increases initially with increasing forward filtration time, reaches a maximum value, and then decreases with further increases in the forward filtration time. The maximum net flux is $\langle J \rangle = 0.0014$ cm/s, occurring at a forward filtration time of $t_f = 40$ s. This represents more than a fourfold improvement over the flux without backpulsing. Figure 9 also shows the predicted net flux from Eq. (8), which follows the trend of the data. A best-fit value of $\beta = 0.19$ is used, found by minimizing the sum of the squared errors between the measured and predicated fluxes.

Figure 10 gives a similar plot of net flux per cycle versus forward filtration time for backpulsing experiments done with a reverse filtration time of only 0.1 s. These experiments were also done using a washed *E. coli* suspension of dry cell weight 1.2 g/L. The maximum net flux is $\langle J \rangle = 0.0039 \pm 0.0005$ cm/s, at

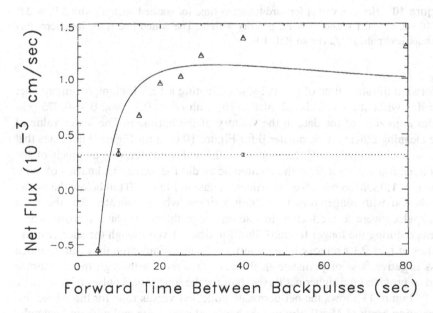

Figure 9 Net flux versus forward filtration time for washed bacteria with $\Delta P_b = \Delta P_f = 10$ psi, $t_b = 1.0$ s, and $c_b = 1.2$ g/L dry weight. The dotted line is the long-term flux without backpulsing, J_s. (From Ref. 10.)

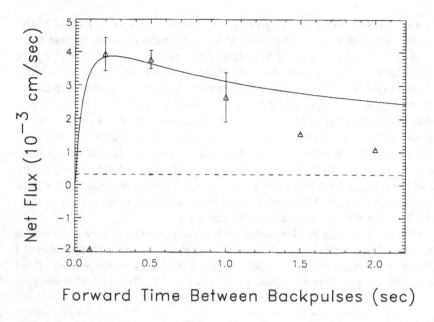

Figure 10   Net flux versus forward filtration time for washed bacteria with $\Delta P_b = \Delta P_f$ = 10 psi, t_b = 0.1 s, and c_b = 1.2 g/L dry weight. The dotted line is the long-term flux without backpulsing, J_s. (From Ref. 10.)

a forward filtration time of t_f = 0.2 s, representing a 12.5-fold improvement over the flux without backpulsing. Equation (9) with t_f^{crit} = 0.1 s and β = 0.075 provides a good fit of the data in the vicinity of the optimum. The lower value of the cleaning efficiency parameter β for Figure 10 than for Figure 9 indicates that the shorter backpulses of duration t_b = 0.1 s did not remove as much of the bacterial fouling layer from the membrane as did the longer backpulses of duration t_b = 1.0 s. Moreover, the experimental data in Figure 10 fall below the model prediction with longer forward filtration times, which indicates that the short backpulses were less effective in removing the thicker foulant deposits which formed during the longer forward filtration times. Even though the shorter backpulses of t_b = 0.1 s removed the bacterial cake less efficiently, they caused much less negative flow of permeate in the reverse direction through the membrane and so provided for higher net fluxes than for the longer backpulses of t_b = 1.0 s.

Figure 11 shows the net permeate collected versus time for the whole fermentation broth of 45 g/L dry weight bacterial cells, with and without backpulsing. Because of the high concentration of foulants, the rate of permeate collection is much lower than for the washed and diluted bacterial cells (compare Figures 8 and 11). Moreover, a severe decline in the rate of permeate collection is ob-

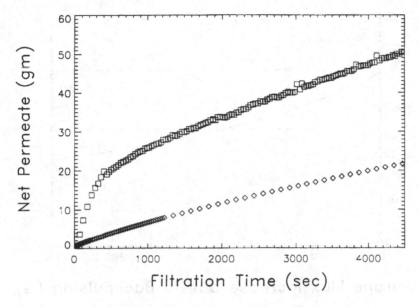

Figure 11 Net permeate collected versus filtration time for bacterial fermentation broth with (□) and without (◇) backpulsing at $\Delta P_b = \Delta P_f = 10$ psi, $t_b = 0.1$ s, $t_f = 3$ s, and $c_b = 45$ g/L dry weight. (From Ref. 10.)

served over the first 1000 s of backpulsing, which indicates irreversible fouling. It was typically found that the net flux for bacterial fermentation broth could be improved about twofold by backpulsing under optimal conditions, which is considerably less than the 10-fold improvements obtained for washed bacterial cells [10]. Nevertheless, even the modest improvement for bacterial fermentation broth makes crossflow microfiltration with backpulsing economically attractive [10].

Figure 12 shows the annual cost of the separation of a soluble extracellular product from the bacterial cells in a 10,000-L fermentation broth by batch concentration followed by diafiltration. The details for the calculations are described by Kuberkar *et al.* [10] and follow the analysis of Zeman and Zydney [11]. Without backpulsing, the cost of cross-flow membrane filtration is sightly below that of rotary vacuum filtration but about 35% higher than for centrifugation. However, the cost of membrane filtration with backpulsing decreases below that of centrifugation when the flux improvement is as little as 50%. When the flux improvement is twofold, as is typical for backpulsing with concentrated fermentation broth [10], the cost of membrane filtration is only 75% of the cost of centrifugation to effect the desired separation.

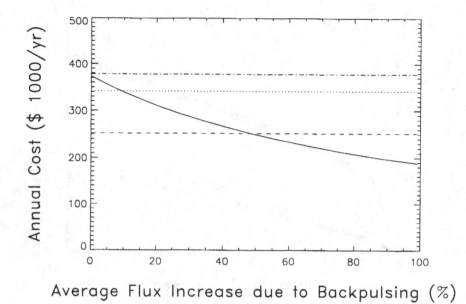

Average Flux Increase due to Backpulsing (%)

Figure 12 Annual cost for separation of a soluble extracellular product from fermentation broth versus the percent increase in flux due to backpulsing for crossflow membrane filtration with backpulsing (solid line). The dashed line is the corresponding cost for centrifugation, the dotted line is the corresponding cost for crossflow membrane filtration without backpulsing, and the dashed-dotted line is the corresponding cost from rotary vacuum filtration. (From Ref. 10.)

IV. SEPARATIONS INVOLVING CELL DEBRIS AND PROTEINS

A. Materials and Methods

Crossflow microfiltration experiments with backpulsing for the recovery of a recombinant protein from bacterial cell debris were performed [36] using the apparatus of Figure 4. The apparatus was operated in a total recycle mode, with the retentate returned to the feed reservoir, the permeate collected on an electronic balance, and the collected permeate continuously replaced in the feed reservoir with a solids-free lysate having the same concentration of solubles as the feed. By adding the regeneration fluid at the same rate as permeation takes place, the concentrations of solids and solubles in the feed lysate remain constant. The membranes used in all backpulsing experiments are the same as the 0.2-μm cellulose acetate membranes used in the first set of yeast experiments. Each membrane was used once for each experiment, and then discarded.

All experiments were performed with *E. coli* provided by Synergen, Inc. (now Amgen, Boulder, CO) which overproduces a 14-kDa, recombinant protein. The concentration of recombinant protein is proportional to the total soluble protein concentration, as measured by absorbance at 280 nm. The cell harvest was concentrated to 35% or more by weight wet solids by centrifugation and then frozen and stored in 150-mL containers at −70°C. Samples were thawed as needed and diluted to 25% wet solids (0.25 g solids/g suspension) with the standard buffer. The cell harvest was then sonicated in three passes through a continuous flow cell (Heat Systems W380, 1/2 inch tip, continuous power at a setting of 10), at a rate of 50 mL min^{-1}. The resulting lysate was stored at 20°C and used within 3 weeks. Clarified lysate for the feed reservoir make-up fluid was generated by centrifugation in an A-641 Sorvall rotor at 40,000 rpm (200,000 g average) for 90 min, which is predicted to remove all particles > 0.16 μm in diameter. The lysate was further diluted to as low as 0.0025 g/g with standard buffer prior to the filtration experiments.

Samples of the permeate and retentate were stored in Eppendorf test tubes and analyzed within 4 hours for protein content. All samples were centrifuged in an Eppendorf microfuge at 4°C for 15 min at ∼ 16,000 g, which is predicted to remove all particles > 0.5 μm in diameter. The samples were then diluted to a total dilution of 1 : 200 (including the operating dilution used in the experiment), and measured for absorbance at 280 nm.

The results with backpulsing [36] are compared to previous results without backpulsing [2]. The experiments without backpulsing used the same cell lysate and flat-filter setup, but surface-modified polyvinylidene fluoride "durapore" membranes (Millipore, Bedford, MA) with 0.2-μm nominal pore diameters were used instead of the cellulose-acetate membranes employed in the backpulsing experiments. Since the very low long-term fluxes obtained indicate deposit-dominated resistance, the membrane material was shown to not have a significant effect on the long-term flux without backpulsing. Experiments without backpulsing were also performed [2] with 0.2-μm polyacrylonitrile MX—2000 asymmetric membranes using a Benchmark rotary filter (Membrex, Fairfield, NJ).

B. Results and Discussion

Figure 13 shows typical transient results for the flux and sieving coefficient without backpulsing [2]. For this flat-filter experiment, the concentration of the bacterial lysate is 2.5 g/L wet solids, the transmembrane pressure is 1 psi, and the nominal shear rate due to the crossflow is 1560 s^{-1}. The sieving coefficient is defined as the protein concentration in the permeate divided by that in the feed. The flux and sieving coefficient decreased with time due to membrane fouling. The clean membrane water flux at $\Delta P_f = 1$ psi is $J_o = 0.01$ cm/s for this membrane, so it is apparent that considerable flux decline occurred during the filtration

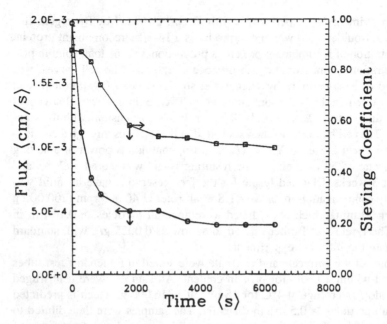

Figure 13 Permeate flux (O) and sieving coefficient (□) versus time for cross-flow microfiltration without backpulsing of a bacterial cell lysate with $c_b = 2.5$ g/L wet solids, $\Delta P_f = 1$ psi, and a nominal shear rate at the membrane surface of $1560 \ s^{-1}$. (From Ref. 2.)

of the cell lysate, even by the time the first measurement was made. Of even greater concern is that the protein transmission decreased from nearly 100% for the clean membrane to < 50% for the fouled membrane, representing considerable loss of the recombinant protein product. The rotary filter gave a similar sieving coefficient and about twofold higher flux at this low-solids concentration, but no improvement at high-solids concentration [2].

In the experiments with backpulsing [36], the net flux decreased rapidly over the first five min of filtration due to irreversible membrane fouling by the cell lysate, but then reached a nearly steady value which is generally much larger than that obtained without backpulsing. Moreover, the net flux increased with increasing transmembrane pressure with backpulsing but not without backpulsing, over the tested range of 1 to 10 psi. This finding indicates that the foulant provides the dominant resistance to filtration without backpulsing, but that the resistance of the partially cleaned membrane plays an important role with backpulsing. Most important, all backpulsing experiments achieved 100% transmission of protein through the membrane [36].

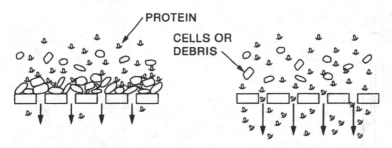

Figure 14 Schematic (not to scale) of forward filtration of cell lysate after a cake has formed (left side) and after the cake has been removed (right side).

The increased permeate flux and protein transmission with backpulsing are thought to be a result of the removal of the fouling cake of cell debris which forms on the membrane surface, as shown schematically in Figure 14. During forward filtration, a fouling layer forms on the membrane surface which may block pores and serve as a dynamic secondary membrane that captures protein molecules which would otherwise pass through the much larger pores of the microfiltration membrane. During reverse filtration, at least a portion of the fouling layer is removed and the trapped protein molecules become resuspended, so that the subsequent permeate and solute fluxes are increased.

Figure 15 shows the long-term net flux with backpulsing at different frequencies and a fixed duration of $t_b = 0.09$ s, with $\Delta P_b = \Delta P_f = 10$ psi, $c_b = 2.5$ g/L wet solids, and a nominal shear rate of $1560 s^{-1}$ for the flat filter. As was seen for suspensions of whole cells, the net flux increases with increasing forward filtration time (decreasing backpulse frequency) to a maximum, and decreases thereafter toward the long-term or steady flux without backpulsing. The maximum net flux occurs at $\sim t_f = 0.3$ s forward time (2.6 s^{-1} frequency) and is > 10-fold greater than the steady flux without backpulsing. A smaller maximum flux enhancement of about threefold was obtained for a larger backpulse duration of $t_b = 0.2$ s, presumably because of the rapid flux decline due to fouling. For $t_b = 0.2$ s, a longer period of forward filtration is required to overcome the larger amount of negative flow which crosses the membrane, as compared to the case $t_b = 0.1$ s [36].

Figure 16 shows the variation in the steady net flux with the nominal shear rate in the flat filter for the bacterial lysate with and without backpulsing. The flux values with backpulsing are five-fold to 10-fold higher than those without backpulsing, but show a weaker dependence on the shear rate. Figure 17 shows the sieving coefficient versus the nominal shear rate for both the flat filter and rotary filter without backpulsing. For both filters, the protein transmission in-

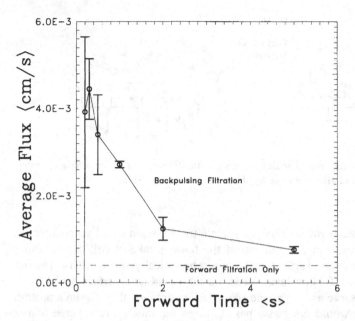

Figure 15 Steady net flux with backpulsing versus duration of forward filtration between backpulses for crossflow microfiltration of bacterial cell lysate in a flat filter with $t_b = 0.09$ s, $\Delta P_b = \Delta P_f = 10$ psi, $c_b = 2.5$ g/L wet solids, and 1560 s^{-1} nominal shear rate. The dashed line is the steady flux without backpulsing, J_s. (From Ref. 36.)

creased from $\sim 40\%$ at low shear rates to $\sim 80\%$ at high shear rates. Moreover, lower sieving coefficients which did not increase with increasing shear rate were observed for more concentrated cell lysates [2]. In contrast, the protein transmission remained at 100% for all experiments with backpulsing [36].

Finally, Figure 18 shows the steady net flux with and without backpulsing versus solids concentration for the flat filter. Although backpulsing provided for a 10-fold enhancement in the net flux for the most dilute lysate, a very steep decline in the backpulse performance with increasing solids concentration is seen, with no improvement over the long-term flux without backpulsing achieved when the solids concentration exceeds 10 g/L wet weight. Increasing the concentration of cell debris leads to rapid formation of thick fouling layers on the membrane surface, which even the most rapid backpulses with the current setup cannot substantially prevent or remove. Nevertheless, 100% protein transmission was obtained with backpulsing, at all concentrations investigated. A possible explanation is that the backpulses create small holes in the fouling layer, through which permeate and protein pass during the periods of forward filtration, but the addi-

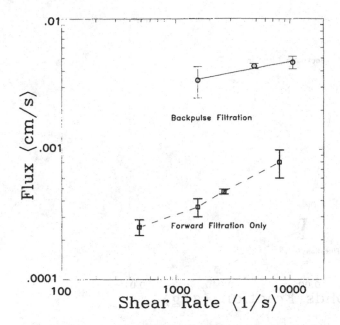

Figure 16 Steady net flux with backpulsing (○) and without backpulsing (□) versus nominal shear rate for crossflow microfiltration of bacterial cell lysate in a flat filter with $\Delta P_b = \Delta P_f = 10$ psi, $t_b = 0.09$ s, $t_f = 0.5$ s, and $c_b = 2.5$ g/L wet solids. (From Ref. 36.)

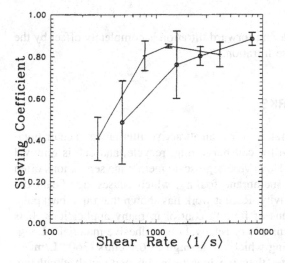

Figure 17 Sieving coefficient versus nominal shear rate for crossflow microfiltration of bacterial cell lysates in flat (○) and rotary (+) filters without backpulsing, with $\Delta P_f = 1$ psi and $c_b = 2.5$ g/L wet solids. (From Ref. 2.)

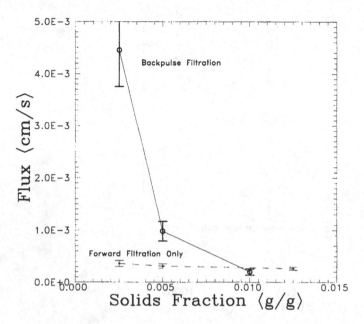

Figure 18   Steady net flux with (○) and without (*) backpulsing versus wet solids fraction for crossflow microfiltration of bacterial cell lysate in a flat filter with $t_b = 0.09$ s, $t_f = 0.3$ s, $\Delta P_b = \Delta P_f = 10$ psi, and 4000 s⁻¹ nominal shear rate. (From Ref. 36.)

tional collection of permeate during forward filtration is completely offset by the loss of permeate during reverse filtration.

V.   CONCLUDING REMARKS

Crossflow membrane microfiltration offers an attractive alternative to centrifugation and conventional filtration for cell harvesting, recycle, and debris removal in biotechnology applications. However, large-scale membrane separations often become uneconomical due to membrane fouling, which causes significant declines in throughput and selectivity. Recent work has shown that rapid backpulsing provides an effective means of fouling control in many applications. It is most effective for dilute suspensions of relatively nonadhesive materials. Long-term net fluxes with backpulsing which are as high as 0.1 cm/s (3600 L/m² − h), and 10-fold or more greater than the long-term flux without backpulsing, have been achieved with washed yeast suspensions of 10 g/L dry weight [35,37]. Similar 10-fold enhancements in net flux were achieved with dilute suspensions of washed bacterial cells [10]. However, due to the adhesive nature of these cells and the large resistance of the bacterial cake which forms on the membrane surface, only a portion of the membrane surface was cleaned and the maximum long-term net flux was only about 2×10^{-3} cm/s (72 L/m² − h) for 10 g/L dry weight. Even greater irreversible fouling occurs for fermentation broth, due to the extracellular material, so the maximum net flux was improved only twofold with backpulsing to 4×10^{-4} cm/s (14 L/m² − h) for a bacterial fermentation broth of 10 g/L dry weight [10]. For a higher concentration of 45 g/L dry weight, the maximum net flux with backpulsing decreased to about 2×10^{-4} cm/s (7 L/m² − h) but is still twofold greater than for crossflow microfiltration of the same fermentation broth without backpulsing [10]. Moreover, the flux improvement with backpulsing makes crossflow membrane microfiltration economically attractive for cell harvesting. The additional capital and energy costs needed to implement the rapid backpulsing are more than offset by the reduction in membrane area and system size afforded by the higher permeate flux.

Membrane fouling is even greater for cell lysates than observed with whole cells. The long-term net flux without backpulsing is $\sim 4 \times 10^{-4}$ cm/s (14 L/m² − h) for a very dilute bacterial cell lysate of 2.5 g/L wet solids, and it decreases gradually with increasing solids concentration [2]. Although backpulsing is able to enhance the net flux by 10-fold under optimum conditions for 2.5 g/L wet solids, the very rapid fouling which occurs at higher concentrations causes backpulsing to be ineffective in flux enhancement for feed concentrations exceeding 10 g/L wet solids [36]. However, rapid backpulsing is able to keep a fraction of the microfiltration membrane surface and pores sufficiently open so that the protein transmission remains at 100%, since most of the forward permeate flow occurs through this cleaned portion of the membrane, whereas the protein transmission typically declines to $\sim 50\%$ in the absence of backpulsing.

Finally, backpulsing conditions must be carefully chosen to optimize performance. From both theory and experiment, the backpulse frequency should be on the order of 0.1 to 1 s^{-1}, with the higher frequencies required when the fouling time constant (τ) is very short due to high concentrations of foulants such as bacteria and debris which form cake layers with high resistance. If the backpulse frequency is too low, then suboptimal performance occurs due to significant fouling during the long periods of forward filtration between backpulses, whereas the net flux can be low or even negative due to the large amount of backpulse fluid used when the frequency is too high. Further work is needed to better predict the optimum backpulse duration, but our experience suggests that 0.1 to 0.2 s is typically sufficient to remove a significant portion of the reversible foulants without causing too much negative flow of fluid in the reverse direction across the membrane.

I close with a caution that rapid backpulsing is not expected to be the panacea which overcomes the problem of membrane fouling in all crossflow microfiltration applications. We have already seen that it has reduced effectiveness for high feed concentrations and adhesive foulants. Moreover, in some applications with multicomponent feeds, the rejected larger particles form a dynamic secondary membrane which captures adhesive smaller particles which would otherwise foul the primary membrane [42–44]. Periodic removal of this secondary membrane by backpulsing would expose the primary membrane to more fouling and might reduce the long-term performance below that without backpulsing.

ACKNOWLEDGEMENTS

The author thanks his former Ph.D. students (Vinod Kuberkar, Charles Parnham, and Sanjeev Redkar) who carried out most of the experiments recorded in this chapter. The work was supported by the National Science Foundation, the U.S. Bureau of Reclamation, the Center for Separations Using Thin Films at the University of Colorado, and the Colorado Institute for Research in Biotechnology. Appreciation is also extended to Amgen, Membrex, and Millipore for providing materials used in the experiments.

REFERENCES

1. Blatt WF, Dravid A, Michaels AS, Nelson L. Solute polarization and cake formation in membrane ultrafiltration: causes, consequences and control techniques. In: JE Flinn, ed. Membrane Science and Technology. New York: Plenum Press, 1970:47.
2. Parnham CS, Davis RH. Protein recovery from cell debris using crossflow microfiltration. Biotechnol Bioeng 47:155–164, 1995.
3. Datar R. Studies on the separation of intracellular soluble enzymes from bacterial cell debris by tangential flow membrane filtration. Biotechnol Lett 7:471–476, 1985.
4. Forman SM, DeBernardez ER, Feldberg RS, Swartz RW. Crossflow filtration for the separation of inclusion bodies from soluble proteins in recombinant *Escherichia coli* cell lysate. J Membr Sci 48:263–279, 1990.
5. Kroner KH, Schütte H, Hustedt H, Kula MR. Crossflow filtration in the downstream processing of enzymes. Proc Biochem 1984; 19:67–74, 1984.
6. Le MS, Spark LB, Ward PS. The separation of aryl acylamindase by cross flow microfiltration and the significance of enzyme/cell debris interaction. J Membr Sci 21:219, 1984.
7. Porter MC. Applications of membranes to enzyme isolation and purification. Biotechnol Bioeng Symp 3:115–144, 1972.
8. Quirk AV, Woodrow JR. Tangential flow filtration—a new method for the separation of bacterial enzymes from cell debris. Biotechnol Lett 5:277–282, 1983.
9. Quirk AV, Woodrow JR. Investigation of the parameters affecting the separation of bacterial enzyme from cell debris by tangential flow filtration. Enz Microb Technol 6:201–206, 1984.

10. Kuberkar VT, Czekaj P, Davis RH. Flux enhancement for membrane filtration of bacterial suspensions using high-frequency backpulsing. Biotech Bioeng 60:77–87, 1998.

11. Zeman LJ, Zydney AL. Microfiltration and Ultrafiltration: Principles and Applications, 1st ed. New York: Marcel Dekker, 1996.

12. Mir L, Michaels SL, Goel V. Crossflow microfiltration: applcations, design and cost. In: Ho WSW, Sirkar KK, eds. Membrane Handbook, 1st ed. New York: Van Nostrand Reinhold, 1992:571–594.

13. Belfort G, Davis RH, Zydney AL. The behavior of suspensions and macromolecular solutions in crossflow microfiltration. J Membr Sci 96:1–58, 1994.

14. Davis RH. Microfiltration: definitions. In: Ho WS, Sirkar K, eds. Membrane Handbook. New York: Van Nostrand Reinhold, 457–460, 1992.

15. Davis RH. Theory of crossflow microfiltration. In: Ho WS, Sirkar K, eds. Membrane Handbook. New York: Van Nostrand Reinhold, 480–505, 1992.

16. Davis RH. Modeling of fouling of crossflow microfiltration membranes. Sep Pur Meth 21:75–126, 1992.

17. Porter MC. Microfiltration. In: Burgay PM, Lonsdale HK, dePinho MN, eds. Synthetic Membranes: Science, Engineering and Applications. Dordrect: D. Reidel, 225–246, 1986.

18. Michaels, SL. Crossflow microfilters: the in and outs. Chem Eng Jan:84–91, 1989.

19. Cheryan M. Ultrafiltration and Microfiltration Handbook. Lancaster: Technomic, 1998.

20. Grandison AS, Lewis MJ. Separation Processes in the Food and Biotechnology Industries: Principles and Applications. Lancaster: Technomic, 1996.

21. Michaels AS. Fifteen years of ultrafiltration: problems and future promises of an adolescent technology. In: Cooper AR, ed. Polymer Science Technology. New York: Plenum Press, 1980.

22. Belfort GT, Baltutis TF, Blatt WF. Automated hollow fiber ultrafiltration: pyrogen removal and phage recovery from water. Poly Sci Technol 13:439, 1980.

23. Harrison RG. Large scale process for the purification of alcohol oxidase. U.S. Patent 4,956,290 (1990).

24. Nipkow A, Zeikus JG, Gerhardt P. Microfiltration cell-recycle pilot system for the continuous thermoanaerobic production of exo-β-amylase. Biotechnol Bioeng 34: 1075–1084, 1989.

25. Matsumoto K, Katsuyama S, Ohya H. Separation of yeast by crossflow filtration with backwashing. J Ferment Technol 65:77–83, 1987.

26. Matsumoto K, Kawahara M, Ohya H. Crossflow filtration of yeast by microporous ceramic membrane with backwashing. J Ferment Technol 66:199–205, 1988.

27. Rodgers VGJ, Sparks RE. Reduction of membrane fouling in the ultrafiltration of binary protein mixtures. AIChE J 37:1517–1528, 1991.

28. Rodgers VGJ, Sparks RE. Effect of transmembrane pressure pulsing on concentration polarization. J Membr Sci 68:149–168, 1992.

29. Rodgers VGJ, Sparks RE. Effect of solution properties on polarization redevelopment and flux in pressure pulsed ultrafiltration. J Membr Sci 78:163–180, 1993.

30. Miller KD, Weitzel S, Rodgers VGJ. Reduction of membrane fouling in the presence of high polarization resistance. J Membr Sci 76:77–83, 1993.

31. Rodgers VGJ, Miller KD. Analysis of steric hindrance reduction in pulsed protein ultrafiltration. J Membr Sci 85:39–58, 1993.
32. Nikolov ND, Mavrov V, Nikolova JD. Ultrafiltration in a tubular membrane under simultaneous action of pulsating pressures in permeate and feed solution. J Membr Sci 83:167–172, 1993.
33. Guerra A, Jonsson G, Rasmussen A, Waagner Nielsen E, Edelsten D. Low crossflow velocity microfiltration of skim milk for the removal of bacterial spores. I Dairy J 7:849–861, 1997.
34. Wenten IG. Mechanisms and control of fouling in crossflow microfiltration. Filtration Separation 32:252–253, 1995.
35. Redkar SG, Davis RH. Enhancement of crossflow microfiltration and performance using high frequency reverse filtration. AIChE J 41:501–508, 1995.
36. Parnham CS, Davis RH. Protein recovery from bacterial cell debris using crossflow microfiltration with backpulsing. J Membr Sci 118:259–268, 1996.
37. Redkar S, Kuberkar V, Davis RH. Modeling of concentration polarization and depolarization with high-frequency backpulsing. J Membr Sci 121:229–242, 1996.
38. Davis RH, Grant DC. Theory of deadend microfiltration. In: Ho WS, Sirkar K, eds. Membrane Handbook. New York: Van Nostrand Reinhold, 461–479, 1992.
39. Redkar SG, Davis RH. Crossflow microfiltration of yeast suspensions in tubular filters. Biotech Prog 9:625, 1993.
40. Mallubhotla H, Belfort G. Semiemperical modeling of crossflow microfiltration with periodic reverse filtration. Ind Eng Chem Res 35:2920–2928, 1996.
41. Tanaka T, Itoh H Itoh K, Nakanishi K, Kume T, Matsumo R. Crossflow filtration of baker's yeast with periodical stopping of permeation flow and bubbling. Biotech Bioeng 47:401–404, 1995.
42. Arora N, Davis RH. Yeast cakes as secondary membranes in deadend microfiltration of bovine serum albumin. J Mem Sci 92:247–256, 1992.
43. Güell C, Czekaj P, Davis RH. Microfiltration of protein mixtures using yeast to reduce membrane fouling. J Membr Sci 155:113–122, 1999.
44. Kuberkar VT, Davis RH. Effects of added yeast on protein transmission and flux in crossflow membrane microfiltration. Biotech Prog 15:472–479, 1999.

5

Effects of Different Membrane Modular Systems on the Performance of Crossflow Filtration of *Pichia pastoris* Suspensions

Vicki L. Schlegel and Michael Meagher
University of Nebraska, Lincoln, Nebraska

I. INTRODUCTION

During the past decade, *Pichia pastoris* expression systems have gained wide acceptance as industrial host strains for the production of heterologous proteins because of several mediating factors. Namely, *P. pastoris* organisms are capable of expressing proteins that are analogous to the primary, secondary, and tertiary structures of their natural protein counterparts [1]. And, similar to other higher-level eukaroytoic systems, *P. pastoris* strains efficiently perform such posttranslational modifications as glycosylation, proteolytic processing, disulfide-bond formation, and proper refolding [2–4]. As a result, biologically active proteins are readily derived from these systems. This characteristic is of particular importance in the industrial setting because costly refolding manipulations and other related downstream operations, which are often necessary when bacterial-based cell lines are used, can be eliminated from the overall process. In conjunction with this attribute, protein recovery and purification steps may be reduced further because of the ability of *P. pastoris* cells to express the recombinant material extracellarly while secreting less endogenous proteins [2,5–8]. *P. pastoris* host strains are also easy to manipulate and can be grown to high cell densities that are comparable to biomasses usually associated with bacterial host organisms, such as *E. coli* [6–9].

189

By combining the advantages inherent to both bacterial and yeast-based systems, *P. pastoris* expression is becoming an attractive alternative for the commercial scale production of heterologous proteins. Yet, relevant to any type of expression system, the method selected to initially recover the cells and/or secreted material from *P. pastoris* fermentation broths can adversely affect protein yields and the efficiency of later purification steps. In the biotechnology industry, centrifugation is the conventional approach for cell harvesting [10]. However, certain economical and operational concerns must be considered, including high maintenance costs, heat production, aerosol formation, protein denaturation, etc., before designing a centrifugation separation method for large-scale productions. Crossflow membrane filtration (CFF) is another versatile tool for recovering cells/proteins from fermentation broths [11,12]. Compared to other harvesting techniques, CFF is usually more effective in separating impurities from the component of interest, in sustaining higher product yields, and in preserving the activity of the biological material. Complementary downstream purification processes also tend to improve when CFF is used as the harvesting tool. Another added benefit of CFF is that the initial investment and maintenance of the necessary equipment are less than those encountered with centrifugation.

Because of these features, CFF techniques are emerging as the method of choice for harvesting yeast-based fermentations within the biopharmaceutical industry, where complete control of a process is crucial and where production of a given product is an expensive endeavor. Nevertheless, it is still imperative that process/product-specific operating conditions be evaluated and defined during the developmental stages of a CFF operation to ensure continual success of that process when used for large-scale purposes. Suitability of CFF for a given process can be determined based upon membrane flux rates, shelf life/cleanability, and protein/cell retention of the membrane. Relative to the cited criteria, selecting a proper membrane/modular system that is compatible with the feed stream is essential for designing a controlled, rugged, and cost-effective CFF separation process. The objective of this chapter is to demonstrate the feasibility of harvesting high cell density *P. pastoris* suspensions by CFF methodologies when appropriate modular systems are chosen.

II. MATERIALS AND METHODS

Various flat-sheet membrane systems were evaluated for their ability to effectively harvest *P. pastoris* fermentation broths. The membrane/modules systems that were tested included a 0.16-μm (pore size) polyethersulfone (PES) membrane module with a membrane surface area of 0.45 m^2 (Pall Filtron Corp., East Hills, NY), a 0.20-μm cross-linked cellulose polymer membrane module, 0.6 m^2 surface area (Sartorius Corp., Edgewood, NY), a 0.20-μm Pellicon

PLCXK regenerated cellulose membrane module, 0.5 m^2 surface area (Millipore Corp., Bedford, MA), a 0.22-µm Prostak GVPP polyvinylidenfluoride membrane module, 0.9 m^2 surface area (Millipore Corp.), a 0.20-µm Supor PES membrane, 0.1665 m^2 surface area (Pall Filtron Corp.), configured into a North Carolina SRT (Cary, NC), Dualport module, a 0.1-µm Koch (Wilmington, MA) membrane, 0.032 m^2 surface area, and a 0.2-µm Koch membrane, 0.032 m^2 surface area.

The last two membranes were cut from larger sheets to specifically fit the North Carolina SRT (NC SRT) Consep module. The procedure of cutting membranes by hand is performed routinely in our laboratories. Modules ranging in channel heights from 0.2 to 1.5 mm can then be tested at a lower initial investment in terms of time and expense. For these experiments, two 0.5-mm channel height plates, one middle 0.96-mm support plate, and two mirror-image shaped membranes were used to assemble the Consep membrane/module system. A Millipore Pellicon holder was employed to secure the Pall Filtron, the Millipore Pellicon, and the Sartorius membrane modules into the CFF skid whereas the Supor/NC SRT, the Koch/Concep, and the Prostak modules were housed within a holder designed specifically for these units.

The experiments were performed by using a skid containing the following CFF components: (1) An insulated stainless-steel 20-L feed tank; (2) an automated rotary lobe pump; (3) the membrane modular system that was tested; (4) pressures gauges located in the feed line (inlet) and in the retentate (outlet) line; and (5) lines/valves that directed the retentate back to the feed tank and the permeate into a collection vessel. A calibrated peristaltic pump (Monistat, New York, NY), was interfaced with the CFF equipment to continuously feed deionized water into the feed tank. Figure 1 depicts the basic arrangement of the individual CFF components used for the purposes described in this paper.

All the membrane modules were pretreated according to the manufacturers' recommended procedures to remove any residual storage chemicals. Clean water flow rate data were then obtained by transferring deionized water to the feed tank and allowing the water to circulate throughout the other components of the CFF system. Permeate flow rates were subsequently determined by collecting the filtrate into a standard graduated cylinder for 1 min. The temperature of the water was monitored with a partial immersion thermometer during this process. Data from the clean water flow rates were ultimately converted into flux rates (in liters, m^{-2}-h^{-1}; LMH).

Fermentation broth composed of the genetically engineered host organism, *P. pastoris* GS115, pHIL-D4, (InVitrogen, San Diego, CA), was prepared via the methods described by Stratton et al. [13]. The *P. pastoris* expression system that served as the feedstream was constructed to express the material of interest intracellularly. The final cell suspension was grown to an O.D.$_{600}$ of 210 with a wet cell weight of $\sim$ 25% and a working volume of 60 L. The fermentation

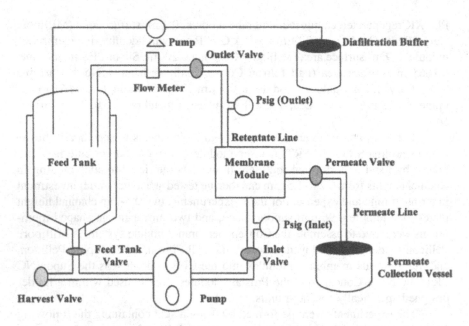

Figure 1   Schematic of the typical CFF system used for harvesting *Pichia pastoris*.

broth was eventually harvested in 6 to 8-L aliquots by using the different membranes systems described previously. The biomass produced from the original
60-L broth was transferred to the CFF recirculation tank, which was chilled below
10°C. The recirculation rate was then adjusted to a value that produced the optimal linear velocity based upon recommendations provided by the manufacturers
or upon the experimental conditions encountered during a process. Unless otherwise noted, 2 L of deionized water was used to wash 1 L of fermentation broth.
Relevant CFF data were collected after every wash cycle, which included the
permeate flow rate, the temperature of the fermentation broth, the recirculation
rate, the pump setting, and the inlet/outlet pressures. In most cases, the suspensions were concentrated immediately by reducing the starting material to a final
volume of 3 to 5 L. Similar to the diafiltration step, appropriate data were collected after a specific volume of permeate was removed from the cells. Permeate
flow data obtained during both the concentration and diafiltration steps were then
converted to flux rates (in LMH).

After the cited harvesting processes were completed, the membranes were
cleaned according to procedures endorsed by the manufacturer. If no instructions
were available, the modules were cleaned as follows. A 4% phosphoric acid solu-

tion was circulated throughout the system for a minimum of 15 min at room temperature. After the system was flushed with DI water, a basic solution of either 1% P53 or 0.1 N sodium hydroxide was used as the cleaning agent. This solution was circulated throughout the components of the CFF system for at least 1 hr at 45 to 50°C. The equipment and the modular unit were thoroughly rinsed with copious amounts of DI water at room temperature. The degree of membrane recovery, in percent, was determined by comparing the clean water data collected after the cleaning procedures were completed with those that were generated prior to exposing the membranes to the fermentation broth.

III. RESULTS AND DISCUSSION

As with the development of any CFF method, the ability of the membrane/modular system to efficiently process the material of interest is of utmost importance. However, yeast-based fermentation broths can pose unique problems because of the high cell densities that are generally encountered with these types of suspensions. It is therefore essential that the compatibility of the feed stream with a given modular unit be ascertained before that CFF system is utilized for routine production purposes. To achieve this end, the performance of each of the cited modules was evaluated by executing the harvesting process at the optimal linear velocities ascribed by the membrane manufacturer. Table 1 summarizes the operating conditions that were required to sustain the linear velocities used for the corresponding membrane systems. It must be noted that the operating parameters shown in the table are averages of the conditions maintained over the entire course of the separation operation. The resulting flux rates were then plotted as a function of the number of liters of solvent exchanged during a given diafiltration step (Figure 2), or as a function of the number of liters of solvent removed during the concentration step (Figure 3).

The graph presented in Figure 2 indicates that the Supor/NC SRT module was much more efficient in washing the fermentation broth than the other modules. More specifically, the LMH values generated by this module ranged between 926 and 933 LMH throughout most of the diafiltration step with the exception of the first several minutes. The dramatic initial surge in LMH, as shown in the Supor/NC SRT graph, occurred because the recirculation was raised in order to increase the linear velocity from 1 m/s to an optimal velocity of 3 m/s. In addition, enough raw data were collected to plot only three points on this graph because the entire process was completed in 10 min.

Similar to the Supor/NC SRT module, the diafiltration step was concluded within a 10-min time frame when the Prostak membrane system was used. Despite the comparable processing times, the flux rates produced by the Prostak

Table 1 Crossflow Filtration Operation Parameters

Membrane/module type	Inlet pressure (psi)	Outlet pressure (psi)	Pump setting (Hz)	Recirculation rate (GPM)	Linear velocity (m/s)
Pall Filtron 0.16-μm PES membrane, 0.45 m²	25	0	21	1.3	1
Sartorius 0.20-μm crosslinked cellulose polymer membrane, 0.6 m²	27	0	20	0.7	0.53
Millipore Pellicon 0.20-μm PLCXK regenerated cellulose membrane, 0.5 m²	27	0	24	0.54	0.41
Supor 0.2-μm PES membrane, 0.1665 m², configured in a NC SRT Duport module	24*	0	46	10.6	3
Millipore Prostak 0.22-μm GVPP PBDF membrane, 0.9 m²	33	2	48	9.48	4
Koch 0.2-μm membrane, 0.032 m², configured in a NC SRT Cocep module	43.5	17	46	9.3	4.7

* Average Operation Settings for the Supor/NC SRT after the recirculation rate was increased.

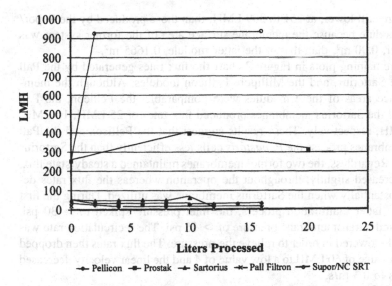

Figure 2 Plot of flux rate (LMH) vs. liters of material processed (or solvent exchanged) for each membrane modular system used to diafilter *Pichia pastoris*.

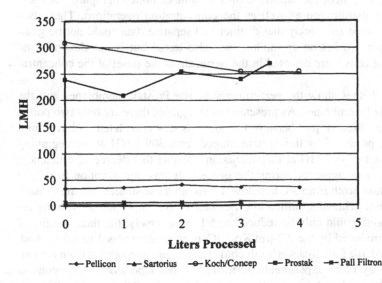

Figure 3 Plot of flux rate (LMH) vs. liters of material processed (or permeate filtered) for each membrane modular system used to concentrate *Pichia pastoris*.

module were still lower, at 251 to 309 LMH, than those produced by the Supor/ NC SRT module because the membrane surface area of the former system was much larger, 0.90 m², than that of the latter module, 0.1665 m².

The remaining plots in Figure 2 chart the flux rates generated by the Pall Filtron, the Sartorius, and the Millipore Pellicon modules. Although the membrane surface areas of these modules were comparable, the Pellicon, the Pall Filtron, and the Sartorius membranes produced flux rates of 25 LMH, 10 LMH, and 32 LMH, respectively. These results suggest that the Pellicon and the Pall Filtron membranes process the *P. pastoris* cells less efficiently than the Sartorius membrane. Regardless, the two former membranes maintained a steady-state flux, or even increased slightly, throughout the operation whereas the flux rates decreased substantially when the Sartorius membrane was utilized. During the first hour of the latter diafiltration process, the inlet pressure spiked to > 30 psi, which created a transmembrane pressure of > 15 psi. The recirculation rate was subsequently lowered in order to relieve the pressure. The flux rates then dropped from a high value of 60 LMH to a low value of 4 and the linear velocity decreased from 1 m/s to 0.3 m/s.

The performances of the referenced membranes, along with a 0.2 μm Koch flat-sheet membrane/module unit, were also evaluated for their ability to concentrate the washed cells, as shown by Figure 3. It must be emphasized, however, that flux rates corresponding to the Supor/NC SRT module are not shown in this figure because the concentration process occurred concurrently with the diafiltration process. The plot of the Supor/NC SRT module exhibited in Figure 2 actually represents the diafiltration as well as the concentration operations. These two processes occurred so quickly that distinct and separate data could not be gathered. A 0.1 μm Koch/Concep module was also tested but is not shown in the figure because cells were detected in the permeate at the onset of the concentration process.

Figure 3 does show the performance of the Prostak membranes and the remaining cited membranes. As presented in this graph, there are only two points comprising the Prostak plot because this process was completed within 1 min. The flux rate produced by this system ranged from 309 LMH at the beginning of the process to 246 LMH at its conclusion. Despite this degree of efficiency, a problem became apparent during the process. It was our intention to reduce the fermentation broth from 8 L to below 4 L for all these studies, but the Prostak module did not achieve specification. Because of the large holdup volume required, the broth could only be reduced to 5 L. Conversely, the final volume of concentrate produced by the 0.2-μm Koch/Concep module was 3 to 3.5 L. And, if so warranted, the fermentation broth could have been concentrated even further, as evidenced by other experiments performed in our laboratory. Analogous to the flux rates produced by the Prostak modules, the 0.2-mm Koch/Consep membrane/module generated an average flux rate of 242 LMH. This rate remained fairly steady throughout the concentration step and even surpassed the Prostak toward the end of the process.

The performances of the other modular units were comparable in regard to concentrating the *P. pastoris* cells, with the notable exception of the Sartorius membrane. As a result of a low flux rate of 2 LMH, approximately 35 min was required to reduce the volume of fermentation broth by only 1 L. Lower volumes were not obtained with this module because the concentration step was terminated in the middle of the experiment. Although the Pall Filtron and the Millipore Pellicon module behaved consistently, the overall performance of these membranes was still quite low compared to the Supor NC/SRT, the Prostak modules, and the Koch/Consep modules.

It is quite evident from the flux rate data that the Supor/NC SRT module was more efficient harvesting *P. pastoris* suspensions than the other membranes modules. To illustrate this point more succinctly, the data generated from modules used to perform both the diafiltration and concentration processes were normalized to total volume of material processed, in L m² of membrane. These relationships were then plotted versus the time required to process the material, as displayed by the bar graph in Figure 4. The Koch/Consep system is not included

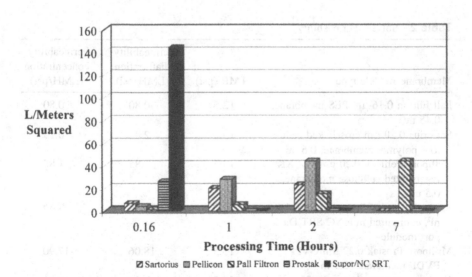

Figure 4 Bar graph of the volume of material processed, in L, per the area of available membrane, in m², vs. processing time (hours) for each modular system used to both diafilter and concentration *Pichia pastoris*.

in the graph because a diafiltration filtration was not conducted with this membrane. A direct examination relative to the other membrane modular systems is therefore not feasible.

Figure 4 clearly shows that the Supor/NC SRT module has the potential of filtering larger volumes of *P. pastoris* broth more efficiently per square meter of membrane surface. In fact, the relationship derived from the Supor/NC SRT flux data suggests that 144 L of *P. pastoris* cells can be processed for every square meter of membrane within a 10-min time period. To even achieve the same processing specifications by the other modules, the membrane surface areas would have to be expanded to 21 m^2, 36 m^2, 117 m^2, and 5.5 m^2 for the Sartorius, the Pellicon, the Pall Filtron, and the Prostak units, respectively. More experimental work is being conducted within our laboratories to substantiate the implications of the available data.

The transmembrane pressure (TMP) produced by the each module was further determined to evaluate the overall performance of the various systems, as described by Michaels et al. [14]. The information listed in Table 2 shows that the TMPs necessary to produce the optimal linear velocities are similar for each module, ranging from 12 psi for the Supor/NC SRT module to 15.5 psi for Prostak modules. Along with the TMP results, Table 2 displays the degree of

Table 2 Solvent Permeability

Membrane/module type	TMP (psi)	Permeability-diafiltration (LMH/psi)	Permeability-concentration (LMH/psi)
Pall Filtron 0.16-μm PES membrane, 0.45 m^2	12.5	0.80	0.80
Sartorius 0.20-μm crosslinked cellulose polymer membrane, 0.6 m^2	13.5	2.3	0.15
Millipore Pellicon 0.20-μm PLCXK regenerated cellulose membrane, 0.5 m^2	13.5	1.85	1.85
Supor 0.2-μm PES membrane, 0.1665 m^2, configured in a NC SRT Duport module	12.0	74.35	NA*
Millipore Prostak 0.22-μm GVPP PVDF membrane, 0.9 m^2	15.5	18.06	17.90
Koch 0.2-μm membrane, 0.032 m^2, configured in a NC SRT Concep module	13.25	NA*	18.26

* Not applicable.

solvent permeability associated with each unit (LMH/TMP). Again the Supor/ NC SRT membrane out performed the other membranes with a solvent permeability of 74.35 LMH/psi.

The high flux rate and high degree of permeability exhibited by the Supor/ NC SRT membrane indicate that fouling of the membrane surface is minimal. Several features pertaining to a membrane system can be ascribed to this phenomenon. Stratton et al. [15] investigated the effects of membrane pore size on the performance of CFF methods for harvesting *E. coli* and *S. cerevisiae*. They found that ultrafiltration membranes, particularly high-molecular-weight cutoff membranes, were better suited for harvesting fermentation broths containing *E. coli*. These results were attributed to the size of the micro-organisms in relation to the pore size of the membrane. Because *E. coli* cells are approximately the same order of magnitude as the pores of microfiltration membranes, the organisms embed directly into the pores of the membranes, which in turn cause blockage and fouling of the membranes during a CFF process. *S. cerevisiae* organisms do not physically block the pores of microfiltration membranes because yeast cells are larger than *E. coli* cells. Although the previously cited investigations certainly agree with the results attained with the Supor 0.2 mm PES membrane/NC SRT module, the relatively low performance of the other microfiltration membranes implies that other contributing factors affect the efficient harvesting of *P. pastoris* by CFF methodologies.

In these studies, the performances of the membranes were evaluated based upon conducting a CFF process at the optimal linear velocities of a particular modular system. The ensuing TMPs were simply an outcome of pressures used to maintain the required velocity. For example, the Prostak module produced a TMP of 15.5 to sustain a steady state linear velocity of 4 m/s. However, the sanctioned TMP industry standard for this module ranges from 3 to 10 psi to prevent membrane fouling. While it is recognized that the degree of fouling is influenced by TMPs, this condition alone cannot explain the high degree of variation between the Supor/NC SRT module and the other systems. The normal relationship between TMP and permeate flow rates is a bell shaped curve. With increasing TMPs, the permeate flux rate also increases until a maximum is attained. After this point, the permeate rate decreases due to formation of a gel layer or to compression of the membrane. Mallubhotla et al. examined the flux rates produced by *S. cerevisiae* suspension broths with changing TMPs [16]. A linear relationship was observed in these experiments but a maximum value was never reached. The flux rates generated by varying concentrations of suspensions did increase as the TMP was varied between 0 and 80 kPa, but only by ~ 0.4 LMH/kPa.

Because the velocity of the retentate fluid is the most important element that controls the formation of a gel layer, TMP optimization should occur only after an appropriate linear velocity has been determined. In most cases, higher linear velocities reduce the overall extent of a gel layer. In our work the linear velocity required for the Prostak module was higher than that of the Supor/NC SRT membrane, yet the performance of the latter system was substantially greater. Verifying the maximum TMP and then performing the tests under these parameters would not offset the influences of the linear velocities and the resulting flux rates to such a degree.

Other work has focused on the effects the composition of the membrane material has on the adsorption of protein at the membrane surface [17–24]. Through their studies of ultrafiltration membranes, Sheldon et al. [17] demonstrated that proteins, specifically bovine serum albumin, tend to unfold at the surfaces of hydrophobic polysulfone membranes. The hydrophobic interior of the protein is then exposed, which thereby interacts with the unfolded hydrophobic portion of native proteins. As a result of this type of contact, a protein layer is formed at the surface of the membrane. The resulting gel layer plugs the membrane pores, which lowers or prevents the flow of material through the filter [22]. The use of hydrophilic membranes may reduce this concentration polarization affect. It was shown that globular proteins accumulated at the surface of a hydrophilic regenerated cellulose membrane but at one-third the binding capacity of the polysulfone membrane [16].

These studies are supported by work performed by Stratton et al. [15]. They found that the hydrophilic membranes out performed other hydrophobic membranes when *S. cerevisiae* and *E. coli* broths were harvested by CFF methodologies. As emphasized by the authors, the hydrophilicity of a membrane is an important factor in the selection of a CFF module for processing protein-based material. Relative to the membranes described in this article, however, the system that performed at the highest efficiency was composed of polyethersulfone, the least hydrophilic material tested. In fact, the more hydrophilic membrane modular systems were not even half as efficient in harvesting *P. pastoris* fermentation broth. It must be noted that the cells used in these experiments were grown on defined media instead of the complex media used by Stratton et al. [15]. This variable may alter the processing capabilities of a particular modular system. The material composition of the membrane may also vary the performance of the process when recovering *P. pastoris* fermentation broth containing product that are expressed extracellarly. Studies are ongoing within our laboratory to determine how the various *P. pastoris* product/process-specific variables, including membrane composition effects, influence the performance of CFF harvesting methods, including membrane composition effects. These results will be reported in a forthcoming article.

While the composition of the membrane is indeed a valid characteristic to consider, the data reported herein imply that the design of the module itself plays an important role in the overall performance of harvesting *P. pastoris* cells by

CFF methods. The membranes comprising the Supor/NC SRT and the Pall Filtron were composed of the same material. Both of these membranes systems processed the same material and were utilized under optimal linear velocities. Yet, the flux rate produced by the former module was much greater than that of the latter membrane. The only significant distinction in the experimental procedures was that the membranes were configured into very different modular designs. Mallubhotla et al. [16] and Kroner et al. [24] also found that the degree of membrane fouling was affected by the design of the spiral filtration modules even when the same type of membrane was used. The modular design of the membrane system could account for the lower flux rates displayed by the Sartorius, Pellicon, and Pall Filtron membranes as well. The design of these modules may promote the formation of biolayers, which would readily restrict the flow of filtrate through the membrane.

The effectiveness of the cleaning procedures, and thus the life of a membrane, is another important parameter that must be contemplated before selecting a membrane system for harvesting *P. pastoris*. When a CFF membrane module cannot be adequately cleaned and subsequently reused, the expense to replace the membranes increases the costs of the overall process. As such, these studies focused on the cleanability of each membrane, also referred to as membrane percent recovery.

Clean-water permeabilities are typically used as a determinant of the cleaning efficacy in CFF systems [14,25,26]. To determine whether the CFF membranes have been adequately cleaned after recovering *P. pastoris* fermentation broths, the clean-water flux rate produced by each of the membranes was measured before use (precleaning) and after cleaning (postcleaning). From these data, the membrane percent recoveries were calculated. A correction factor was then applied to the data to account for the differences in the water viscosity as a result of fluctuation in temperatures of the precleaning and postcleaning water [14]. The membrane percent recovery was ultimately determined based on the corrected values, as shown in Table 3.

The membrane percent recoveries achieved by the Pellicon system (102%) and by the Supor/NC SRT module (97%) were much higher than those produced by the other membranes. These membranes could be reused if so needed for harvesting additional batches of *P. pastoris* fermentation broths because they recovered within the normal acceptable specifications recommended by most manufacturers and pharmaceutical industries—i.e., at least 80% recovery [14]. Still, the procedures used to clean each of the membranes systems were based on the manufacturers' instructions. The percent recovery of the other membranes may improve with the development of more defined cleaning procedures that are applicable to the modular system as well as to the chemical and physical characteristics specific to *P. pastoris* broths.

Table 3 Clean-Water Flux Data

Membrane/module type	Flux before (LMH)	Clean-water temp (°C) for flux before	Flux after (LMH)	Clean-water temp (°C) for flux after	Membrane recovery	
					Uncorr.	Corr.
Pall Filtron 0.16-μm PES membrane, 0.45 m²	612	22	186	22	30%	30%
Sartorius 0.20-μm crosslinked cellulose polymer membrane, 0.6 m²	299	23	NA*	NA	NA	NA
Millipore Pellicon 0.20-μm PLCXK regenerated cellulose membrane, 0.5 m²	431	23	419	19	97%	102%
Supor 0.2-μm PES membrane, 0.1665 m², configured in a NC SRT Du-port module	1510	23	1402	21	93%	97%
Millipore Prostak 0.22-μm GVPP PVDF membrane, 0.9 m²	572	24	73	23	7.8%	13%
Koch 0.2-μm membrane, 0.032 m², configured in a NC SRT Cocep module	778	18	628	25	81%	68%

* Not applicable: Postcleaning flux data were not obtained because this membrane was so plugged that cleaning could not be accomplished.

IV. CONCLUSIONS

When harvesting *Pichia pastoris* suspensions by CFF techniques, it is important to use a an efficient membrane modular system that does not pass cells into the permeate and that can be reused time and again. In this study, the performances of several types of microfiltration modules were evaluated in terms of accomplishing the defined objectives. Through this preliminary work, it was determined that the Supor 0.2-mm PES membrane configured into a Dualport NC SRT module out performed all the other modules. However, in order to optimize the harvesting process of *P. pastoris* further, and thereby reduce time and expense while maintaining the integrity of the product, other factors must be analyzed, including channel heights, membrane composition, composition of the feed stream, etc. [27]. Nonetheless, these studies do show that *P. pastoris* fermentation broths can be effectively and efficiently harvested via CFF when the proper modular system is selected.

REFERENCES

1. Higgins DR, Cregg JM. Introduction to *Pichia Pastoris*. In: DR Higgins, JM Cregg, eds. Pichia Protocols. Totowa, NJ: Humana, 1998:1–15.
2. Tscopp JF, Sverlow G, Kosson R, Craig W, Grinna L. High level secretion of glycosylated invertase in methylotropic yeast, *P. pastoris*. Bio/Technology 5:1305–1308, 1989.
3. Smith PM, Suphioglu D, Griffith, IJ, Theriault, K, Knox RB, Singh, MB. Cloning and expression in yeast *P. pastoris* of a biologically active form of Cynd 1, the major allergen of Bermuda grass pollen. J Allergy Clin Immunol 93:331–343, 1996.
4. Peng YC, Acheson, NH. Production of active polyomavirus large T antigein in yeast *P. pastoris*. Virus Res 49:41–47. 1997.
5. Bar KA, Hopkins SA, Sreekrishna, K. Protocol for efficient secretion of HAS developed from *P. pastoris*. Pharm Eng 12:48–51, 1992.
6. Diggon ME, Tschoop JF, Grinna L, Lair SV, Craig WS, Velicelebi GR, Thill G. Secretion of heterologous proteins from methylotropic yeast *P. pastoris*. Dev Ind Microbiol 29:59–65, 1988.
7. Larouch Y, Strme V, DeMuetter J, Messens J, Lauwerys M. High level secretion and very efficient isotopic labeling of tick anticoagulent peptid (TAP) expressed in the methyltropic yeast *P. pastoris* Bio/Technology, 12:1119–1124, 1994.
8. Cregg JM, Tschopp JF, Stillman C, Siegel R, Akong M, Craig WS, Buckholz RG, Madden KR, Kellais PA, Davis GR, Smiley BL, Cruze J, Rorregrossa R, Velicelibi G., Thill, GP. High Expression and efficient assembly of hepatitis B surface antigen in methyltropic yeast, *P. pastoris* Bio/Technology 5:479–485, 1987.
9. Siegel RS, Brierley, RA. Methyltropic yeast *P. pastoris* produce in high cell density fermentations with high cell yields as a vehicle for recombinant protein production. Biotechnol Bioeng 34:403–404, 1989.

10. Titchener, HNJ, Gritsis D, Mannweiler K, Olbrich R., Gardiner SAM, Fish NM, Hoare M. Integrated process design for producing and recovering proteins from inclusion bodies. BioPharm 7:34–38, 1991.
11. Forman SM, Debernardez ER, Feldberg RS, Swartz RW. Crossflow filtration for the separation of inclusion bodies from soluble proteins in recombinant *E. coli* lysate. J Membrane Sci 48;263–279, 1990.
12. Bailey FJ, Warf RT, Maigetter RZ. Harvesting recombinant microbial cells using crossflow filtration. Enz Microb Technol 12:647–652, 1990.
13. Stratton J, Chiruvolu V, Meagher M. High Cell-Density Fermentation. In: Higgins DR, Cregg, JM, eds. Pichia Protocols. Totowa, NJ: Humana, 95–106.
14. Michaels SL, Clean-water permeability as a determinant of cleaning efficacy in tangential-flow filtration. BioPharm Oct.: 38–45, 1994.
15. Stratton J, Meagher M. Effect of membrane pore size and chemistry on crossflow filtration of *E. coli* and *S cerevisiae*: simultaneous evaluation of different membranes using a versatile flat-sheet membrane module. Bioseparation 4:255–262, 1994.
16. Mallubhotla H, Nunes E, Belfort G. Microfiltration of yeast suspensions with self-cleaning spiral vortices: possibilities for a new membrane module design. Biotechnol Bioeng 48:375–385, 1995.
17. Sheldon MJ, Reed IM, Hawes CR. The fine-structure of ultrafiltration membranes. J Membrane Sci 62:87–102, 1991.
18. Pitt AM. The nonspecific protein binding of polymeric microporous membranes. J Parenter Sci Technol 41:110–113, 1987.
19. Sarry C, Sucker H. Adsorption of proteins on microporous membrane filters. Part 1. Pharm Technol 17(1):72–82, 1993.
20. Sarry C, Sucker H. Adsorption of proteins of microporous membrane filters, Part 2. Pharm Technol 17(10):30–44, 1993.
21. Palecek SP, Zydney AL. Intermolecular electrostatic interactions and their effect of flux and protein deposition during protein filtration. Biotechnol Prog 10(2):207–213, 1994.
22. Kim KJ, Fane AG, Fell, CJD, Joy, DC. Fouling mechanisms of membranes during protein ultrafiltration. J Membrane Sci 68:79–91, 1992.
23. Brose DJ, Waibel P. Adsorption of proteins to commercial microfiltration capsules. BioPharm Jan.: 36–39, 1996.
24. Kroner KH, Nissen N, Ziegler H. Improved dynamic filtration of microbial suspensions. Bio/Technology 5:921–926, 1987.
25. Parenteral Drug Association. Industrial perspective on validation of tangential flow filtration in biopharmaceutical applications. J Parenteral Sci and Technol Technical Report No. 15. 46:S3-S13, 1992.
26. Gyure DC. Set realistic goals for crossflow filtration. Chem Eng Prog Nov.: 66, 1992.
27. Meagher M., Stratton J., Schlegel V. Unpublished results.

6

Filter Applications in Product Recovery Processes

Paul K. Ng, Alfred C. Dadson Jr., G. Michael Connell, and Bernard P. Bautista
Bayer Corporation, Berkeley, California

I. INTRODUCTION

In large-scale manufacture of proteins, filter applications that utilize micro- and ultrafiltration membranes play a key role in many processing stages. The capability of membrane filtration such as ultrafiltration has been well recognized and generally secured as a full-scale industrial process for many years [1,2]. The unit operation offers benefits such as low energy requirements, continuous flow, ease of operation, reproducibility, scalability, and gentleness to the product. The filtration system can be divided, on the basis of membrane geometry, into four commonly known modules: parallel plate, spiral, hollow-fiber, and tubular. Much of the methodology on these systems has been evaluated [2] and reduced to practice in the plasma fractionation industry [3]. Previous work using mostly hollow-fiber systems has not only established and refined the technique but has also elucidated factors affecting process recovery and process efficiency. Once the membrane with the proper pore size is determined, differences in product recovery among the different filtration modules are generally not substantial. The selection of one type over another for a specific application may be dependent on other factors. For example, one system could be more favorable than the others in regard to size, cost, process time, and ease of availability. There are a large number of commercial units available; therefore we decided to limit our present investigation to the parallel-plate module which allows scaleup by adding modules in parallel. This chapter will discuss our experience using parallel plates in the following areas:

Tangential flow cell separation of tissue culture fluid
Tangential flow filtration and dead end filtration as applied to tissue culture fluid
Concentration of cell-free filtrate
Ultrafiltration/diafiltration in downstream processing
Nanofiltration

II. TANGENTIAL FLOW CELL SEPARATION OF TISSUE CULTURE FLUID

Clarification of tissue culture fluid removes DNA-containing cells and cell debris. The operation could be achieved with centrifugation which carries the potential for heat generation, aerosol formation, and labor-intensive assembly. In the present study, we investigated a plate-and-frame unit for tangential flow cell separation. Tissue culture fluid derived from large scaled fermentors was used. The system, Sartocon II, was purchased (Sartorius Corp., Bohemia, NY). The membrane was a 0.45-μm microfilter with an area of 0.6 m². Recirculation was achieved with a Watson-Marlowe 701 S/R manual control peristaltic pump (Bacon Technical Industries, Concord, MA). The module, pump, and pressure gauges were mounted on a mobile trolley, which was stationed conveniently close to the fermentors. Figure 1 depicts the equipment arrangement in the experiment.

Permeate flux over a range of cell concentrations was obtained as shown in Table 1. For complete rejection of cells, the film theory [4] relates the permeate flux J to the solute concentration as follows:

$$J = k \ln(C_w/C_b) \tag{1}$$

where k is the local mass transfer coefficient, C_w is the concentration at the membrane wall and C_b is the concentration in the bulk solution. The mass transfer coefficient calculated [5] by best fit using the nonlinear least-squares iterative procedure is 2.5 L/m²-min (standard error is 25% of the parameter value). This compares with 0.4 L/m²-min for bacterial cells at cell concentration of <1% dry weight [6]. The higher mass transfer seen with mammalian cells (20 μm in size) versus bacterial cells (<10 μm in size) could be explained by (1) lower osmotic pressure from mammalian cells as predicted by the Van't Hoff equation for an ideal solution, and (2) larger cells enhance the momentum which dislodges the cells accumulating on the surface.

Previously, it was demonstrated that an aseptic concentration of living microbial cells by crossflow filtration preserves cell viability [7]. However, cell lysis

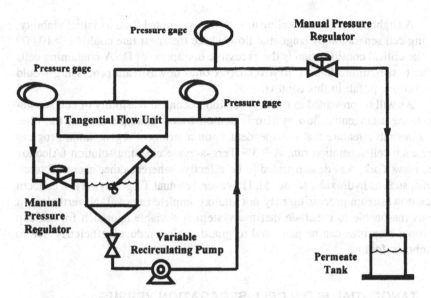

Figure 1 Tangential flow separation of fermentor fluid. Sartocon II system obtained from Sartorius Corp., Bohemia, NY. The membrane is a 0.45-μ microfilter with a surface area of 0.6 m².

was apparent across crossflow filtration of an erythrocyte suspension [8]. In the current study, samples from baby hamster kidney cell culture were collected for cell counts with transmembrane pressure (TMP) of 1, 3.5, 4, and 5 psi.

TMP is defined as follows:

$$\text{TMP} = \frac{P_F + P_R}{2} - P_P \tag{2}$$

where P_F = feed pressure, P_R = retentate pressure, and P_P = permeate pressure.

Table 1 Effects of Cell Concentrations on Permeate Flux

Cell concentration × 1000/mL	Flux L/min m²
92	1.93
148	1.77
157	1.37
208	1.00
212	0.73

A slight, insignificant decline in viability was noted from its initial viability. During cell separation by tangential flow where the shear rate could be $>10,000$ s^{-1}, the critical consideration is the excessive disruption of DNA-containing cells in the tissue culture fluid. Carryover DNA onto downstream processing would be a serious pitfall in this context.

As will be presented in the next section, technical feasibility of cell separation using a tangential flow system is demonstrated. Multiple uses of the filter are a desirable feature that is dependent upon a stringent regeneration program after each cell separation run. A 0.3% Terg-a-zyme cleaning solution (Alconox Inc., New York) was demonstrated to be effective whereas other chemical treatments, such as hydroxide, failed [5]. However, residual Terg-a-zyme is a concern since downstream processing may not attain complete removal. Nevertheless, it seems reasonable to conclude that the system is a viable approach for scaleup if proper measures can be instituted to guard against reduced efficiency due to membrane fouling.

III. TANGENTIAL FLOW CELL SEPARATION VERSUS DEAD-END FILTRATION

Figure 2 shows the filtration profiles of tangential flow filtration using the Sartorius device versus dead-end filtration using a 0.45-μ absolute filter (triangles) or a 0.2-μ nominal prefilter plus a 0.45-μ absolute filter (circles). Each set of data was obtained from an experiment that was started with a similar amount of tissue culture fluid. Experience with dead-end filters showed that the filtration rate of water or buffer was always proportional to the pressure differential. With a fluid-containing suspension of cells, however, although increasing the initial pressure differential would result in initial faster filtration, deterioration of the rate invariably developed rapidly; the quantity filtered at higher pressure was always less than that at low pressure. Accordingly, dead-end filtration was started with a minimum pressure of <1 psi. The pressure drop monitored over the course of the experiment under a constant pump setting indicated a slow rise to a final value of 6.5 psi. As shown in Figure 2, the nonlinear relationship between the filtrate collected and the filtration time resulted from the backpressure generated by gradual blocking of the pores in the filter. Comparable data were obtained from a 3500-L batch where tissue culture fluid was filtered with a prefilter and a 0.45-μ filter in series (Figure 3). Gradual increases in pressure differential across the prefilter and the absolute filter was apparent. As predicted by the small-scale experiment, the slope of the filtrate weight to time relationship decreased over time.

With the tangential flow system, in contrast, the transmembrane pressure of 4 psi remained constant and the filtrate weight increased proportionately with

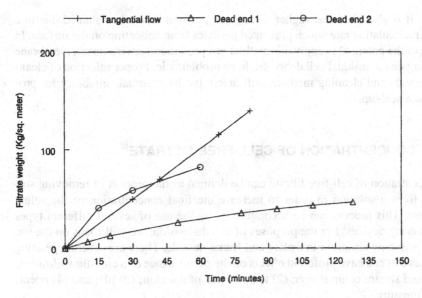

Figure 2 Cell separation of fermentor fluid by tangential flow filtration or dead-end filtration. Dead-end 1 (triangles) is the flow profile using a 0.45-μ absolute filter. Dead-end 2 (circles) is the flow profile using a 0.2-μ nominal filter plus a 0.45-μ absolute filter.

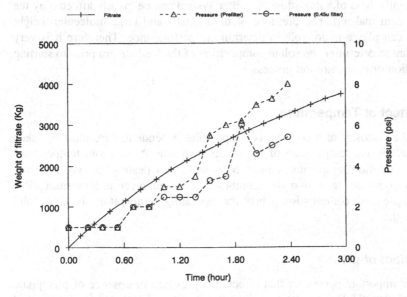

Figure 3 Dead-end filtration (prefilter + 0.45-μ filter) of a 3000 to 4000-kg batch. Weight of filtrate and pressure across the filters were monitored over time.

time. It is also clear that higher throughput per unit area was obtained due to a high recirculation rate which prevented particles from collecting on the surface. In spite of the potential to apply this method in a production environment, membrane fouling due to residual cell debris could be problematic. Proper selection of cleaning agents and cleaning methods will determine its eventual suitability for production scaleup.

IV. CONCENTRATION OF CELL-FREE FILTRATE

Concentration of cell-free filtrate can be defined as the process of removing solvent from a solution in order to increase the final concentration of the solute present. This process can be accomplished by the use of several different types of filtering devices. For the purposes of this discussion, we will focus on the use of a crossflow ultrafilter in a plate-and-frame module. There are several operating factors that can have profound effects on the performance of a crossflow filtration: (1) feed stream composition; (2) temperature of solution; (3) pH; and (4) operation pressure.

A. Effect of Feed Stream Composition

The retentiveness of a crossflow ultrafilter system can be greatly affected by the feed stream makeup. The presence of both small- and large-molecular-weight solutes can play a major role in determining performance. Therefore it is very important to determine the solute composition of the feed stream prior to starting a filtration or concentration process.

B. Effect of Temperature

Since the viscosity of a solution is temperature dependent, care must be taken not to allow the temperature of the solution to rise above room temperature. Warmer solution temperatures have been shown to destroy gel layer formation by compressing it into the membrane. The gel layer is less susceptible to collapse and compression when the operating temperature is kept cold, ~4°C [9].

C. Effect of pH

Another important parameter that affects the presence or absence of precipitate is the solution pH. During ultrafiltration, care must be taken not to reach or exceed the isoelectric point of the protein or solute being processed. Reaching the isoelectric point will cause the solute to precipitate out of solution and eventually lead to membrane fouling.

D. Effect of Operation Pressure

Since pressure is a very important operating parameter in any filtration, extreme caution must be taken in selecting the correct operating pressure. In crossflow filtration, TMP is the main driving force and its determination for different solutions and membrane type or pore sizes must be carefully determined through experimentation. For the data presented in this chapter, after a series of concentration experiments, we determined that the optimal TMP for our system was 25 psi.

The ultrafiltration system from which the data presented were obtained is a Sartoflo 30 Skid (Sartorius Corp., Bohemia, NY) with 36 Pellicon II Membranes. The total surface area was 18 m^2 and the MWCO of the membrane was 50 kDa. Operating at a pressure <25 psi extended overall processing time in some cases by >2 hours. On the other hand, pressures >25 psi led to a rapid increase in gel layer formation thus affecting process efficiency. By using a TMP of 25 psi, we were able to minimize processing time and maximize filter membrane efficiency in all of the products tested.

The cell-free filtrate containing the solute debris is brought to the surface of the membrane at the beginning of the ultrafiltration process when the membrane is porous by convection. The flow of the filtrate passes tangentially through the pore and leaves the feed stream into the permeate stream, where it is discarded to the drain. As the filtrate passes tangentially, solute tends to back-diffuse via Brownian motion to the flowing aqueous stream. The flow stream also acts to sweep the material brought to the membrane surface by inertial lift back into the retentate flow [2,9,10].

As the ultrafiltration concentration progresses, a localized higher concentration of solute occurs at the membrane surface due to permeate removal from the main stream and liquid friction at the membrane surface. The continuation of this solute accumulation results from concentration polarization, which is the primary cause of membrane fouling.

As the condition of the filter begins to deteriorate, one might suggest that manipulation of the Hagen-Poiseuille equation [11] will lead to better or improved flux rate. The equation states that liquid flux is directly proportional to membrane pressure and inversely proportional to liquid viscosity.

$$J = \varepsilon r^2 \Delta p / 8 \eta \Delta x \tag{3}$$

where J = liquid flux, ε = membrane porosity, r = mean pore radius, Δp = transmembrane pressure, Δx = pore length, and η = liquid viscosity. This equation can be interpreted to indicate that increasing the pressure will increase the flux. Increasing the transmembrane pressure (TMP) is not always the best approach to achieving an efficient concentration process. A very high TMP (above optimum) will result in premature filter clogging. Because premature fouling is undesirable, how quickly or slowly the gel layer forms must be carefully monitored and controlled. Gel layer formation [2] is due to several factors, namely: the rate of solute back diffusion; the amount of shear generated by the moving liquid; and the rate of solute to membrane contacts.

Figure 4 shows TMP data from four runs of a typical 3000 to 4000 L size concentration batch. Figure 5 depicts the retentate flow rate for the same runs while Figure 6 represents the permeate flow rate over the duration of the volume reduction. To decrease the gel layer buildup during concentration, we employ a deliberate fluid management technique. This involves intermittently opening and closing the permeate valve at the beginning of the ultrafiltration process. Closing the permeate valve during concentration for 5 to 10 min, increases the amount of crossflow across the membrane surfaces and thus increases the sweeping action. This action tends to maximize the mass-transfer coefficient and maintain permeate flux at near maximum for longer periods.

Analysis of Figures 4 and 5 indicates that toward the end of the concentration process, the gel layer increases, filter efficiency declines, and TMP rises sharply. At the same time, the retentate flow rate increases while the permeate flow rate decreases. The conditions discussed in the figures above occur due to the resistive nature of the gel layer to flow once it is formed. As the filtration concentration progresses, more and more solute diffuses back into the feed

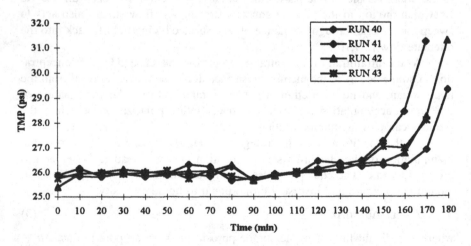

Figure 4   Transmembrane pressure vs. time for ultrafiltration of clarified fermentor fluids containing an 85-kDa protein. Data obtained from four consecutive runs: 40, 41, 42, and 43.

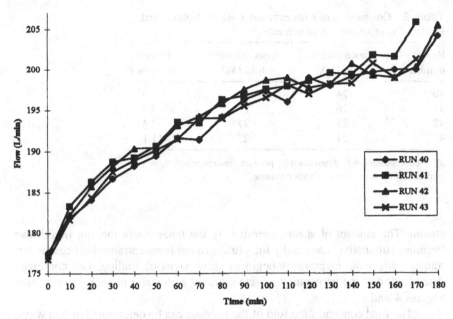

Figure 5 Retentate flow vs. time for ultrafiltration of clarified fermentor fluids containing an 85-kDa protein. Data obtained from four consecutive runs: 40, 41, 42, and 43.

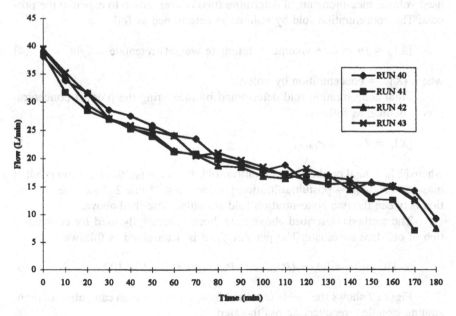

Figure 6 Permeate flow vs. time for ultrafiltration of clarified fermentor fluids containing an 85-kDa protein. Data obtained from four consecutive runs: 40, 41, 42, and 43.

Table 2 Comparison of Concentration Fold by Volume and
Product Concentration Measurements

Run number	Concentration fold (X_p)	Concentration fold (X_V)	Percent difference
40	24	26	7.6
41	26	27	3.7
42	25	27	7.4
43	24	27	11.1

X_p = concentration fold determined by product concentration; X_V = concentration fold determined by UFTCF volume.

stream. The amount of shear generated by the tangentially moving liquid also becomes cumulative. Eventually the ultrafiltration (concentration) process is terminated when the membranes begin to show signs of fouling, i.e., continued deterioration of the permeate flow rate coupled with a rise in back pressure (see Figures 4 and 6).

The final concentration fold of the product can be determined in two ways, either by volume or by product concentration. Short of measuring the product concentration online, which determines the final concentration, we frequently used volume measurements to determine final concentration to expedite the process. The concentration fold by volume is determined as follows:

$$[X]_V = (\text{permeate volume} + \text{retentate weight})/\text{retentate weight} \qquad (4)$$

where $[X]_V$ = concentration by volume.

Final concentration fold determined by measuring the product concentration is defined as follows:

$$[X]_P = P_{\text{POST-UF}}/P_{\text{PRE-UF}} \qquad (5)$$

where $[X]_P$ = final product concentration fold, $P_{\text{PRE-UF}}$ = preultrafiltration product mass, and $P_{\text{POST-UF}}$ = postultrafiltration product mass. Table 2 shows the correlation between the two concentration fold formulae described above.

The methods described above have been successfully used for concentration of cell-free supernate. The percent yield is determined as follows:

$$\% \text{ UF yield} = [(1 - (P_{\text{PRE-UF}} - P_{\text{POST-UF}}))/P_{\text{PRE-UF}}] \times 100 \qquad (6)$$

Figure 7 shows the yields obtained from a fermentation campaign, demonstrating excellent recovery across the step.

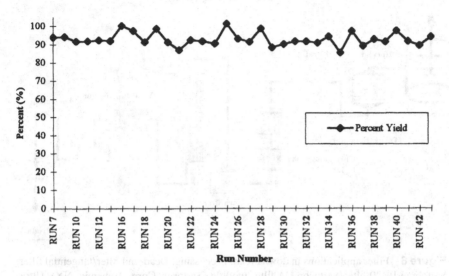

Figure 7 Product yield across ultrafiltration of clarified fermentor fluids containing an 85-kDa protein. Feed stock weight: 3000 to 4000 kg. Titer of product before and after ultrafiltration was measured by ELISA.

V. ULTRAFILTRATION/DIAFILTRATION IN DOWNSTREAM PROCESSING

Use of ultrafilters in downstream recovery processes can be grouped into two principal categories: concentration—removal of solvent and microsolutes; and desalting—salt exchange by removal or exchange of micorsolutes (e.g., buffer ions, small peptides, small molecules, etc.).

A general flow diagram of a purification train is shown schematically in Figure 8. As can be seen, there are various use options of ultrafiltration. Ultrafilter 1 is used for concentration prior to chromatography, ultrafilter 2 is used for concentration before size exclusion, and ultrafilter 3 is used for concentration and desalting just before the final formulation step. It is instructive to discuss several aspects of ultrafiltration in downstream processing.

A. Regeneration of Ultrafilters

It is desirable to regenerate all downstream processing equipment with a cleaning-in-place (CIP) method in which cleaning reagents are recirculated through the equipment. A sequence of cleaning solutions using sodium hydroxide, WFI, phosphoric acid and WFI has been successfully developed for ultrafilters in product

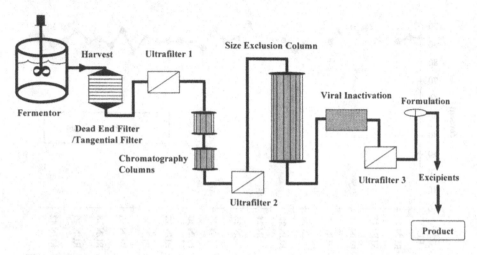

Figure 8 Filter applications in downstream processing. Dead-end filter/tangential filter: Sartobran-PH20 filter/Sartocon CA filter module (Sartorius Corp., Bohemia, NY). Ultrafilter 1: Pall Filtron Centrasette UF cassette (Pall Filtron Corp., Northborough, MA) or Millipore spiral-wound membrane cartridge (Millipore Corp., Bedford, MA). Ultrafilter 2: Pall Filtron Centrasette UF cassette (Pall Filtron Corp., Northborough, MA) or Millipore spiral-wound membrane cartridge (Millipore Corp., Bedford, MA). Ultrafilter 3: Pall Filtron Centrasette UF cassette (Pall Filtron Corp., Northborough, MA).

recovery processes. The procedure has been validated for cleaning by challenging the system with medium solution. The content of the WFI rinse after each CIP cycle was determined. Results from rinse samples submitted for chemical analysis are shown in Table 3. The CIP procedure was shown to effectively and repeatedly clean and sanitize the system and to result in no residual chemicals in the ultrafilters, tubings, or associated pipings.

Table 3 CIP Chemical Analysis

Test	Actual results	Limit
pH	6.13	5.0–7.0
Conductivity, μmho/cm at 25 ± 1°C	1.2	USP WFI (stage 2) ≤ 2.1
Total organic carbon	103 ppb	≤500 ppb
A280	<0.001 AU	≤0.005 AU
Sodium	<0.5 ppm	≤0.5 ppm
Phosphate	<1 ppm	≤1 ppm

Microbial load of the system was measured by taking plate counts of the WFI rinse samples before and after ultrafiltration for bacterial contamination. In seven consecutive runs, only one batch showed a significant detection of plate counts. It exceeded an internal limit of 10 cfu/100 mL. This resulted in an immediate investigation that identified an assignable cause. Subsequent runs with our cleaning regimen demonstrated microbial load of <10 cfu/100 mL.

B. Product Recovery Across Ultrafiltration

Studies of tangential flow system for ultrafiltration of antibody prior to size exclusion chromatography in a purification train were undertaken with a small-scale cassette system (Pellicon-2) manufactured by Millipore (Biomax, MWCO = 10,000). A summary of the results obtained for 10 runs is given in Table 4. The forgoing experiments were expanded to large-scale concentration of the same antibody using an equivalent cassette system (Omega Centrasette MWCO = 10,000) purchased for cGMP studies (Pall Filtron Corp., Northborough, MA). The results on product recovery of 11 clinical batches compare favorably with experimental data reported in the small-scale experiments. Our findings demonstrated that the operation could be scaled up without any compromise in product recovery, thereby providing data for process design in a production environment. Correspondingly, another study has shown that a 400-fold linear scaleup directly to production was attained without the additional effort and expense of a pilot-scale test [12].

C. Contaminant Removal

Due to its extended configuration and its susceptibility to chain scission, DNA contains molecules of widely different molecular masses. Because of this, it was of interest to check if DNA clearance occurred across ultrafiltration. Pre- and postconcentrate DNA in an antibody solution was measured by dot blot hybrid-

Table 4 Product Recovery of Small-Scale Runs and Full-Size Runs

Ultrafilter	Antibody processed (g)	Surface area (m^2)	Step recovery (%)	Number of batches
Millipore Biomax polyether sulfone	5	0.1	88.0 ± 5.7	10
Filtron Omega Centrasette polyether sulfone	400	5.52	85.0 ± 6.5	11

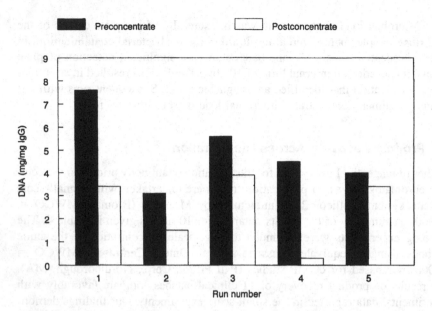

Figure 9 DNA clearance across ultrafiltration. IgG-containing solution was concentrated 4 to 7 fold using a Pall Filtron cassette system with 30,000-MWCO membranes (Pall Filtron Corp., Northborough, MA). DNA of the IgG solution was measured before and after ultrafiltration by dot blot hybridization.

ization in five experimental runs, as shown in Figure 9. When the results were expressed in mg DNA/mg IgG, it was shown that a reduction of <1 order of magnitude was achieved. This correlated with DNA spike experiments where DNA was resolved into two main peaks across a size exclusion column. One contained the high- and the other the low-molecular-mass fractions [13]. Similarly, attempts were made to investigate removal of cellular contaminants by ultrafiltration. In the experiment depicted in Figure 10, cellular contaminants in a protein solution derived from Chinese hamster ovary (CHO) cell culture were monitored over the course of diafiltration against the formulation buffer. At each of seven diafiltration volumes, the retentate was assayed for CHO protein by ELISA. It was evident that an incremental decrease of approximately twofold was achieved in two or three volumes, and no further decrease was apparent in subsequent volumes. Taken together, it is apparent that the process of concentration or diafiltration, when compared to purification by chromatography, provides a subsidiary benefit by fractionally reducing contaminants such as DNA or cellular proteins.

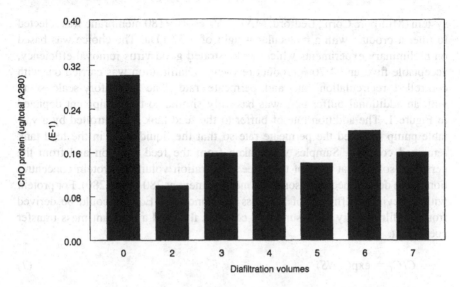

Figure 10 Chinese hamster ovarian cell contaminant reduction across diafiltration. Solutions containing an 85-kDa protein were diafiltered in a Pall Filtron cassette system with 30,000-MWCO membranes (Pall Filtron Corp., Northborough, MA). Product samples were taken pre- and postdiafiltration and assayed for host cell contaminant by ELISA.

VI. NANOFILTRATION

Mammalian cell culture has become the method of choice for the production of complex proteins that require posttranscriptional modifications. Associated with the proteins is a potential risk of contamination by virus or virus-like particles. Thus, manufacturers have included procedures of virus removal/inactivation in the purification train. One such procedure is virus removal by nanofiltration developed to specifically remove viruses on the basis of size difference between proteins and viruses. It has an advantage over chemical treatment [14,15] for reducing the virus load in biological products since it does not entail the use of toxic reagents that subsequently have to be eliminated from the proteins. The procedure also ensures the least disturbance to the therapeutically active protein composition. It makes use of tangential flow across a nanofilter with nominal cutoff values of 70 and 160 kDa and 15 to 75 nm [16–18]. In essence, if the step is operated under the proper conditions, the membrane should retain viruses while the product of interest passes through. The permeate could then be processed or formulated further. Systems are available commercially from several manufacturers, so we decided to direct our present investigation to the Viresolve system (Millipore Corp., Bedford, MA). A Viresolve/180 membrane was selected to filter a product with a molecular weight of ~57 kDa. The choice was based on preliminary experiments which demonstrated good virus removal efficiency, acceptable flux, and >70% product recovery. Diafiltration was carried out with controlled recirculation rate and permeate rate. The laboratory-scale setup with an additional buffer tank was basically similar to the equipment depicted in Figure 1. The addition rate of buffer to the feed tank as controlled by a variable pump matched the permeate rate so that the liquid level in the feed tank remained constant. Samples were taken from the feed solution and from the permeate solution at each of the three diafiltration volumes. Protein concentrations were determined by absorbance measurement at 280 nm (A280). For protein with a sieving coefficient of S across the nanofilter, Eq. (7) could be derived from the film theory [4] assuming a constant flux and a constant mass transfer coefficient.

$$C/C_i = \exp(-NS) \tag{7}$$

$$S = C_p/C_i \tag{8}$$

where C = the concentration of retentate at any time, C_i = the initial concentration of retentate at time zero, C_p = the permeate concentration at any time, and N = the number of diafiltration volumes. For mass balance of solute after the first volume:

$$C_i V_i = CV + C_{p1}V \tag{9}$$

where C_{p1} = the permeate concentration in the first diafiltration volume. Substituting Eqs. [8] and [9] into Eq. [7], the percent of solute recovered in the permeate after the first volume could be expressed as

$$C_{p1}/C_i = 1 - \exp(-S) \tag{10}$$

Similarly, the percent of solute recovered in the permeate after the second volume or third volume could be expressed as

$$C_{p1}/C_i + C_{p2}/C_i = 1 - \exp(-2S) \tag{11}$$

$$C_{p1}/C_i + C_{p2}/C_i + C_{p3}/C_i = 1 - \exp(-3S) \tag{12}$$

where C_{p2} = the permeate concentration in the second volume and C_{p3} = the permeate concentration in the third volume.

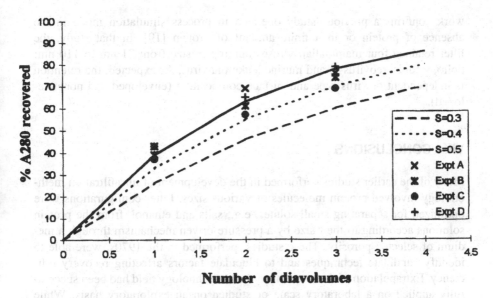

Figure 11 Percent A280 recovered in permeate vs. diafiltration volumes. The feed stock contained a 57-kDa protein. It was adjusted to a protein concentration of 0.2% to 0.5%, sterile filtered, and diafiltered in a system with a Viresolve-180 filter (surface area = 0.03 m^2). Transmembrane pressure was maintained under 5 psi during the course of the experiment.

Figure 11 shows the experimental and predicted percent solute (A280) recovered in the permeate when sieving coefficients of 0.3, 0.4, and 0.5 were applied to the 57-kDa protein. The data are in general agreement with a sieving coefficient of 0.4 to 0. 5. It also demonstrates that Eqs. (10), (11), and (12) could be used for predicting the performance after each diafiltration volume.

This discussion would not be complete without a comment on some pitfalls of the step. While it is desirable to have a high feed concentration so that the amount of diafiltration volume is minimized, it must be counterbalanced by the reduction in permeate flow rate. Therefore, determination of an optimum feed concentration is required and in our case, 0.2% to 0.5% gave consistent flow characteristics and product recovery. Higher protein concentration resulted in rapid loss in permeate flow rate, presumably due to irreversible fouling of the membrane.

Virus clearance work has been carried out (P. Blair/M. Connell, personal communication, Bayer Corp., Berkeley, CA, 1998). The data demonstrate that the nanofilter predictably and reproducibly removes model mammalian virus. Our work confirms a previous study operated in process simulation mode in the absence of protein or in a finite amount of protein [19]. In that study, the filter retained four mammalian viruses ranging in size from 31 nm to 110 nm: polio, sindbis, reovirus-3, and murine leukemia virus. As expected, the retention is independent of virus type and surface composition (enveloped and nonenveloped).

VII. CONCLUSIONS

Many of the earlier studies performed in the development of ultrafiltration methodology involved protein molecules of various sizes. Filter configurations were optimized for separating small solutes, e.g., salts and ethanol, from the protein solutions according to their size by a pressure-driven mechanism through a medium of selected porosity. These studies, performed in the 1970s, were able to identify candidate techniques and to elucidate factors affecting recovery efficiency. Extrapolation of these results to the biotechnology field had been successfully applied on a laboratory scale or studied on an exploratory basis. While tangential flow filtration of tissue culture fluid is a viable approach for removing cells or cell debris, dead-end filtration also offers ease of adaptability to scaleup. After cell separation, ultrafiltration is a well-accepted technique for concentration of the product. As a general prerequisite for upstream or downstream processing in a cGMP environment, cleanliness of an ultrafiltration system is maintained by a stringent cleaning regimen. The WFI rinse samples are routinely subjected to a battery of chemical/microbial analyses. Satisfactory regeneration of the ultrafilter aside, several factors can influence the success of the step. Among these factors are feed stream composition, temperature, pH, pressure, viscosity, and filter medium type. The effects of these factors must be determined by conducting repeated concentration experiments to determine the optimum operating parameters. Finally, by employing effective fluid management techniques, operating factors such as TMP and temperature can be manipulated to result in a very efficient concentration process.

   Biologicals prepared from tissue culture carry a safety risk due to a potential of viral contaminants in the initial material. It is for this reason that any purification train must have viral inactivation/clearance step(s) to ensure a virus-safe product. One method under evaluation is nanofiltration, which discriminates viruses from proteins of relatively smaller size. The feed liquid contacts one side of a membrane, which selectively permeates the product of interest. Theoretical calculation shows that actual protein recovery at each of the diafiltration volumes corresponds to a rejection coefficient of 0.4 to 0.5 for a marker protein of 57 kDa when it is subjected to filtration across a Viresolve/180 filter. Once the feasibility study and virus clearance validation studies are completed, nanofiltration could then be considered for scaleup. If the economics of the step are sufficiently addressed, one can expect that nanofiltration will become an alternative to other virus treatment methods and can be used efficiently in product recovery processes to provide clearance of both enveloped and nonenveloped viruses.

ACKNOWLEDGMENTS

The authors wish to acknowledge the GMP fermentation and purification groups on many phases of the project, and QA staff on various assays. Review of this paper by Drs. Paul Wu and D.Q. Wang is greatly appreciated.

REFERENCES

1. Harrison RG, ed. Protein Purification Process Engineering. New York: Marcel Dekker, 1993.
2. Cheryan M. Ultrafiltration Handbook. Lancaster, PA: Technomic Publishing, 1986.
3. Mitra G, Ng PK. Practice of ultrafiltration-diafiltration in the plasma fractionation industry. In: WC McGregor, ed. Membrane Separations in Biotechnology. New York: Marcel Dekker, 1993:115–134.
4. deFillippi RP, Goldsmith RL. Application and theory of membrane processes for biological and other macromolecular solutions. In: FE Finn, ed. Membrane Science and Technology. New York: Plenum, 1970:47–97.
5. Ng PK, Obegi IA. Tangential flow cell separation from mammalian cell culture. Separation Sci Technol 25:799–807, 1990.
6. Le MS, Atkinson T. Crossflow microfiltration for recovery of intracellular products. Process Biochem 20:26–31, 1985.
7. Mourot P, Lafrance M, Oliver M. Aseptic concentration of living microbial cells by crossflow filtration. Process Biochem 2:3–8, 1989.
8. Zydney AL, Colton CK. Continuous flow membrane plasmapheresis: theoretical models for flux and hemolysis prediction. Trans Am Soc Artif Intern Organs 28:408–412, 1982.
9. Dosmar M, Brose D. Crossflow ultrafiltration. In: TH Meltzer, MW Jornitz, eds. Filtration in the Biopharmaceutical Industry. New York: Marcel Dekker, 16:493–532, 1996.
10. Wolber P, Dosmar M, Banks J. Cell harvesting scaleup: parallel-leaf crossflow microfiltration methods. Biopharm 1:38–45, 1988.
11. Bird RB, Stewart WE, Lightfoot EN. Transport Phenomena. New York: John Wiley, 1960:160.
12. van Reis R, Goodrich EM, Yson CL, Frautschy LN, Dzengeleski S, Lutz H. Linear scale ultrafiltration. Biotechnol Bioeng 55:737–746, 1997.
13. Ng P, Mitra G. Removal of DNA contaminants from therapeutic protein preparations. J Chromatogr 658:459–463, 1994.

14. Horowitz B, Wiebe ME, Lippin A, Stryker MH. Inactivation of viruses in labile blood derivatives. I. Disruption of lipid-enveloped viruses by tri(n-buytl) phosphate detergent combinations. Transfusion 25:516–522, 1985.
15. Highsmith F, Xue H, Chen L, Benade L, Owens J, Shanbrom, Drohan W. Iodine-mediated inactivation of lipid- and nonlipid-enveloped viruses in human antithrombin III concentrate. Blood 86:791–796, 1995.
16. DiLeo AJ, Allegrezza AE Jr. Builder SE. High resolution removal of virus from protein solutions using a membrane of unique structure. Biotechnology 10:182–188, 1992.
17. Hamammoto Y, Harada S, Kobayashi S, Yamaguchi K, Iijima H, Manabe S, Tsurumi T, Aizawa H, Yamamoto N. A novel method for removal of human immunodeficiency virus: filtration with porous polymeric membrane. Vox Sang 56:230–236, 1989.
18. Burnouf-Radosevich M, Appourchaux P, Huart JJ, Burnouf T. Nanofiltration, a new specific virus elimination method applied to high-purity factor IX and factor XI concentrates. Vox Sang 67:132–148, 1994.
19. Viresolve membrane validation guide. Millipore Corp., Bedford, MA, 1992.

7

Ultrafiltration Membranes in the Vaccine Industry

Peter Schu and Gautam Mitra
SmithKline Beecham Biologicals, Rixensart, Belgium

I. INTRODUCTION

Large-scale industrial downstream processes of biotechnological products have utilized ultrafiltration/diafiltration processes since the mid-1970s, when conventional separation techniques, such as salt precipitation, centrifugation, evaporation, lyophilization, and membrane dialysis were supplemented/replaced with alternative industrial-grade ultrafiltration membranes. Two primary objectives have been addressed: concentration of biological products from dilute solutions (ultrafiltration), and removal of small molecules via constant-volume buffer exchange (diafiltration). The availability of membranes in the molecular-weight cutoff range of 500 to 1,000,000 Daltons (Da) and manufactured with a wide range of polymers provides a reasonable degree of choice to the process engineer. Industrial-grade ultrafiltration membranes are amenable to cleaning in place for reuse, the process of which is extensively validated.

With the advent of hepatitis B vaccines in the 1980s the vaccine industry has undergone a renaissance with a plethora of new products. Almost all of the downstream processes utilize membrane processes in one form or the other. Vaccine antigens are usually derived from three possible sources:

1. Sub-unit vaccines containing recombinant proteins expressed in prokaryotes or eukaryotes (e.g., hepatitis B surface antigen expressed in yeast, lipoprotein Osp A (vaccine for lyme disease) expressed in *E. coli*, etc.).

2. Infectious or toxinogenic strains of bacteria producing either proteins or capsular polysaccharides as antigens which are subsequently inactivated (e.g., *Bordetella pertussis*–derived antigen proteins for whooping cough vaccine;

225

Haemophilus influenca–derived capsular polysaccharide for *Haemophilus* type b vaccine, etc.).

  3. Inactivated or live-attenuated viral vaccines (e.g., hepatitis A, polio (oral as well as injectable), measels, mumps, rubella, etc.).

  In this report we discuss ultrafiltration in the manufacture of two vaccine products: hepatitis B surface antigen and acellular pertussis antigens. For the former we describe a front-end ultrafiltration process which achieves the goal of concentration as well as removal of contaminant proteins and lipids. For the latter we describe a diafiltration process which needs to be run aseptically because the globular crosslinked protein cannot be sterile-filtered through an absolute 0.22-μm membrane.

II. ULTRAFILTRATION DURING PROCESSING OF HEPATITIS B SURFACE ANTIGEN VACCINE

A. Hepatitis B Surface Antigen Vaccine (Engerix B)

Engerix B, SmithKline Beecham's vaccine against hepatitis B infection, contains the hepatitis B surface antigen as immunogen. It is the first recombinant vaccine to receive regulatory approval in Europe. Since then, it has been introduced in about 130 countries worldwide for application.

B. Manufacturing of Hepatitis B Surface Antigen

The hepatitis B surface antigen is a recombinant 24-kDa protein expressed in *Saccharomyces cerevisiae* and accumulated in the vacuole of the recombinant yeast cells during growth. Its expression is promotor controlled and regulated via the physiological status of the yeast cells during fermentation. The antigen accumulation occurs in parallel with the growth of the cells. The fermentation scale for the growth of the yeast cells is approximately 1600 L at the end of the fed-batch process, and about 170 kg of yeast cell mass are contained in the fermentation broth upon harvest.

  Harvest of the recombinant antigen from the yeast cells involves centrifugation, cell disruption, several precipitation steps, clarification, and finally an ultrafiltration and concentration step, before the concentrated antigen solution is subjected to antigen purification and sterile filtration in further processing. In Figure 1 a schematic flow diagram indicating the individual processing steps during manufacture of the purified bulk antigen is provided.

  As indicated above, multiple process operations are necessary to get hold of the recombinant antigen in soluble form. These processing operations eliminate important process impurities. Precipitation's occurring upstream of the ultrafiltration/diafiltration involves the processing of significant volumes of 2000 to 3000 L. During these initial process operations the recombinant antigen is

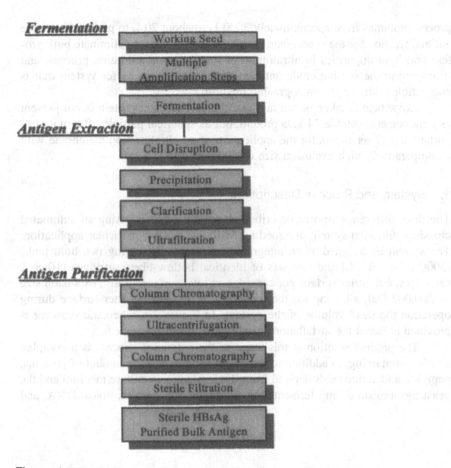

Figure 1 Flow scheme of the manufacturing process of Engerix B bulk.

transformed from its monomer soluble form into spherical particulate form. These particles can be observed through electron microscopy. Their mean diameter is about 26 nm and they have a mean density of 1.2 g/cc. This spherical form is maintained throughout the manufacturing process to be present in the final vaccine. The purified bulk antigen is released via quality control tests and adsorbed onto aluminum hydroxide as final vaccine, Engerix B.

C. Ultrafiltration/Diafiltration Procedure

A key element in the production of the recombinant surface antigen is the diafiltration and ultrafiltration of the crude antigen preparation at the end of the extraction process. The purpose of the ultrafiltration/diafiltration is threefold: to reduce process volumes from approximately 3000 L to about 20 L to prepare the crude antigen fraction for the subsequent chromatography steps; to eliminate bulk protein and lipid impurities in ultrafiltrate of the antigen diafiltration process; and to transform the soluble crude antigen preparation into a buffer system that is compatible with the chromatographic medium.

Advantage is taken of the fact that the recombinant protein is not present as a monomeric soluble 24-kDa protein, but as spherical particles of a quite important size. This allows for the application of an ultrafiltration membrane with a comparatively high exclusion size of 500,000 Da.

1. System and Process Description

The three different purposes described above are achieved using an automated crossflow filtration system designed by Millipore for this particular application. The system is designed as an integrated two-step unit serving two hold tanks (3000 L and 300 L) and two sets of identical hollow-fiber cartridges (10 + 2 cartridges, 6 m² filter surface per cartridge = total surface 72 m² , exclusion size = 500,000 Da), allowing for the reduction of the applied filter surface during operation the dead volume of the system. In Figure 2 a schematic drawing is provided to detail the installations established to run the system.

The product solution at this stage of the extraction process is a complex solution containing, in addition to the target antigen, a large amount of proteins, peptides, and amino acids derived from the fermentation culture medium and the yeast metabolism during fermentation. In addition it contains lipids, DNA, and

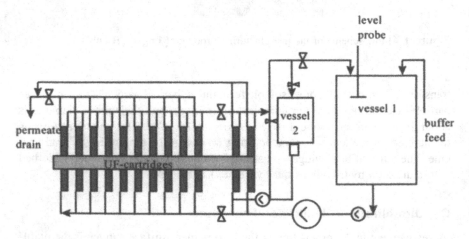

Figure 2 Ultrafiltration system.

sugars from the yeast cell's disruption process which is performed to liberate the intracellularly trapped antigen.

Following the clarification filtration after the second precipitation, the clear supernatant containing the soluble antigen is collected in the large UF-hold tank (vessel 1) to make up a total volume of ~ 3000 L. The product solution is then circulated in the crossflow system. Prior to the introduction of the product solution into the system, the membranes have been preconditioned by circulating one tank volume of 0.1 % Tween 20 solution. The purpose of this preconditioning is to avoid adherence of the different components contained in the product solution to the membrane surface.

The pool is then subjected to diafiltration against deionized water containing 0.5 M NaCl. The solution is passed through all 12 hollow-fiber cartridges in parallel at a constant inlet pressure of 2 bar and an outlet counterpressure of 0.5 bar. After 10 min of recirculation the permeate valve is opened and the permeate solution is allowed to drain. Fresh buffer is fed into the system at a constant rate matching exactly the volume drained on the permeate side. This is controlled by the computerized system controller as function of the volume level maintained constant in vessel 1 (compare Figure 3: circuit 1). Diafiltration is continued until five initial volumes of buffer solution have been washed through. The accumulated wash volume therefore is > 12,000 L.

Subsequently, concentration is started. During about 2 hours of operation at a constant transmembrane pressure of about 0.8 bar the volume of the product-containing solution is reduced to ~ 100 L in the vessel, corresponding to a total volume of about 200 L (including the volume contained in the cartridges and the piping). At this stage the second product flow circuit between the smaller hold tank and a limited number of membrane cartridges (2) is established (compare

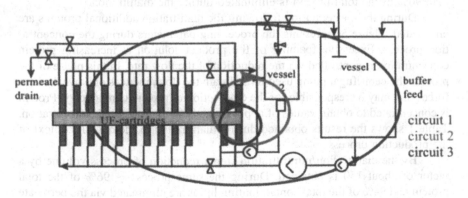

Figure 3   Different product circuits established.

Figure 3: circuit 2). Total filter surface is reduced to 16% of the initial surface. Product is transferred into the smaller hold tank and concentration is continued. As concentration continues at a constant pressure differential and decreasing permeate flow rate, retentate protein concentration continues to increase rapidly. Final protein concentration shall then be 40 to 50 mg/mL. In the very last phase of the concentration process (solution volume $\sim$ 30 L), when the hold tank (compare Figure 3: vessel 2) has been emptied completely, an additional short circuit is established to bypass the hold tank to establish a circulation between cartridges and pump only (compare Figure 3: circuit 3) and concentration is continued at low transmembrane pressure and low flow rates to avoid excessive foaming. The concentrated product solution is then collected directly from the system into a mobile container to be transported to the purification suite for further processing. The final volume obtained is $\sim$ 20 L, representing a total concentration factor of 150.

2.   Ultrafiltration/Diafiltration: Elimination of Impurities

As indicated previously, the reduction of the processing volume is the first essential goal achieved by this ultrafiltration step. A second task is the elimination of bulk impurity proteins from the product solution. Total protein concentration in the product-containing solution after cell disruption is $\sim$ 35 mg/mL, amounting to a total protein content of $\sim$ 56 kg. This amount is then subsequently reduced stepwise by the two precipitations indicated in Figure 1 by a factor of 7; total protein concentrations are approximately 3 to 4 mg/mL. Prior to ultrafiltration/diafiltration, lipids, liberated during cell disruption are present at 2–3 mg/mL.

Figure 4 shows the decrease in protein, determined via optical density measurement at 280 nm, in the permeate in as well as in the retentate. The permeate flow rate increases steadily while pressure conditions are maintained stable. More than 90% of the total protein is eliminated during the diafiltration.

During the concentration following the diafiltration additional proteins are eliminated. Figure 5 represents the processing parameters during the concentration process. Excessive foaming of the product solution at increasing protein concentration is controlled via the reduction of the flow rate. In this phase of the process the centrifugal pump used to maintain the flow, induces only weak shear forces that may be responsible for the destruction of protein components. Protein is concentrated to obtain values of approximately 45 mg/mL final concentration. Table 1 shows the results obtained during diafiltration in the overall context of the production process.

By the diafiltration/ultrafiltration step a reduction of process volume by a factor of about 150 is obtained. During the same process $\sim$ 96% of the total protein and 94% of the total contaminating lipids are eliminated via the permeate drain. At the same time the total quantity of active protein is retained and the

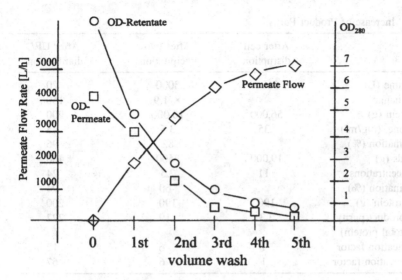

Figure 4 Elimination of protein during diafiltration.

special purity of the antigenic protein is increased by a factor of 22. No further purity of that is achieved up to this point; the processing step is a diafiltration: purification of the recombinant surface antigen achieved during diafiltration is following the step.

3 Sanitation and Maintenance of the Crossflow Equipment

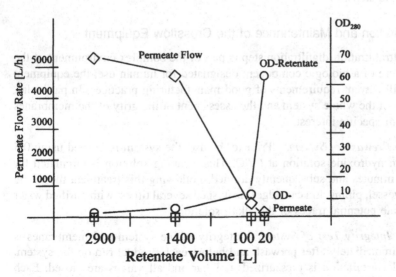

Figure 5 Process parameters and results during concentration.

Table 1 Increase of Product Purity

	After cell disruption	After two precipitations	After UF/ diafiltration
Total volume (L)	1600	3000	20
Volume change		× 1.9	150
Total protein (g)	56,000	10,000	900
Protein conc. (mg/ml)	35	3.5	45
Step elimination (%)		82	96
Total lipids (g)	19,000	7500	480
Lipid concentrations	11	2.5	24
Step elimination (%)		60	94
Product protein (g)	190	190	200
Relative product purity (mg/g total protein)	3.3	19	222
Step purification factor	1	6	12
Total purification factor	1	6	67

specific purity of the antigenic protein is increased by a factor of 12. The total purification factor achieved up to this particular processing step is ~ 67. Further purification of the recombinant antigen is achieved during the subsequent chromatographic steps.

3. Sanitation and Maintenance of the Crossflow Equipment

As this ultrafiltration/diafiltration step is part of the complex environment of the manufacture of a biologic component designated for human use, the equipment must fulfill current requirements of good manufacturing practices. In particular, sanitation of the whole system and the assessment of integrity of the membranes used are of specific interest.

(a) Prewash of System.   Prior to the use, the system is flushed using 0.5 M sodium hydroxide solution at 65°C. This cleaning solution is circulated for about 15 minutes and subsequently drained. Following this treatment the whole system (vessel, piping and cartridges) is flushed several times, with purified water to eliminate potential traces of the alkali solution.

(b) Integrity Test of System.   Integrity of the system and membranes is assessed immediately after prewash and before each product run on the system. The whole installation is pressurized to 1 bar and all valves are closed. Each individual cartridge is inspected visually for the appearance of air bubbles along

the hollow fibers. The pressure is maintained for 10 min and the drop during this period should not exceed 100 mbar. If so, the whole system is checked for leaks.

(c) Post-wash of System. After use of the system, it is emptied to the drain and the whole installation is rinsed by recirculating large amounts of purified water. After the use of the system, it is cleaned using 0.5 M sodium hydroxide solution at 65°C. The whole system (vessel, piping, and cartridges) is flushed several times, with purified water to eliminate potential traces of the alkali solution. During system validation the last rinse water has been assessed for absence of any traces of proteins, alkali, or endotoxins. After rinse the cartridges are kept under purified water until the pre-wash before the next run is started.

(d) Cycle Use of the Cartridges. The hollow-fiber ultrafiltration cartridges are maintained for a maximum of 50 product runs. At the end of each cleaning procedure after a production run, the water flow rate of the whole system is measured under defined conditions to confirm the reestablishment of consistent starting conditions for the next run. If the visible inspection of the cartridges after each run indicates an integrity problem of an individual membrane cartridge, this cartridge is immediately replaced by a new cartridge.

D. Summary

As described above, the ultrafiltration/diafiltration of the crude product solution generated during the manufacturing process of Engerix B bulk represents an essential element in the manufacturing process; > 95% of protein impurities and > 90% of lipid impurities can be eliminated. The product concentration and buffer conditions established during diafiltration and ultrafiltration allow for the further processing of the product solution in subsequent chromatographic steps. The technical layout of the installations allows for a consistent reproduction of the manufacturing and sanitation operations.

III. DIAFILTRATION OF ACELLULAR PERTUSSIS ANTIGENS UNDER ASEPTIC CONDITIONS

A. Acellular Pertussis Antigens

Whooping cough is the disease caused by *Bordetella pertussis* infection. Vaccination against the disease during the past few years has been done with a new generation of pertussis vaccines. These acellular pertussis vaccines contain the purified and inactivated antigens of *Bordetella pertussis* instead of the whole inactivated bacterial cells. Infanrix, SmithKline Beecham's trivalent vaccine against diphtheria, tetanus, and pertussis infection contains three different purified

B. pertussis proteins: pertussis toxin (PT), filamentous hemagglutinin (FHA), and outer membrane protein pertactin (69K protein) as active antigens.

B. Manufacture of Acellular Pertussis Antigens and Infanrix

The different acellular pertussis antigens are produced during fermentation in 800-L fermenters from the cultivation of *Bordetella pertussis* as host organism. *Bordetella pertussis* is a Gram-negative, pathogenic bacterium which exhibits poor or even no growth on most common culture media. Its cultivation requires the application of quite specific cultivation techniques. The fact that *B. pertussis* is pathogenic to humans necessitates at the same time-specific safety measures to prevent its release to the environment or the potential contamination of the manufacturing staff.

During growth in the main fermentor, the bacterium secretes two of the three desired antigens into the culture medium: pertussis toxin and the filamentous hemagglutinin. The third antigen (69K protein) remains bound to the outer surface of the bacterial membrane. The three antigens are subsequently separately purified to homogeneity via several chromatographic steps. A schematic manufacturing process flow is provided in Figure 6. In a subsequent process operation the active pertussis toxin needs to be inactivated (toxin → toxoid) to be used as an antigenic component in the vaccine Infanrix. This inactivation involves a treatment using formaldehyde and glutaraldehyde. FHA and 69K-protein are also formaldehyde treated to inactivate potential traces of PT.

Following the inactivation procedure the detoxified/inactivated antigens are diafiltrated to remove excessive amounts of glutaraldehyde or formaldehyde. Subsequently the antigens are adsorbed onto aluminum hydroxide to await further processing in vaccine formulation. During formulation preadsorbed purified diphtheria and tetanus antigens are added to the acellular pertussis antigens to make the final vaccine bulk which is then distributed into the final vaccine containers, such as prefilled syringes or ampules.

C. Diafiltration Procedure

During inactivation of PT the active protein is chemically crosslinked via intra-molecular and intermolecular aldehyde bridges. After detoxification/inactivation excessive formaldehyde and glutaraldehyde must be removed completely from the antigen preparation. As the crosslinking yields globular antigenic structures several micrometers in size, the inactivated toxoid is no longer filterable via standard sterile filtration membranes with a mean pore size of 0.22 μm. Therefore, all process operations of inactivation involving the addition of the inactivation agents, the incubation during detoxification, and the removal of excessive agents

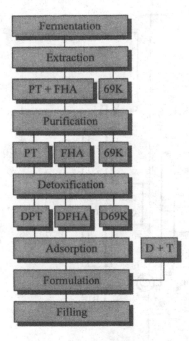

Figure 6 Manufacturing scheme of acellular Pertussis antigens.

via membrane diafiltration have to be performed under aseptic conditions. The aseptic diafiltration process uses individual modular amicon-polysulfone ultrafiltration membranes of 10 kDa exclusion size which are sterilized by formalin treatment.

1. System and Process Description

In this specific type of application of a diafiltration process, the "biological" processing conditions (the assurance of sterility of the process) outweigh the relevance of the physical processing conditions. As a general prerequisite, the environmental conditions of the operations are well defined to allow for the application of aseptic operation conditions. The diafiltration takes place in a manufacturing suite designed to meet the room classification criteria of a class B environment. In addition, any critical process operations that may bear the risk of exposing the product to the environment are handled within a class A environment. These process operations involve the establishment of sanitary connections between containers or the diafiltration membranes, sampling of the system, or any manipulations that interfere with the integrity of the system.

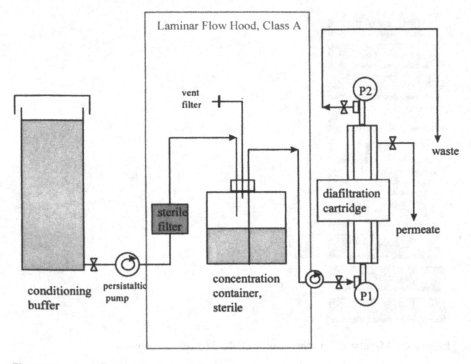

Figure 7   Preconditioning of the diafiltration system.

The aseptic diafiltration process can be divided into three different operational segments:

1. Preconditioning of the membranes
2. Introduction of the product antigen and concentration
3. Diafiltration of the product antigen

Any subsequent operations involve the sanitation and sterilization of the system and storage of the system under aseptic conditions. In routine manufacturing practice this sequence of operations has been performed as such for a maximum of 20 runs before the diafiltration membrane is discarded and replaced by a new one.

(a) Preconditioning of the Membranes.   In Figure 7 a schematic drawing of the system during the preconditioning phase is provided. During this phase of the process, preceding the introduction of the product into the system, the sanitized and sterile cartridges are equilibrated with an appropriate sterile buffer to prepare the system for the reception of the detoxified antigen. About 20 L of

conditioning buffer, which was prepared using water for injection, is fed into the system via flexible tubes using a peristaltic pump. The nonsterile conditioning buffer contained in the feed tank is sterilized by passing it through a sterilizing filter into the concentration container. The conditioning buffer is eliminated from the system via the permeate drain and the via the retentate drain at the same time.

   (b) Introduction of the Product Antigen and Concentration   The purified and detoxified bulk antigen is contained in a sterile bulk container, from which it can easily be transferred into the sterile concentration container which was previously used during the conditioning of the diafiltration membrane (Fig. 8). The product antigen is then concentrated via the ultrafiltration membrane following reception from the bulk antigen container. During this concentration phase a counterpressure of 0.8 to 1.0 bar is maintained on the outlet valve of the retentate (P2). Permeate is drained without further analysis. Concentration is continued until the whole content of the bulk antigen container has been transferred and a theoretical concentration of 500 (g antigen/mL is achieved. A mean permeate flow rate of ~ 500 mL/min is obtained.

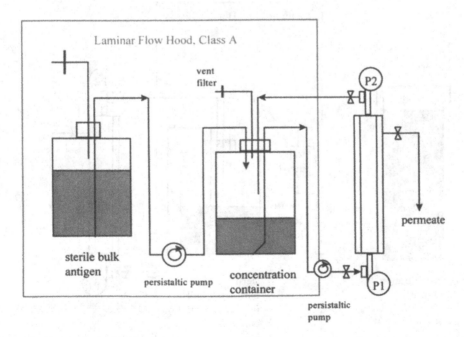

Figure 8   Concentration of the product antigen.

  (c) Diafiltration of the Product Antigen. Following the concentration
phase the concentrated antigen (8 to 10 L) is diafiltered using 10 volumes of an
inline, filter-sterilized sodium chloride solution. A schematic drawing of the system is provided in Figure 9. Processing conditions applied are identical to those
applied during concentration. A transmembrane pressure of 0.5 bar is maintained
during diafiltration. Permeate is drained at a constant flow rate of ~ 500 mL/
min. The aim of this diafiltration is the elimination of any residual formaldehyde
and glutaraldehyde added during the detoxification process.

  At the end of the diafiltration, the system (flexible tubes and cartridge) is
emptied into the concentration container. The pool is sampled for in-process and
sterility analysis, deconnected from the system, closed, and transferred into a
storage area to await further processing.

2. Validation of Sanitation and Sterilization of the System

As the sterility assurance of the aseptic process is critical with regard to the use
of the antigens in human vaccines, the sanitation of the system and the assurance

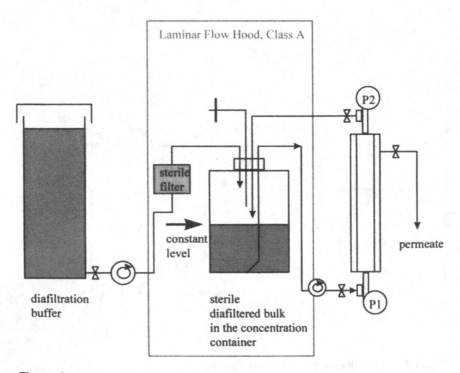

Figure 9   Diafiltration of the product antigen.

of sterility are key parameters. Following the operation of the system the membrane cartridge is sanitized applying a sequence of different solutions: diafiltration buffer for rinsing; 0.5 M NaOH solution for 1st sanitation; 0.5 M NaOH solution for 2nd sanitation; and 1% formaldehyde solution for sterilization.

Subsequently, the system is maintained under 1% formaldehyde solution until the next use. This sanitation and sterilization aspect has been subject of validation studies applying challenge with endotoxin solutions and contaminating bacteria suspensions. The data of these validation studies are presented hereafter.

(a) Validation of the Elimination of Bacterial Contaminants. As indicated earlier, the diafiltration process must be maintained as an aseptic process. The product antigen is provided as a sterile solution and the buffer added is inline sterile-filtered. Sterility is assessed at the end of each run.

To validate the sterility assurance of the process operations and at the same time the capability of a complete elimination of bacterial contaminants by the sanitation and sterilization procedures, a challenge validation protocol was designed. The system was contaminated, by applying a concentrated suspension of live *Bacillus subtilis* cells. This cell suspension, contained in a volume of 5 L, was circulated among the system for about 15 min, as if it were product solution. Following this circulation, the system was drained as established for recovery of the diafiltrated antigen solution. Subsequently the sequence of sanitation and sterilization applying the 1% formaldehyde solution was performed. At the end of this sanitation-and-sterilization procedure, the system was flushed with conditioning buffer and the startup of a subsequent product run was simulated. In this particular case a solution of tryptic soy broth (TSB), a common bacterial growth medium, was circulated in the system as if it were antigen product. This growth medium was then sampled and incubated accordingly to assess the absence of live bacteria. The results of this validation are summarized in Table 2.

The results showed that by the application of the established sanitation and sterilization procedures even massive bacterial contamination of greater than 10^9 bacteria were efficiently eliminated. The growth potential of these bacteria after introduction into the system was assessed in a positive control. The same was assessed for the TSB medium used for the determination of the sterility assurance.

(b) Validation of the Elimination of Endotoxin. Endotoxin represents a major potential contaminant in pharmaceutical products. These biological structures derived mainly from the outer membranes of gram-negative bacteria are composed of complex lipopolysaccharides which may cause nonspecific, generally weak toxic reactions such as fever, redness, or swelling at the injection site. But when contained in larger amounts in pharmaceutical preparations, the reaction may be severe. Their elimination from biological products, which are derived from the cultivation of bacteria, through appropriate purification procedures represents a major task of the manufacturing process. This task is rendered even more difficult as these endotoxins are chemically and thermally very stable.

Table 2 Sterility Assurance and Elimination of Bacteria by System Sterilization

	B. subtilis cells in system after contamination (bacteria/mL)	Bacterial growth on culture medium	TSB medium promotes bacterial growth	Sterility of TSB medium after passage
Validation run 1	10×10^9	positive	positive	pass
Validation run 2	1.3×10^9	positive	positive	pass
Validation run 3	4.9×10^{12}	positive	positive	pass

As described above, the sanitation and sterilization procedures in place are appropriate to eliminate a potentially contaminating bacterium from the diafiltration system. Endotoxins may, however, still be present, so to validate the efficacy of the sanitation procedure for the elimination of potential endotoxin contaminants a challenge protocol was established. In each of three independent runs, the system was intentionally contaminated by connecting a 5-L flask in place of the product container, containing an endotoxin solution prepared from a culture supernatant of a *Neisseria meningitis* culture (Endotoxin concentration of 1.3×10^6 EU/mL) to yield a system concentration of ~ 4000 EU/mL. This suspension was then circulated for 15 min in the system, while the permeate drain remained closed. Subsequently the system was emptied as established for the recovery of a diafiltrated antigen product, and the membranes and the tube were subjected to the sanitation procedure established. At the end of this sanitation procedure the concentration of endotoxins in the conditioning buffer was determined. The results of these studies are presented in Table 3.

The validation aimed for a reduction of the endotoxin load of a minimum factor of 1,000. As shown in Table 3, the sanitation resulted in a complete elimination of the endotoxin introduced into the system, corresponding to a reduction factor of 100,000.

Table 3 Elimination of Endotoxin by Sanitation

	Endotoxins in system after contamination (EU/mL)	Endotoxins in system after sanitation (EU/mL)
Validation run 1	4.1×10^3	<0.05
Validation run 2	5.1×10^3	<0.05
Validation run 3	4.9×10^3	<0.05

D. Summary

The validations described above show that the diafiltration system may well be maintained as an aseptic system, and at the same time, by the application of appropriate sanitation and sterilization procedures, live bacterial contaminants, as well their metabolites or bacterial membrane components may be completely eliminated from the system.

IV. CONCLUSIONS

The transformation of the hepatitis B surface antigen to a spherical particulate of a mean diameter of 26 nm during the initial processing steps allows for the selection of comparatively large molecular cutoff (500 kDa) hollow-fiber membranes. Consequently, large amounts of low-molecular-weight contaminants (lipids and proteins) can be eliminated during a diafiltration and a subsequent concentration step on these membranes. Elimination of > 90% of contaminating lipids and proteins, resulting in a step purification factor of 12, is achieved by this membrane filtration. Furthermore, established process conditions and sanitation procedures allow for the repetitive use of the membranes over a large number of manufacturing cycles maintaining their appropriate performance.

In a final processing step of the preparation of *Bordetella pertussis* antigens, an aseptic diafiltration process has been established to allow for the complete elimination of chemicals introduced during detoxification of these antigens. This diafiltration process, using 20-kDa membranes, has been validated to prove sterile operation conditions and adequate sanitation procedures. Regeneration and consistent reuse of these diafiltration membranes over multiple manufacturing cycles was established as a relevant quality criterion. The sanitation and regeneration procedures allowed for a complete removal of potential contaminating bacteria, as well as for a complete removal of contaminating endotoxin.

The membrane uses described here demonstrate that in a complex manufacturing environment of vaccines for human use industrial-grade ultrafiltration membranes are capable of fulfilling all necessary processing requirements under validation control and good manufacturing practices.

8

Protein Purification by Affinity Ultrafiltration

Igor Yu. Galaev and Bo Mattiasson
Lund University, Lund, Sweden

I. INTRODUCTION

Tremendous success in molecular biology and gene engineering has made it possible to express practically any given amino acid sequence and to synthesize practically any desired protein. That success has overshadowed the fact that to be used commercially the expressed protein should be purified to a sufficient degree depending on its final application. The higher the required purity of the protein is the more expensive purification procedure is, mounting to 80% of the overall production costs [1]. The efficiency and economic liability of the purification process determine often the fate of the protein product on the market. Purification of a protein from a fermentation broth generally involves a combination of techniques resolving the proteins according to their size, charge, hydrophobicity, or ability to bind some particular substances (affinity ligands, chelated metal ions, HS-containing substances). These techniques have traditionally been optimized individually rather than as a part of the integrated, continuous process. Moreover, a lot of the development and refining of the purification protocol has been carried out on a laboratory scale with little consideration given to the economy of scaling up.

In order to cut down on costs and improve product recovery in the process consisting of several purification stages, it is critical to design purification techniques that could combine high-resolution power, high-throughput capacity, and process impure feed streams.

Thus, the development of the successful techniques should address two main points. The first imperative requirement is the selection of the protein of interest from the rest of the proteins and its preferential, ideally complete, accumulation into one phase, while the rest of proteins accumulate in the other phase. The second step is the mechanical separation of these two phases arranged in a way allowing high flow rates and the ability to process turbid solutions with minimal mechanical entrapment of impurities in the phase enriched with target protein. Let us address each of these requirements in turn.

II. SELECTIVE PARTITIONING OF THE TARGET PROTEIN—USE OF AFFINITY LIGANDS

Many proteins possess intrinsic ability to bind some particular substances. For example, many enzymes have strong affinity toward their substrates or substrate-similar molecules; protein inhibitors that selectively bind proteins to which they are targeted and antibody-antigen pairs present nearly perfect selectivity of binding. All these types of interactions combined under the name of affinity interactions are successfully used for selective binding of the target protein from the crude extract [2]. Whenever the natural binding force of the protein is insufficient for the purification process, the target protein could be modified genetically via fusing different affinity tails or tags in order to facilitate the purification process [3]. Thus, affinity ligands coupled to a matrix present an efficient tool capable of recognizing the target protein and binding it, resulting in the enrichment of the matrix compared to the aqueous phase containing crude extract.

III. SEPARATION OF TWO PHASES, ENRICHED WITH AND DEPLETED OF TARGET PROTEIN

The solution of the second problem, separation of enriched phase from the depleted crude extract, was traditionally realized in the format of the column chromatography. The ligand-bearing matrix is represented by chromatographic adsorbent and packed into a column with a porous bottom. The starting mixture is pumped through the column, providing enough contact time for binding of the target protein by adsorbent to take place. The column format suffers a few drawbacks. High flow rates through the column result in significant flow resistance of the column and require high pressures to maintain desired flow rates. Moreover, the solution applied on a column should be completely free of particulate material. Even minute "haze" frequently appearing in concentrated protein solutions could be deadly for the performance of the column, which works as a depth filter accumulating particulate material between the beads of the adsorbent. The final result is a complete blocking of the flow through the column. Expanded bed chromatography has emerged as an attempt to use particulate-containing feeds in chromatography [4]. On the other hand, new techniques were developed where the separation of enriched and depleted phases was carried out in other than column format. In affinity precipitation the target protein is accumulated in the polymer precipitate; the latter is separated by centrifugation or filtration [5]. In affinity partitioning in aqueous polymer two-phase systems, target protein is enriched in one of the liquid phases; the other one is separated using separating funnel or other techniques for liquid-liquid separation [6].

Ultrafiltration separates molecules on the basis of their size and shape. For maximum separation efficiency, there should be 10-fold difference in the sizes of the species to be separated [7]. Thus, to make ultrafiltration applicable for protein purification, the target protein should be somehow included in a complex of at least 10 times larger than the impurities, which have usually the same size range as target protein. This could be achieved by binding the target protein to a macroligand, a soluble polymer, or a microparticle-bearing affinity ligand specific toward the target protein. On the other hand, the macroligand should not be too big and precipitate on gravity. It should remain in a soluble or suspended state enabling easy mass transfer and avoiding any aggregation. The high-molecular-weight soluble polymers or microparticles (latexes) seems to be objects of choice for affinity ultrafiltration.

IV. ULTRAFILTRATION

Membrane filtration in general and ultrafiltration in particular is an established technology in biochemical and pharmaceutical industry. Ultrafiltration membranes have pore sizes of 0.001 to 0.1 μm. Because of the low or negligible osmotic pressure of macromolecules, ultrafiltration operates at relatively low pressures, typically 2 to 5.5 bar. Ultrafiltration is an efficient process for removing macromolecules, colloids, emulsified oil, endotoxins, pyrogens, viruses, and bacteria. It is widely used in the food industry, for treating industrial wastewater, and processing of drinking water. In 1996, the market for ultrafiltration membranes and systems was nearly $530 million [7]. Developments in polymer chemistry and module construction have resulted in the availability of an array of ultrafiltration membranes of high chemical and physical stability combined with high fluxes, for different scales of operation. Hydrophobic membranes can be modified to increase flux and reduce fouling. The modification techniques includes the following: reacting of the base polymer with hydrophilic pendant groups before casting; entrapping a hydrophilic moiety before casting; blending polymer before casting; modifying membrane surface or charge; grafting the ceramic membrane surface [7].

Filtration is most effective when operated as a tangential flow (also termed crossflow filtration) such that the flow of the feed is parallel to the membrane surface and is recirculated rapidly enough to prevent concentration polarization.

V. AFFINITY ULTRAFILTRATION

A. Concept

In order to combine the high-volume processing capability of membrane filtration and at the same time increase its specificity, the concept of affinity ultrafiltration was developed. This technique is based on a principle that target protein and impurities, when free in solution, pass through the ultrafiltration membrane, whereas when the target protein is bound to a macroligand, the resulting complex is retained by the membrane. The impurities are washed out after the binding of the target protein to the macroligand. When elution buffer is introduced, the target protein dissociates from the macroligand and is transported in a purified form through the membrane with the flux (Figure 1) [8–14].

This process can be operated as a batch, semibatch, or continuous process. The respective sizes of the target protein and macroligand alongside with the

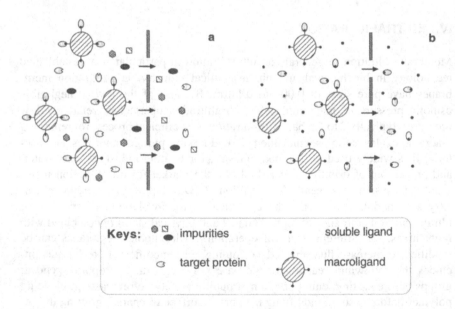

Figure 1 Principle of affinity ultrafiltration: (a) binding of the target protein and washing out impurities; (b) dissociation of the complex target protein-macroligand mediated by free ligand and transport of the purified target protein through the membrane.

cutoff of the membrane determine the efficiency of the purification procedure. The membrane should prevent the leakage of macroligand and, at the same time, should be highly permeable to the free protein and the impurities. Therefore, the study of the individual behavior of the macroligand and target protein prior to the affinity step is highly recommended [15].

The choice of the membrane is dictated not only by its cutoff properties but also by the chemical nature of the membrane material. Macroligand could adsorb or deposit as a layer on its surface resulting in reduced filtration rates. Membrane could adsorb target protein as was demonstrated during isolation of urokinase by affinity ultrafiltration using copolymer of N-acryloyl-*m*-aminobenzamidine and acrylamide as a macroligand [8]. Urokinase, with a molecular weight of 31,000, was not detected in permeates through polysulfone Millipore membranes with cutoff 10,000 and 100,000. However, the enzyme passed with 95% recovery through the 100,000-cutoff Amicon YM membrane made of cellulose acetate, suggesting that urokinase was bound to polysulfone membrane.

Protein-protein interactions in the crude could also contribute to the complicated ultrafiltration patterns. During urokinase purification from human urine, high-molecular-weight material present in the crude formed a complex with the enzyme, preventing its passage through 100,000-cutoff membrane. The addition of 2 M NaCl resulted in dissociation of the complex and 70% recovery of urokinase in the permeate [8].

B. Binding

As all protein purification techniques based on selective binding of the target protein, affinity ultrafiltration protocol is divided into four steps: binding, washing, elution, and regeneration of the adsorbent, in this case, macroligand.

Binding of the target protein to the macroligand could be realized either (1) by mixing crude extract with macroligand in the chamber at the same side of the membrane, or (2) by pumping crude extract and macroligand solution/ suspension at the opposite sides of the membrane, letting the target protein to diffuse through the membrane to the macroligand-containing chamber where the binding takes place.

The first alternative is more straightforward and constructively simple, but it requires preliminary clarification of the feed. Most of the studies reported so far on affinity ultrafiltration were done in this mode. The second mode is a more general approach and could be directly integrated with, for example, a fermentation broth or complex crude extracts. Here the membrane forms a protective barrier for the ligand against particulate material present in the crude extract [9]. α-Amylase from *Bacillus licheniformis* was purified with 95% recovery by the latter approach using crosslinked starch slurry as the affinity adsorbent [16].

The ligand content in the macroligand affects the affinity ultrafiltration procedure. The binding capacity initially increases with the increase in ligand content of the macroligand, but reaches a plateau at a high ligand content as ligands become sterically inaccessible for protein binding. A similar phenomenon has been observed during affinity chromatography [9].

C. Washing

Washing of the complex target protein-macroligand is usually performed in the diafiltration mode such that the volume of the retentate is maintained constant by adding the washing buffer at a rate equal to the filtration rate [9].

To avoid leakage of the target protein from the complex with macroligand, affinity ultrafiltration requires higher binding strength of the target protein to the macroligand than affinity chromatography. The protein molecule dissociated from the chromatographic matrix during washing has high chances to bind again due to the high volumetric (in the chromatographic column) concentration of the ligands. The protein molecule encounters many interactions on its way through the column, binds to the matrix once again, and remains inside the column. The total volumetric concentration of the ligands in affinity ultrafiltration is lower than in affinity chromatography, and the protein molecule dissociated from the macroligand has much lower chances to meet other ligands when drawn by the flux through the membrane. Thus, it is desirable to use high-affinity ligands for affinity ultrafiltration. According to Ling and Mattiasson [10], if the binding between the protein and macroligand is assumed to be in equilibrium, the binding constant should be $>10^6 \, M^{-1}$. However, for affinity crossflow filtration of human serum albumin using Cibacron Blue agarose as a macroligand, Herak and Merrill [17] showed that release of the target protein during the washing step has very slow kinetics and essentially proceeds under nonequilibrium conditions. This observation opens an opportunity for using affinity ligands with somewhat lower binding strength than $10^6 \, M^{-1}$. One more factor could be also favorable during washing of bound multisubunit proteins, namely, multisite attachment. High local concentration of the ligands in macroligand facilitates multisite binding of these proteins, especially when soluble macroligands are used. Flexible polymer backbone of the macroligand can adapt in solution conformation that allows simultaneous binding of a few ligands to a multisubunit protein. The effect of polymer flexibility on multisite polymer attachment to the protein was demonstrated recently in affinity precipitation using Cu-loaded copolymers of N-isopropylacrylamide and 1-vinylimidazole [18].

D. Elution

The elution of bound protein is achieved by changing the buffer so that the binding interaction is unfavorable and the target protein is released from the complex with macroligand. As in other affinity techniques, elution is usually achieved either by use of a competitive affinity ligand or by altering pH or ionic strength. The first alternative is attractive, since it is performed under conditions that preserve the biological activity of the target protein. The drawback might be the high cost of using competitive affinity ligand. Nonspecific elution at extreme pH or high salt concentrations is cheap but may be harmful to the target protein. Extreme pH values or high salt concentrations could also adversely affect the properties of the ultrafiltration membrane or macroligand itself.

Affinity ultrafiltration is operated in the mode close to complete mixing, and there is no buffer front as in affinity chromatography. Affinity between the protein and the macroligand changes gradually as the elution buffer replaces the washing buffer. The important aspect of this step is to avoid dilution of the released target protein with excessively large volumes of buffer [9]. The precise selection of elution conditions is very much dependent on the system concerned.

The eluate containing the purified product is very dilute after affinity ultrafiltration. The additional ultrafiltration step could be useful resulting in concentration of the product and at the same time removal of competing ligand when the latter was used for protein elution.

E. Regeneration

After the elution step, the free macroligand should be regenerated and reequilibrated in the binding buffer. Regeneration may include washing out proteins, that were nonspecifically bound to the macroligand and were not eluted with the elution buffer. High salt concentrations or extreme pH values used for washing out nonspecifically bound proteins could be detrimental for both macroligand and the membrane. Hence, the macroligand should be developed with regard to minimal nonspecific interactions with the proteins present in the crude extract. The reequilibration of the macroligand is achieved in a separate membrane filtration step where the elution buffer or washing solution is replaced with the fresh binding buffer by diafiltration.

VI. CHOICE OF MACROLIGAND

The choice of macroligand is critical for the applicability of affinity ultrafiltration. The general requirements to the macroligand are the following:

1. Selective binding of target protein with minimal nonspecific interaction with the other proteins present in the crude extract
2. High capacity, i.e., binding large amounts of target protein per gram of the macroligand
3. Chemical stability, i.e., ability to withstand many cycles without leak-

age of the ligand and chemical destruction of a soluble polymer or microparticle

4. Mechanical stability, of microparticles, i.e., ability to maintain their size and shape under high shear forces resulted in the liquid flow during pumping

5. Meeting legal requirements, e.g., being nontoxic, nonpyrogenic, etc.

6. Cost efficiency

By and large, finding commercially available macroligand is difficult. This limitation is overcome by "tailoring" the macroligand; that is, a ligand molecule is coupled either to a water-soluble polymer or a water-insoluble carrier that does not itself bind proteins.

A. Water-Soluble Macroligands

The advantages of water-soluble macroligands lie in the high yields obtained from ligand coupling, the rapid control of the amount immobilized, and easy complexation (no mass-transfer limitations) with the target protein. High binding capacities per gram of the macroligand could be achieved. Protein binding takes place under equilibrium conditions, resulting in quantitative complexation [9].

Macroligands based on soluble polymers have much smaller size than macroligands based on insoluble carriers. When used in affinity ultrafiltration, soluble macroligands require much smaller pores, resulting in much lower flux rates; accordingly, a longer processing time or a larger membrane surface is needed. To provide sufficient resolution in affinity ultrafiltration, the size of the complex of target protein with the macroligand should differ significantly from the size of protein impurities. Most of the soluble proteins have molecular weight <100,000, so the polymer carrier should be of the molecular weight >100,000, preferably in the range of 1,000,000. Aqueous solutions of polymers of that high molecular weight are often very viscous even at relatively low concentration due to the entanglement of macromolecular chains. Both reversible and irreversible fouling can be a significant problem in ultrafiltration of polymeric solutions. On the other hand, soluble polymer is not likely to be degraded by attrition.

The requirement of molecular weight in the range of 10^5–10^6 limits the choice of water-soluble polymer carriers to only a few options—namely, polysaccharide-based polymers (dextran, starch) and poly(acrylamide) based polymers (Table 1). The chemistry of attaching ligands to carbohydrate polymers is well established, and high-molecular-weight dextran was used as ligand carrier in a number of studies [19–22]. Unfortunately, separation of trypsin and α-chymotrypsin using trypsin specific conjugate, dextran with coupled soybean trypsin inhibitor, faced significant problems due to the nonspecific adsorption of the proteins to the polymer carrier.

Table 1 Macroligands Used in Affinity Ultrafiltration

Carrier	Size	Ligand	Target protein	References
Soluble macroligand				
Poly(acrylamide)	MW >100,000	m-amino-benzamidine	trypsin	15,55
Poly(acrylamide)	MW >100,000	m-amino-benzamidine	urokinase	8
Dextran	MW 2,000,000	protein A	IgG	19
Dextran	MW 2,000,000	estradiol	$\Delta_{5 \to 4}$3-ostosteroid isomerase	22
Dextran	MW 2,000,000	p-amino-benzamidine	trypsin	21
Dextran	MW 2,000,000	soybean trypsin inhibitor	trypsin	20
Insoluble macroligand				
Silica nano particles	0.012 μm	Cibacron Blue	alcohol dehydrogenase, lactate dehydrogenase	10
Liposomes	0.027 μm	biotin	avidin	29,30
Polystyrene core-shell latex particles	0.25–0.4 μm	Cibacron Blue	bovine serum albumin	32
Liposomes	0.06 μm	p-amino-benzamidine	trypsin	31
Cell parts of *Streptococcus*	1 μm	surface protein G	bovine IgG	56
Saccharomyces cerevisiae cells	5 μm	surface carbohydrate groups	concanavalin A	12
Polystyrene core-shell microspheres	2.0–7.7 μm	endopectin-lyase	polyacid, Eudragit 100–55	35
Starch granules	1–10 μm	Cibacron Blue	alcohol dehydrogenase	13
Superose beads	11–15 μm	Cibacron Blue	human serum albumin	57
Superose beads	11–15 μm	Cibacron Blue	lysozyme	57
Sepharose beads	45–165 μm	Cibacron Blue	human serum albumin	17,57
Sepharose beads	45–165 μm	Cibacron Blue	lysozyme	57
Sepharose beads	45–165 μm	concanavalin A	horseradish peroxidase	57
Agarose beads	45–165 μm	protein A	IgG	57
Agarose beads	not stated	p-aminobenzyl-1-thio-β-D-galactopyranosid	β-galactosidase	28
Sepharose beads	45–165 μm	heparin	lactoferrin	56
Sepharose beads	45–165 μm	protein G	bovine IgG	56

Poly(acrylamide)-based macroligands proved to be more successful for trypsin purification [8,15,23]. The poly(acrylamide)-based macroligands are synthesized by copolymerization of acrylamide with a monomer, N-acryloyl-*m*-aminobenzamidine, which has a benzamidine group specifically binding to the trypsin active site. The mechanism of trypsin inhibition by macroligand seems to be competitive with binding constant dependent on the *m*-aminobenzamidine content in the macroligand. At high *m*-aminobenzamidine content the inhibition capacity of the polymer decreased. The optimal benzamidine content in the macroligand, maintaining a balance between the capacity of the macroligand and strength of binding was around 17 mol%. The macroligand with molecular weight $>10^5$ (fractionated using ultrafiltration through a membrane with 100,000 cutoff) was used for the separation of a model trypsin/α-chymotrypsin mixture. At optimal conditions, 98% pure trypsin with 90% recovery was obtained in a batch process [24]. Trypsin binding to the polymer proceeds very fast, eliminating the need of preliminary incubation of the crude extract with the macroligand. With the reuse of the same polymer in a continuous process and using benzamidine as the eluent, trypsin was purified from crude porcine pancreatic extract with 97% purity and 77% yield. The decrease in polymer concentration from 0.75% to 0.5% resulted in decrease in trypsin retention from 90% to 54%, suggesting that bound trypsin became more susceptible to the shear forces with the decrease in polymer concentration.

It was shown important to maintain the pH value around 8 during affinity ultrafiltration. Both trypsin and α-chymotrypsin passed slowly through the membrane at low or high pH values. Hence, acidic conditions used for trypsin elution from benzamidine-containing chromatographic matrices could not be used for elution during affinity ultrafiltration. Benzamidine (50 mM) or arginine (1 M) solutions were used instead for trypsin elution [25]. The polymer solution is stable for months and trypsin when bound to macroligand is stabilized against autolysis [26].

The same soluble macroligand was used for urokinase purification from a model mixture containing peroxidase and urokinase and from crude human urine [8]. The bound urokinase could be eluted easily using a buffer solution containing 100 mM benzamidine and 100 mM NaCl. Purification of urokinase from crude human urine was complicated by the fact that when this material was ultrafiltered through a 100, 000-cutoff membrane, only 5% of the urokinase was detected in the filtrate. Apparently, there was an aggregation between urokinase and some high-molecular-weight materials to form a complex. When saline solution (2.0 M NaCl) was added to the system at the same rate as the filtrate, about 70% of urokinase was detected in the filtrate. The overall yield of the enzyme in affinity ultrafiltration procedure was 49% and sevenfold purification was achieved.

The recent advances in polymer chemistry resulted in the appearance of new water-soluble polymers. Some of them could be of interest as potential carriers in affinity ultrafiltration. Dendrimers are a new class of polymers, which emerged recently. They have hyperbranched treelike structures of a very rigid nature [27]. Some unique properties of the dendrimers make them attractive as carriers for macroligand production in affinity ultrafiltration. The rigidity of the structure improves the selectivity of the ultrafiltration; no "wormlike" squeezing of the linear polymer through membrane pores could take place. High and easily regulated density of reactive groups at the outer surface of the dendrimers combined with the rigidity allows tuning spatial adjustment of ligands to provide all the advantages of multisite binding of the target protein. Finally, whereas the viscosity of the linear polymer increases logarithmically with the molecular weight, viscosities of dendrimers reach plateaus as the molecular weight increases, making it possible to create high volumetric ligand concentration without macroligand solution being too viscous. With new polymers available, one could expect the revival of the interest to affinity ultrafiltration as well as to other techniques of protein purification based on the target protein binding to a soluble macroligand.

B. Water-Insoluble Macroligands

By definition, insoluble macroligands are much larger than soluble ones, providing higher resolution force between the complex target protein-macroligand and impurities. Particles as large as those used in affinity chromatography could be applicable for ultrafiltration. Agarose and starch particles have been found to be relatively easy to pump through the cross-filtration device with minimal fouling of the membranes [17]. However, particle degradation occurred with time. Particles of a rather rigid highly crosslinked gel, Sepharose CL-6B, were found to be fractured 5% after 24 h of circulation while 30%–40% particles were damaged in just 3 h for a soft gel, Affi-Gel Blue [17]. Particles of a commercially available affinity gel, p-aminobenzyl-1-thio-β-D-galactopyranoside (PABTG) agarose, were used for purification of β-galactosidase from *E. coli* homogenate by continuous affinity recycle extraction (CARE) process (28). When dealing with chromatographic materials as ligand carriers, it is no longer necessary to operate with ultrafiltration membranes. The introduction of microfilters gave much better fluxes and thus allowed larger volumes to be processed per unit time.

High volumetric concentration of the ligand is advantageous for the affinity ultrafiltration process, increasing total capacity of the system and concentration of the target product in the eluate. When the ligands are coupled to insoluble carrier, the available surface of the particles sets a limit on the amount of ligand. The larger the particles, the less the surface area they have. Thus to increase surface area one should use either small particles or large porous particles, e.g., agarose-gel particles. In the later case the binding and/or desorption kinetics could be slow due to the diffusional limitations resulting in the obligatory incubation of the macroligand with a crude extract in a separate vessel for binding to take place or a very dilute eluate.

One of the first studies on affinity ultrafiltration with small insoluble macroligand employed heat-killed cell of *Saccharomyces cerevisiae* (mean diameter, 5 μm) as a macroligand for the purification of concanavalin A (molecular weight 71,000) from jack bean extract. The target protein is specific for carbohydrates and binds to the available carbohydrate ligands exposed on the surface of the yeast cells. The mixture of the cells and crude extract was pumped through a hollow-fiber unit with 1,000,000-cutoff at a flow rate of 500 mL/min. High glucose concentration (150 g/L) was used to dissociate the affinity complex. The glucose was partially removed by passing the permeate from the dissociation step through a membrane with 35,000 cutoff, while the yeast cell suspension was washed from the glucose in the regeneration step. Concanavalin A in a highly purified form was obtained with an overall yield of 70% [12].

Alternatively, small particles (diameter of 12 nm and a surface area of 200 m^2/g) of fumed silica with immobilized Cibacron Blue were used for purification of alcohol dehydrogenase from yeast and lactate dehydrogenase from porcine heart [10]. The ligand density was found to be 34 μmol/g, which is equivalent to 0.2 μmol/m^2 and comparable to the density of 0.33 μmol/m^2 for an NAD derivative coupled to silica beads and used in HPLC columns.

Due to their small size, the silica particles formed a stable colloid suspension. The binding kinetics was similar to those obtained for binding in free solution, with binding and dissociation occurring almost instantaneously. Thus the process could be run continuously without the risk of incomplete binding.

Although filtration rate has been greatly improved as compared to previous work, it was still the limiting step in the process. The recovery in the elution step has been shown to follow an exponentially decreasing curve as expected for unbound protein. An alternative to increasing the filtration rate is to decrease the working volume. In this case a higher percentage of particles could be used [13].

Insoluble macroligand should not be necessarily based on a solid carrier. The use of affinity-modified liposomes or vesicles formed by phospholipids for affinity ultrafiltration seems to be an interesting approach [29–31]. Unimolecular liposomes with an average diameter 200 to 700 Å possess an extremely large surface area; e.g., 300-Å-radius liposomes have a specific surface area of 100 m^2/mL of liposomes which in terms of area per carrier volume is approximately the same as for porous silica. The hydrophilic surface of the liposomes minimizes the nonspecific protein adsorption and results in a rather strong repulsion between liposomes. Hence, the membrane polarization and plugging during ultrafiltration are reduced.

Liposomes have historically been used to study the structure and function of biological membranes and as biocompatible drug carriers. The chemistry of liposome modification is well developed. Affinity ligands could be covalently attached to the liposome surface, either coupling the molecule to suitable functional groups on the outer surface, or incorporating derivatized phospholipids into the membrane during liposome formations. Liposomes are sufficiently stable for this type of modification.

The first approach, a surface modification, was used for purification of trypsin. Liposomes comprised of dimyristoyl phosphatidyl choline, cholesterol, and dimyristoyl phosphatidyl ethanolamine with amino groups at the surface were modified with diglycolic anhydride and p-aminobenzamidine was coupled to the terminal carboxy groups via carbodi-imide procedure. The equilibrium binding constant between trypsin and immobilized ligand was shown to be dependent on the ligand density of the liposome surface. The ligand density on the liposome can be carefully controlled by simply varying the amounts of phospholipids during liposome preparation. Bound trypsin was eluted from the liposomes by the trypsin inhibitor benzamidine. Trypsin was purified from a trypsin/chymotrypsin mixture and from porcine pancreatic extract. A recovery yield from the crude mixture of 68% was obtained with a trypsin purity of 98%. Loss of trypsin during washing was thought to be due to the relatively low equilibrium binding constant of 1.7 $\times$ 10^4 M^{-1} between trypsin and immobilized ligand. The affinity-modified liposomes were stable in the complex mixture and retained their trypsin binding capacity after multiple adsorption/elution cycles over a 1-month period [31].

The second approach, an incorporation of a modified phospholipid into liposome during formation, was used for avidin purification from avidin model mixtures with lysozyme or cytochrome c, and also from its naturally occurring source, hen egg white. Biotinylated dimyristoyl phosphatidylethanolamine was synthesized by coupling biotinyl-N-hydroxysuccinimide ester to phosphatidylethanolamine. The liposomes comprised of dimyristoyl phosphatidylcholine and biotinylated dimyristoyl phosphatidylethanolamine. Only ~1% of the total biotin in the system was accessible for avidin binding. The equilibrium binding constant was determined to be 6 $\times$ 10^7 M^{-1} as compared to 10^{15} M^{-1} for the free biotin-avidin in solution. The drastic decrease in binding efficiency was attributed to the steric hindrance between the avidin molecule and the liposomes as the biotin was immobilized directly on the liposome surface; however, the binding strength was sufficient for purification purposes. Crude hen egg white extract was subjected to ion exchange chromatography and the eluate was processed in affinity crossflow ultrafiltration system resulting in 96% pure avidin with 80% recovery of its initial activity [30].

Polystyrene core-shell latex particles of submicron size were prepared by a seeded emulsion polymerization and composed of a hard polystyrene core surrounded by a hydrophilic shell consisted of crosslinked glycidyl methacrylate, acrylamide, and vinyl sulfonate [32]. Triazine dye, Cibacron Blue, was used as an affinity ligand and coupled directly to the latexes after the hydrolysis of surface epoxide groups. The degree of coupling was controlled by the ionic strength of the reaction medium. The adsorption of bovine serum albumin to the latex was proportional to the ligand density and reached 70% of the maximal within 1 min. Desorption was accomplished using 2 M NaCl. Complete desorption required up to 5 min. A relatively slow binding/desorption kinetics prevented using continuous crossflow systems. Centrifugal filtration with 0.2-μm pore diameter membrane was used instead. The functionalized core shell particles show a high colloidal stability and a low nonspecific adsorption.

Water-soluble polymers are often used as stabilizers of colloid systems [33]. The polymer-stabilized latex could be used as affinity latex when a target protein has affinity for the polymer stabilizer. For example, concanavalin A binds efficiently to poly(methyl methacrylate) latex synthesized using dextran as the polymer stabilizer [34]. The functional microspheres (2.0 to 7.7 μm diameter) for the reversible binding of endopectin-lyase were prepared by free-radical dispersion polymerization of styrene using polyacid Eudragit L 100–55 (1 : 1 statistical copolymer of methacrylic acid with methylmethacrylate) as the polymer stabilizer [35]. The enzyme binds to the microspheres at pH 4.7 and is eluted quantitatively and fast at pH 6.6. The separation of the microspheres was carried out using fast filtering system with 1.2-μm pore size. When bound to the microspheres, the enzyme retains about 60% activity, indicating that an active site is not involved directly in the binding process. Binding to the microspheres improves storage stability of the enzyme.

The examples presented undoubtedly show that affinity ultrafiltration could be successfully used for protein purification, provided the proper macroligand is developed, while the selection of ultrafiltration system presents no problem as many of them are available at the market. The recent achievements in the chemistry of microparticles resulted in the development of many macroligands capable of reversible binding/elution of a large variety of proteins. These macroligands have not been yet used in ultrafiltration mode but there are no plausible restrictions, why they could not be used for affinity ultrafiltration of proteins. The macroligands, which are suitable for but have not been used in ultrafiltration mode, are reviewed in the following section.

C. Potential Macroligands for Affinity Ultrafiltration

A straightforward way of macroligand synthesis is a direct covalent coupling of the ligand to appropriate microspheres or latexes. There are right now commercially available latexes with a variety of properties. For example, Bang Laboratories Inc (Fishers, IN) produces microspheres with sizes from 20 nm to 1 mm, carrying hydroxyl, sulfonate, amino, carboxy groups, or preactivated with chloromethyl, epoxy, and aldehyde surface groups. Preactivated latexes require only mixing with a ligand in a buffer of the optimum pH for covalent coupling. Parking area, i.e., area occupied by a single active group, varies from a few to a few hundred square Å, making possible the choice of the optimal ligand density of the macroligand and/or the surface properties with minimal nonspecific protein adsorption.

On top of the numerous latexes available for the synthesis of macroligands, quite a few tailor-made latexes with different ligands were developed. DNA oligomers were immobilized by coupling to epoxy-groups of the latexes comprised of styrene copolymer with glycidyl methacrylate crosslinked with divinylbenzene. The affinity latex was used for purification of DNA-binding transcription factor ATF/E4TF3 from HeLa cell crude nuclear extracts [36]. The latex particles with bound protein were separated by centrifugation and elution was achieved with 1.0 M KCl. At least eight polypeptides with molecular masses 116,000, 80,000, 65,000, 60,000, 55,000, 47,000, 45,000, and 43,000 were copurified. Affinity-purified polypeptides stimulated transcription *in vitro* from a promoter, in which ATF/E4TF3-binding sites were present [37]. Protein kinase II was later shown to copurify with ATF/E4TF3 factor [38]. The enzyme does not bind to the DNA latex but only to the ATF moiety of the ATF/E4TF3 factor.

Latex particles with a single stranded DNA were produced by coupling double-stranded DNA. More than 80% of the double-stranded DNA was coupled through the single-stranded protruding ends. The particles were then treated with alkali or heated to denature double-stranded DNA. The resulting single-stranded DNA particles allowed selective and efficient isolation of complementary RNA from total cellular RNA [39], or rapid and sensitive detection of specific genome DNA, separating normal and point mutant DNA [40,41].

TATA-binding protein from yeast was immobilized via spacer on the latex of the same structure by converting epoxy groups to amino, coupling ethyleneglycol diglycidylethter followed by its hydrolysis, and activating of hydroxy groups formed using tosyl chloride. The affinity latex was used for the purification of human transcription factor from an enriched protein fraction of HeLa cell nuclear extracts [42]. A common transacting factor, Ad4-binding protein, was purified from nuclear extracts of bovine adrenal cortex when polymerized Ad4 sequences were coupled to this latex [43]. A peptide Gly-Arg-Gly-Asp-Ser was coupled to the same latex via ethyleneglycol diglycidylethter spacer, and the affinity latex was used for purification of receptor proteins from octylglucoside extracts of human platelets. Western blotting with monoclonal antihuman platelet GP IIIa antibody indicated that the purified proteins were GP IIb/IIIa [44]. Arg-Gly-Asp-Ser-carrying latex was used not only for purification of the respective receptor from cell membrane, but also for cell activation [45].

Carboxylated polystyrene/poly(acrylamide) latexes were used for immobilization of bovine serum albumin (BSA) and purification of anti-BSA antibodies [46]. Centrifugation was used to separate latex particles from the supernatant in all the above-mentioned purification procedures.

Recently, Kawaguchi and coworkers [47] proposed a versatile technique for affinity latex production. Hydrophilic monodisperse latex particles are composed of acrylamide, methacrylic acid, methylenebisacrylamide (crosslinker), and nitrophenyl acrylate. The nitrophenyl acrylate presents a reactive moiety allowing a variety of chemical reactions for ligand coupling as well as direct coupling of biopolymers (protein, peptide, DNA) via amino groups (Figure 2).

The latexes are inherently stable; i.e., the latex suspension does not precipitate on gravity. High centrifugal forces are required to separate latex from a supernatant. The latexes composed of styrene, N-isopropylacrylamide and glycidyl methacrylate, or methacrylic acid showed thermosensitivity originated from the thermosensitive nature of N-isopropylacrylamide; that is, at the temperatures above critical, the latex particles flocculate and could be separated from the supernatant by precipitation on gravity or low-speed centrifugation. BSA was covalently immobilized on these latexes and used for purification of anti-BSA antibodies from antiserum. The increase in temperature to 40°C resulted in immediate flocculation of the latex with bound antibodies and easy separation from the supernatant [48].

Another approach for mechanical separation of latex particles from a supernatant was proposed by Kondo et al. [49]. Affinity latex was prepared by immobilizing human γ-globulin onto carboxylated latex composed of polystyrene and poly(acrylamide). A fusion protein (ZZB1B2) of IgG and albumin-binding domains was expressed in *Escherichia coli* and purified by the affinity latex. In a polyethylene glycol (PEG)/potassium phosphate aqueous two-phase system affinity latex with bound recombinant protein was partitioned into the PEG-rich phase, while cells and cell debris were displaced into the salt-rich bottom phase.

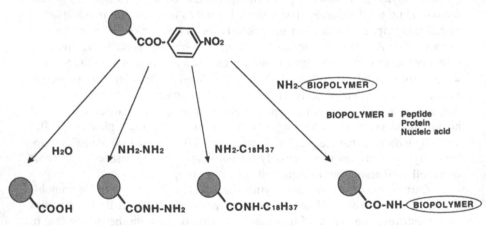

Figure 2 Ligand coupling to nitrophenyl acrylate-containing latex.

The separation of microspheres from supernatant could be achieved using magnetic properties of the microspheres produced of conventional Fe_3O_4 pigment covered with poly(vinyl alcohol). Immobilized m-aminophenylboronic acid allows specific binding of glycated Hb [50]. High density of the affinity particles formed by poly(vinyl alcohol)-coated perfluorocarbon emulsions allows separation by gravity but requires special construction of adsorber with uprising flow to keep the particles suspended. The continuous purification of malate dehydrogenase from unclarified homogenate of *Saccharomyces cerevisiae* was achieved when a triazine dye, Procion Red HE-7B, was immobilized on the perfluorocarbon particles [51].

The above-mentioned examples do not exhaust all the possible ways of latex synthesis and latex derivatization applicable for the production of affinity macroligands for protein purification. Apart of protein purification, affinity latexes are used widely for analytical purposes, especially in immunoanalysis [52], and less frequently, for immobilization of enzymes [53]. The synthetic procedures and ligand coupling to the latexes used for analytical purposes are reviewed by Kawaguchi [54].

Membranes are available in many materials and with almost any pore size. The choice of membrane is thus not a limitation. What has been limiting the development of affinity ultrafiltration has rather been lack of suitable ligand carriers and laborious and cumbersome coupling procedures. The recent development in polymer chemistry has dramatically changed that picture.

VII. CONCLUSION

Affinity ultrafiltration has not found yet a wide commercial application. It remains mainly the object of academic study. The commercial availability of ultrafiltration equipment of different designs and productivities combined with a variety of macroligands, which are already developed or could be easily synthesized using available carriers, make affinity ultrafiltration an attractive alternative to other protein purification techniques. The possibilities of quickly processing large solution volumes and easy scaleup that make the process automatic, and the feasibility of operating the same unit for purification of different proteins when using different membranes/macroligands constitute the obvious advantages of affinity ultrafiltration and justify hopes for its coming wide application in protein purification.

REFERENCES

1. Walsh G, Headon D, eds. Protein Biotechnology. Chichester: John Wiley and Sons, 1994.

2. Jones C, Patel A, Griffin S, Martin J, Young P, O'Donnell K, Silverman C, Porter T, Chaiken I. Current trends in molecular recognition and bioseparation. J Chromatogr A 707:3–22, 1995.

3. Nygren P-Å, Stål S, Uhlén M. Engineering proteins to facilitate bioprocessing. Trends Biotechnol 12:184–188, 1994.

4. Chase HA, Draeger NM. Affinity purification of proteins using expanded beds. J Chromatogr 597:129–145, 1992.

5. Gupta MN, Mattiasson B. Affinity precipitation. In: G Street, ed. Highly Selective Separations in Biotechnology. London: Blackie Academic & Professional, 1994:7–33.

6. Johansson G, Tjerneld F. Affinity partitioning. In: G Street, ed. Highly Selective Separations in Biotechnology. London: Blackie Academic & Professional, 1994:55–85.

7. Singh R. Industrial membrane separations. Chemtech 4:33–44, 1998.

8. Male KB, Nguen AL, Luong JHT. Isolation of urokinase by affinity ultrafiltration. Biotechnol Bioeng 35:87–93, 1990.

9. Kaul R, Mattiasson B. Affinity ultrafiltration for protein purification. In: T Ngo, ed. Molecular Interactions in Bioseparations. New York: Plenum Press, 1993:487–498.

10. Ling TGI, Mattiasson B. Membrane filtration affinity purification (MFAP) of dehydrogenases using Cibacron Blue. Biotechnol Bioeng 34:1321–1325, 1989.

11. Mattiasson B, Ramstorp M. Ultrafiltration affinity purification. Ann NY Acad Sci 413:307–310, 1983.

12. Mattiasson B, Ramstorp M. Ultrafiltration affinity purification. Isolation of concanavalin A from seeds of Canavalia ensiformis. J Chromatogr 283:323–330, 1984.

13. Mattiasson B, Ling TGI. Ultrafiltration affinity purification. A process for large-scale biospecific separations. In: WC McGregor, ed. Membrane Separations in Biotechnology. New York: Marcel Dekker, 1986:99–114.

14. Mattiasson B, Ling TGI, Nilsson JL. Ultrafiltration affinity purification. In: IM Chaiken, M Wilchek, I Parikh, eds. Affinity Chromatography and Biological Recognition. Orlando: Academic Press, 1986:223–227.

15. Luong JHT, Male KB, Nguen AL. Synthesis and characterization of a water-soluble affinity polymer for trypsin purification. Biotechnol Bioeng 31:439–446, 1988.

16. Somers W, Van't Riet K. Downstream processing of α-amylase by affinity interactions. In: C Christiansen, L Munck, J Willadsen, eds. Proceeding from 5th European Congress on Biotechnology, Vol. 2, Copenhagen, July 8–13, 1990. Copenhagen: Munksgaard International Publishers, 1990:774–777.

17. Herak DC, Merrill EW. Affinity cross-flow filtration: experimental and modeling work using the system of HSA and Cibacron Blue-agarose. Biotechnol Progr 5:9–17, 1989.

18. Kumar A, Galaev IY, Mattiasson B. Affinity precipitation of alpha-amylase inhibitor from wheat meal by metal chelate affinity binding using Cu(II)-loaded copolymers of 1-vinylimidazole with N-isopropylacrylamide. Biotechnol Bioeng 59:619–624, 1998.

19. Nylén U. Swedish patent application SE 8006102; European Patent Application, publication number 0,046,915 (1980).

20. Choe TB, Masse P, Verdier A. Separation of trypsin from trypsin α-chymotrypsin mixture by affinity ultrafiltration. Biotechnol Lett 8:163–168, 1986.

21. Adamski-Medda D, Trong NQ, Dellacherie E. Biospecific ultrafiltration: a promising purification technique for proteins? J Membr Sci 9:337–342, 1981.

22. Hubert P, Dellacherie E. Use of water soluble biospecific polymers for the purification of proteins. J Chromatogr 184:325–333, 1980.

23. Sigmundsson K, Filippusson H. Synthesis and characterization of an acrylamide-based water soluble affinity polymer for trypsin purification. Polymer Int 14:355–362, 1996.

24. Luong JHT, Nguen AL. Novel separations based on affinity interactions. Adv Biochem Eng Biotechnol 47:138–158, 1992.

25. Luong JHT, Male KB, Nguen AL. A continuous affinity ultrafiltration process for trypsin purification. Biotechnol Bioeng 31:516–520, 1988.

26. Luong JHT, Male KB, Nguen AL, Mulchandani A. Affinity ultrafiltration for purifying specialty chemicals. In: J Gavora, DF Gerson, J Luong, A Storer, JH Woodley, eds. Biotechnology Research and Applications. New York: Elsevier, 1988:79–93.

27. Stinson SC. Delving into dendrimers. Chem Eng News Sept 22:28–30, 1997.

28. Pungor E, Afeyan NB, Gordon NF, Cooney CL. Continuous affinity-recycle extraction: A novel protein technique. Bio/Technology 5:604–608, 1987.

29. Powers JD, Kilpatrick PK, Carbonell RG. Protein purification by affinity binding to unilamellar vesicles. Biotechnol Bioeng 33:173–182, 1989.

30. Powers JD, Kilpatrick PK, Carbonell RG. Purification of avidin from egg whites using affinity-modified unilamellar liposomes. In: DA Butterfield, ed. Progress in Clinical and Biological Research. New York: Alan R. Liss, 1989:173–182.

31. Powers JD, Kilpatrick PK, Carbonell RG. Trypsin purification by affinity binding to small unilamellar liposomes. Biotechnol Bioeng 36:506–519, 1990.

32. Gebben B, van Houwelingen GDB, Zhang W, vanden Boomgaard T, Smolders CA. Protein separation using affinity binding. 1. Polystyrene core-shell latex as ligand carrier. Colloids Surfaces B: Biointerfaces 3:75–84, 1994.

33. Sato T, Ruch R, eds. Stabilization of Colloidal Dispersions by Polymer Adsorption. New York: Marcell Dekker, 1980.

34. Chern CS, Lee CK, Tsai YJ. Dextran stabilized poly(methyl methacrylate) latex particles and their potential application for affinity purification of lectins. Colloid Polym Sci 275:841–849, 1997.

35. Dinella C, Doria M, Laus M, Lanzarini G. Reversible adsorption of endopectin-lyase to tailor-made core-shell microspheres prepared by dispersion polymerization. Biotechnol Appl Biochem 23:133–140, 1996.

36. Kawaguchi H, Akira A, Ohtsuka Y, Watanabe H, Wada T, Hiroshi H. Purification of DNA-binding transcrition factors by their selective adsorption on the affinity latex particles. Nucleic Acid Res 17:6229–6240, 1989.

37. Inomata Y, Kawaguchi H, Hiramoto M, Wada T, Handa H. Direct purification of multiple ATF/E4TF3 polypeptides from HeLa cell crude nuclear extracts using DANA affinity latex particles. Anal Biochem 206:109–114, 1992.

38. Wada T, Takagi T, Yamaguchi Y, Kawase H, Hiramoto M, Ferdous A, Takayama M, Lee KAW, Hurst HC, Handa H. Copurification of casein kinase II with transcriptionfactor ATF/E4TF3. Nucleic Acids Res 24:876–884, 1996.

39. Imai T, Sumi Y, Hatakeyama M, Fujimoto K, Kawaguchi H, Hayashida N, Shiozaki K, Terada K, Yajima H, Handa H. Selective isolation of DNA or RNA using single-stranded DNA affinity latex particles. J Colloid Interface Sci 177:245–249, 1996.

40. Hatakeyama M, Iwato S, Fujimoto K, Handa H, Kawaguchi H. DNA-carrying latex particles for DNA diagnosis. 1. Separation of normal and point mutant DNA ac-

cording to the difference in hybridization efficiency. Colloids Surfaces B: Biointerfaces 10:161–169, 1998.

41. Hatakeyama M, Nakamura K, Iwato S, Fujimoto K, Handa H, Kawaguchi H. DNA-carrying latex particles for DNA diagnosis. 2. Distinction of normal and point mutant DNA using S1 nuclease. Colloids Surfaces B: Biointerfaces 10:171–178, 1998.

42. Shiroya T, Hatakeyama M, Fujimoto K, Watanabe H, Wada T, Imai T, Kawaguchi H, Handa H. Purification of transcription factor IIA using yeast tata-binding protein affinity latex particles. Int J Bio-Chromatogr 1:191–198, 1995.

43. Morohashi K, Honda S, Inomata Y, Handa H, Omura T. A common transacting factor, Ad4-binding protein, to the promoters of steroidogenic P-450s. J Biol Chem 267:17913–17919, 1992.

44. Inomata Y, Kasuya Y, Fujimoto K, Handa H, Kawaguchi H. Purification of membrane receptors with peptide-carrying affinity latex particles. Colloid Surfaces B: Biointerfaces 4:231–241, 1995.

45. Kawaguchi H. Functionalization of polymer particles—focusing on affinity latices. Macromol Symp 101:501–508, 1996.

46. Kondo A, Yamasaki R, Higashitani K. Affinity purification of antibodies using immunomicrospheres. J Ferment Bioeng 74:226–229, 1992.

47. Kawaguchi H, Fujimoto K, Nakazawa Y, Sakagawa M, Ariyoshi Y, Shidara M, Okazaki H, Ebisawa Y. Modification and functionalization of hydrogel microspheres. Colloids Surfaces A: Physicochem Eng Aspects 109:147–154, 1996.

48. Kondo A, Kaneko T, Higashitani K. Development and application of thermo-sensitive immunomicrospheres for antibody purification. Biotechnol Bioeng 44:1–6, 1994.

49. Kondo A, Kaneko T, Higashitani K. Purification of fusion proteins using affinity microspheres in aqueous two-phase systems. Appl Microbiol Biotechnol 40:365–369, 1993.

50. Mueller-Schulte D, Brunner H. Novel magnetic microspheres on the basis of poly(vinyl alcohol) as affinity medium for quantitative detection of glycated Hb. J Chromatogr A 711:53–60, 1995

51. Owen RO, McCreath GE, Chase HA. A new approach to continious counter-current protein chromatography: direct purification of malate dehydrogenase from a *Saccharomyces cerevisiae* homogenate as a model system. Biotechnol Bioeng 53:427–441, 1997.

52. Bastos-Gonzales D, Ortega-Vinuesa JL, DeLasNieves FJ, Hidalgo-Alvarez R. Carboxylated latexes for covalent coupling antibodies. J Colloid Interface Sci 176:232–239, 1995.

53. Yunus WMZW, Salleh AB, Basri M, Ampon K, Razak CNA. Preparartion and immobilization of lipase on to poly(methyl acrylate–methyl methacrylate–divinylbenzene) beads for lipid hydrolysis. Biotechnol Appl Biochem 24:19–23, 1996.

54. Kawaguchi H. Polymer materials for analysis and bioseparation. 5.1. Microspheres for diagnosis and bioseparation. In: T Tsuruta, ed. Biomedical Application of Polymeric Materials. Boca Raton, FL: CRC, 1993:299–324.

55. Male KB, Luong JHT, Nguen AL. Studies on the application of a newly synthesized polymer for trypsin purification. Enzyme Microb Technol 9:374–378, 1987.

56. Chen J-P. Novel affinity based process for protein purification. J Ferment Bioeng 70:199–209, 1990.

57. Herak DC, Merrill EW. Affinity cross-flow filtration: some new aspects. Biotechnol Progr 6:33–40, 1989.

9

Process Development for the Isolation of a Recombinant Immunofusion Protein Using a Membrane Adsorber

William K. Wang
SmithKline Beecham Pharmaceuticals, King of Prussia, Pennsylvania

Harold G. Monbouquette
University of California, Los Angeles, California

W. Courtney McGregor
XOMA Corporation, Santa Monica, California

I. INTRODUCTION

In moving protein purification processes from the bench to production scale, the focus must shift increasingly to practical issues related to economics. More rapid and efficient bioseparation processes that operate at low pressure provide savings in both capital and operating costs. Also, faster protein product isolation early in the purification process can improve product yield and cost-effectiveness by limiting proteolysis and denaturation. Toward these processing goals, significant advances have been made in analytical to preparative level column chromatography through the use of more size-monodisperse and smaller packings which minimize dispersion and reduce diffusion times to interior binding sites within gel particles. However, as column methods are scaled, constant performance typically comes at an increasingly alarming cost due to the high price of monodisperse, small bead packings, and the high pressure drop across even shallow beds of these particles [1–3]. Alternatives to standard packings and column geometries have been proposed and marketed in recent years including perfusion chromatography [4] and membrane chromatography [5–7] which hold promise for improved scalability and economic viability.

The primary focus in the development of alternative geometry adsorptive systems has been on the reduction or elimination of diffusional mass-transfer resistance to solute binding. Bead packings have been redesigned with macropores that enable some convective transport to the particle interior [4], or pores have been eliminated altogether [8]. In the most radical departure from convention, planar [2,9–28], wound (for radial flow) [5,10,29–33], and hollow-fiber [1,6,33–38] membrane adsorptive systems have been designed and evaluated. In these systems, ion-exchange [5,9,10,12,14–25,27–33,37–39], hydrophobic [12,14,20,21], or affinity [2,6,11,13,18,26,34–36,40–49] ligands are attached to polymeric membranes or fibrous mats, and transport of solutes to binding sites occurs primarily by convection [2,6,11,35,40]. Elimination of diffusional resistance usually leaves a system controlled by much faster binding kinetics, thereby enabling adsorptive separation of proteins in typically one-tenth the time common for packed columns [2,5,17,28]. Many membrane configurations exhibit high binding capacities, similar in magnitude to packed columns [5,12–15,18,21,29]. Using stepwise elution, proteins can rapidly be concentrated 10-fold or more at recoveries of 85% to 100% [5]. Using stacks of sufficient numbers of membranes and gradient elution methods, analytical separations of proteins can be achieved equivalent to column chromatography [16,20]. The linear scalability of the membrane systems further adds to the attractiveness of the technology [5,18,24, 28,32,39]. More detailed discussion regarding membrane adsorption chromatography can be found in review articles by Thommes and Kula [50], Charcosset [51], and Zeng and Ruckenstein [52].

In this study, ion-exchange columns and membrane adsorbers were first evaluated in parallel for recovery and concentration of cytochrome c [53]. Based on these results, the isolation of a 51-kDa targeted immunofusion (TIF) protein produced extracellularly by recombinant $E.$ $coli$ subsequently was pursued to the pilot scale using a recently developed membrane adsorber product of Sartorius (Goettingen, Germany). The proprietary membranes used in the Sartorius units were constructed from regenerated crosslinked cellulose with a pore size range of 3 to 5 μm. The sulfonic acid active groups were located in the pores with an optimized functional group density. TIF consists of a CD5 binding moiety and a toxin moiety. Such engineered immunofusion molecules are candidates for treatment of immune disorders and malignancies. Initial rapid isolation of the protein from the culture medium is critical, as TIF undergoes proteolytic degradation in the culture broth. The need for a rapid, cost-effective technique to concentrate and purify recombinant protein makes this an ideal test case for industrial application of membrane adsorber technology. Cost-effectiveness of membrane adsorbers will likely be dependent on membrane reusability which, in this evaluation, required repeated membrane exposure to a cycle of use, washing, sanitization in NaOH solution, and reequilibration in buffer.

II. MATERIALS AND METHODS

A. Determination of Cytochrome c Loading Capacities for Different Chromatography Media

The cytochrome c (from horse heart; Sigma) loading capacities were determined for four different media (Table 1). All studies were done at room temperature using a peristaltic pump (Masterflex 7520–25, 7016–20 pump head). All solutions were clarified with a 0.45-μm filter unit (Nalgene) prior to application to the membrane adsorbers or columns. Each unit was equilibrated in 10 mM NaPhos, pH 7.0, until the eluate pH and conductivity matched that of the equilibration buffer. A 10 mM NaPhos, pH 7.0, 0.2 mg/mL cytochrome c solution was loaded onto each unit. Loading continued until the permeate absorbance (A_{280}) reached a level equivalent to 80% to 90% of the original loading solution. The units were washed with 10 mM NaPhos, pH 7.0, until the A_{280} returned to baseline. Cytochrome c was eluted from the units with 10 mM NaPhos, 1 M KCl, pH 7.0.

B. Protein Assays

All cytochrome c fractions were assayed using the Bradford protein assay kit (Bio-Rad). TIF concentrations were determined by a quantitative HPLC assay using gradient elution from an MA7S strong cation-exchange column on a Hewlett-Packard 1050 HPLC system.

C. SDS-PAGE

Protein samples were analyzed by sodium dodecyl sulfate–polyacrylamide gel electrophoresis (SDS-PAGE) by the method of Laemmli [54].

Table 1 Characteristics of Ion-Exchange Units

Unit	Type	Active group	ID (mm)	Length (mm)	Bed Volume (mL)
SP MemSep 1000	Membrane	Sulfopropyl	19	5	1.4
Resource S	Column	Methyl sulfonate	6.4	30	1.0
Poros SP	Column	Sulfopropyl	4.6	50	0.8
S15	Membrane	Sulfonic acid	25.2	0.6	0.3

D. TIF Isolation Using S15 and S100 Membrane Adsorbers

The S15 and S100 membrane adsorbers (15 and 100 cm^2 of surface area; Sartorius) were equilibrated in 10 mM NaPhos, 0.02 M NaCl, pH 7.0, until the eluate pH and conductivity matched that of the equilibration buffer. TIF loading samples were diluted to below 0.02 M NaCl with WFI prior to loading. Solutions were clarified by filtration through a 0.45-μm filter unit (Nalgene). Following loading, the membrane adsorbers were washed with 10 mM NaPhos, 0.02 M NaCl, pH 7.0, until the A_{280} baseline had stabilized. To minimize the elution volume, the TIF was eluted from the units with 10 mM NaPhos, 0.6 M NaCl, pH 7.0, using a syringe. The membranes were stripped with 10 mM NaPhos, 2 M NaCl, pH 7.0; sanitized in 1.0 M NaOH, 0.5 M NaCl; and stored in 50 mM NaOH after each TIF run.

E. TIF Isolation Using an S11K Membrane Adsorber

For large-scale TIF purification, a membrane adsorber with an effective membrane surface area of 11,200 cm^2 was constructed using a stainless-steel filter holder (Sartorius model SM16277) and twenty 29.3-cm-diameter S membrane adsorbers (Sartorius). At the end of a fermentation, the TIF-containing broth was clarified through a microfiltration unit (Microgon Kros Flo II, 0.2 μm), but not concentrated or buffer-exchanged. Approximately 20 L of broth was diluted 10-fold with WFI prior to loading with a Masterflex (77300-00, H-07529-10 pump head) or Watson-Marlow 710U/R peristaltic pump. The pH of the starting material was 6.2 to 6.4, and the conductivity was 22 to 26 mS/cm. Following dilution, the sample load conductivity was between 3.1 and 3.9 mS/cm, and the pH remained the same. The S11K unit was equilibrated in 10 mM NaPhos, 0.02 M NaCl, pH 7.0, until the eluate pH and conductivity matched that of the equilibration buffer. Following loading, the unit was washed with at least 65 L (>200 times the holdup volume) of equilibration buffer until well after the A_{280} baseline had returned to zero. The sample was then eluted with 10 mM NaPhos, 0.6 M NaCl, pH 7.0. Following elution, the S11K was stripped and sanitized with 10 L of 1.0 M NaOH, 0.5 M NaCl. The membrane adsorber then was flushed and stored in 50 mM NaOH before a subsequent TIF run.

III. RESULTS AND DISCUSSION

A. Cytochrome c Loading Capacities and Evaluation of Adsorptive Media

At harvest time, TIF must be recovered quickly to stem proteolytic degradation. Four ion-exchange adsorption units were identified as candidates for rapid TIF isolation. The units included two membrane adsorbers (SP-MemSep 1000, Millipore; and S15 membrane adsorber, Sartorius) and two columns prepacked with media designed for increased flow rate (Resource S, Pharmacia; and P series Poros SP, PerSeptive Biosystems). Each of these units incorporates strong cation exchange sulfonate ligands and they are available in ready-to-use packages (Table 1).

The 12.4-kDa-protein cytochrome c was used to compare directly the performance of these units. At least two runs were carried out for each unit. The cytochrome c dynamic loading capacities and the maximum linear velocities (i.e., volumetric fluxes) for each of the units are shown in Table 2. The dynamic loading capacity was defined as the maximum amount of protein that could be bound at a given flow rate to the membrane adsorber before breakthrough per adsorber unit volume. Note that (with the exception of the S15 membrane adsorber, where no maximum flow rate was determined) the maximum velocity was attained at the maximum operating pressure of the peristaltic pump and tubing used, not necessarily at the performance limitation of the adsorber unit. As shown in Table 2, the S15 dynamic capacity for cytochrome c (115 to 124 mg/mL) was almost three times that of the next best unit (Resource S, 39 mg/mL). Further, the velocity could have been increased beyond 10.0 cm/min for the S15 without reaching the maximum operating pressure of the pump, although the capture of cytochrome c decreased slightly (5%) with the increase in velocity from 4 to 10 cm/min. The second highest linear velocity of 5.9 cm/min was achieved with the Poros SP column, although the loading capacity of the column was significantly lower than the other units at 4.4 mg/mL. In terms of volumetric flow rates, the S15 membrane adsorber, with the largest cross-sectional area, was able to process 10 times more fluid volume than any of the other units per unit time.

Table 2 Cytochrome c Dynamic Loading Capacities

Unit	Velocity (cm/min)	Loading capacity (mg/mL)	Number of loads
SP-MemSep 1000	1.1	28	2
Resource S	5.5	39	2
Poros SP	5.9	4.4	2
S15	2.0	124	3
S15	4.0	121	3
S15	10.0	115	3

Reproduced with permission from Biopharm, Vol. 8, Number 5, June 1995, page 54. Copyright by Advanstar Communications Inc. Advanstar Communications Inc. retains all rights to this article.

The breakthrough curves for the S15 membrane adsorber were reproducible over repeated runs, indicating that performance is not diminished with use in the short term. In addition, the breakthrough curve was steep, the result of maximum loading being achieved with minimal flowthrough losses. The Resource S and Poros SP had comparable, steep breakthrough curves. The SP-MemSep 1000 had a shallower breakthrough curve, with more protein lost during loading. Given these results, the S15 membrane adsorber was selected for development of a rapid TIF isolation process at low pressure and high volumetric throughput.

B. TIF Isolation Using Membrane Adsorbers

Purified TIF at 0.47 mg/mL was loaded at 1.0 cm/min (5 mL/min) onto an S15 membrane adsorber equilibrated with 10 mM NaPhos, 0.02 M NaCl, pH 7.0. Six 6-mL elutions at sequentially increased salt concentrations of 0.15, 0.3, 0.6, 1.0, 1.5, and 2.0 M NaCl in 10 mM NaPhos, pH 7.0, were performed with a syringe. Approximately 94% of the 6-mg TIF load was recovered in the 0.15 to 0.6 M NaCl fractions. Of a second 3.7-mg loading of purified TIF on the same membrane adsorber, 99% was recovered in 6 mL of a 0.6 M NaCl eluate. The SDS-PAGE gel of Figure 1 shows little or no 51-kDa TIF in the permeate during

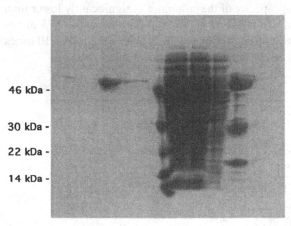

Figure 1 A 14% Coomassie Blue–stained SDS-PAGE gel illustrating the performance and separation achieved for a purified TIF load (lanes 1–4) and for a crude preparation (lanes 6–10). The gel shows protein bands in the loaded solutions (lanes 1 and 6), the flowthrough permeate and wash (lanes 2, 7, and 8), the 0.6 M NaCl eluate (lanes 3 and 9), and the 2.0 M NaCl strip solutions (lanes 4 and 10). The bands in lane 5 are molecular-weight standards.

loading (i.e., the flowthrough) or in the subsequent washes with the buffer to remove loosely bound protein; however, TIF appears highly concentrated in the eluate (see Figure 1, lane 3).

After the pH and ionic strength conditions for TIF binding and elution were established, the S15 membrane adsorber was evaluated with crude TIF prepared from the recombinant *E. coli* fermentation broth. The broth was clarified by microfiltration. Several fermentation lots were concentrated and buffer-exchanged using a diafiltration unit. In a representative run, this crude preparation was loaded at 1.2 cm/min. In this buffer-exchanged preparation, the salt concentration was 10 mM sodium phosphate at pH 7.0 and did not include 20 mM NaCl as used above. Although a number of additional proteins were present, of course, in this crude preparation, TIF clearly is concentrated in the 0.6 M NaCl eluate (see Figure 1, lane 9). A relatively small amount of the 51-kDa protein is evident in the 2.0 M NaCl strip solution.

In a representative run, a crude TIF harvest (lot A) which had been clarified with a 0.45-μm filter, buffer-exchanged, and concentrated was loaded at a velocity of 3.6 cm/min (18 mL/min). Approximately 81% of the protein was recovered with a 25-fold increase in concentration (Table 3). According to SDS-PAGE, the major protein species in the eluate was TIF (Figure 2). The relatively low estimated purity (53%) reflects the fact that the sample had been concentrated and diafiltered prior to loading, during which time TIF degradation products accumulated.

The high binding capacity of these membrane adsorbers accounts for much of the performance of the system for rapid TIF isolation. In a separate study, the TIF loading capacity of the S15 membrane adsorber was determined. A purified TIF solution was loaded until the A_{280} trace of the flowthrough solution increased above baseline, indicating breakthrough. Following washing with the equilibration buffer, 13.5 mg of TIF was eluted at 0.6 M NaCl. The S membrane adsorber, therefore, had an dynamic loading capacity of 0.9 mg TIF/cm². Using an S100

Table 3 TIF Isolation with the S Membrane Adsorber from the Bench to Pilot Scale

Sample lot	Membrane adsorber	Load (mg TIF)	Eluate (mg TIF)	% TIF recovery	Concentration factor	% TIF purity
A	S15	10.5	8.5	81	25	53
B	S100	81.8	56.6	69	24	73
2	S11K	962	895	93	9	73
4	S11K	624	490	79	10	73
5	S11K	552	664	120	10	89

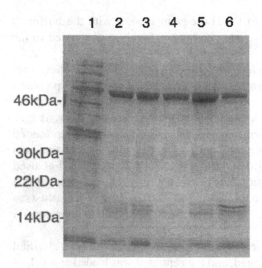

Figure 2 TIF eluates from the S membrane adsorbers. Aliquots of TIF eluate samples were examined by 14% SDS-PAGE gel electrophoresis under reducing conditions, stained with Coomassie Blue. Lane 1 corresponds to a typical crude preparation loaded on the membrane adsorbers. Lanes 2 through 4 show TIF isolated with the S11K membrane adsorber. Lanes 5 and 6 contain the S100 and S15 membrane adsorber eluates, respectively. The major protein species in each sample is the 51-kDa TIF band. The protein patterns for each sample are comparable regardless of the membrane adsorber used or the TIF lot applied. Reproduced with permission from Biopharm, Vol. 8, Number 5, June 1995, page 55. Copyright by Advanstar Communications Inc. Advanstar Communications Inc. retains all rights to this article.

membrane adsorber the equilibrium loading capacity was determined. A purified TIF solution was loaded until the A_{280} trace of the eluate rose to the level of the inlet, indicating membrane saturation. Following washing with the equilibration buffer, an average equilibrium loading capacity for the membrane adsorber was calculated to be 1.0 mg TIF/cm^2.

The same operating conditions were used to scaleup from the S15 membrane adsorber to the S100 unit. The two units differ in the following respect: the S100 unit contains a total of 100 cm^2 of membrane surface area in a five-membrane stack, while the S15 unit harbors a total of 15 cm^2 of membrane surface area in a three-membrane stack. A lower velocity of 3 cm/min (60 mL/min) was employed throughout the equilibration, loading, and washing steps with the S100 unit, reflecting limitations of the pumping equipment rather than the adsorber unit.

To simplify the TIF recovery process and to improve yield, the sample (lot B) loaded on the S100 membrane adsorber was clarified, diluted 10-fold with

WFI (to reduce NaCl concentration), and pH-adjusted to 7.0, but not concentrated or buffer-exchanged. In this way, time for product degradation was minimized. The salt concentration in the loaded TIF sample was about 0.015 M NaCl. After loading and washing, the TIF was eluted with 60 mL by syringe to minimize elution volume.

HPLC assays and SDS-PAGE of the eluate samples (Table 3 and Figure 2) indicated that the membrane adsorber TIF isolation was scaled successfully from the S15 to the S100. The SDS-PAGE protein pattern was similar to that observed earlier with the S15 unit, and the purity of the eluate was estimated at 73%. There was no detectable TIF contained in the flowthrough or wash fractions. Note that the HPLC assay may be limited by the low TIF concentrations relative to the high levels of protein impurities in the loading samples, and that the lower recovery of 69% for the S100 may be in error by as much as 20%.

C. Pilot-Scale Isolation of TIF Using a Membrane Adsorber

For large-scale TIF isolation, an S11K membrane adsorber which had a total usable surface area of 11,200 cm^2 was tested. This S11K unit consisted of a 293-mm-diameter stainless-steel filter holder into which a stack of 20 S membrane adsorber sheets had been inserted. Each membrane had an effective surface area of 560 cm^2. Although this membrane stack provided excess loading capacity, usage of 20 membrane layers was suggested by the manufacturer to minimize the possibility of leakage from the filter holder during operation. In the initial studies, the system was operated at a conservative 2 cm/min velocity (1.1 L/min). A later run (TIF lot 5) was carried out successfully at an increased velocity of 5.4 cm/min (3.0 L/min).

The performance of the same S11K membrane stack was evaluated three times with TIF from three different fermentation lots. A total volume of about 20 L of clarified fermentation broth was diluted 10-fold and loaded on the S11K. In an effort to further improve the efficiency of the process, the crude TIF was not pH-adjusted from 6.2–6.4 to 7.0. Following washing with equilibration buffer, the TIF was eluted. The elution profile consistently showed a sharp, symmetrical elution peak for each of the three lots (Figure 3). A 10-fold concentration of TIF was obtained in the eluate. The concentration factors were not as high with the S11K membrane adsorber as with the S15 and S100 units, since elution was performed with a pump rather than a syringe and there was significant excess capacity.

The quantitative results for the S15, S100, and S11K systems are summarized in Table 3. For all lots, there was essentially no TIF detected in the flowthrough or wash fractions assayed. The 120% recovery for lot 5 reflects the uncertainty of the HPLC quantitation assay. The SDS-PAGE protein patterns for the bench to pilot-scale isolations essentially were identical (Figure 2), demonstrating that similar performance was achieved within the operating conditions

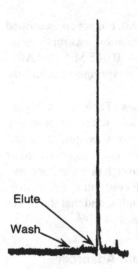

Figure 3 Process chromatograph for the isolation of TIF lot 4 by the S11K membrane adsorber showing the wash and elution steps. The S11K membrane adsorber was loaded at 2 cm/min (1.14 L/min), washed with 10 mM NaPhos, 0.02 M NaCl, pH 7.0, and 10 mM NaPhos, 0.03 M NaCl, pH 7.0, and eluted with 10 mM NaPhos, 0.6 M NaCl, pH 7.0. Reproduced with permission from Biopharm, Vol. 8, Number 5, June 1995, page 56. Copyright by Advanstar Communications Inc. Advanstar Communications Inc. retains all rights to this article.

of this study regardless of adsorber size. All eluate samples contain predominantly TIF. For the streamlined procedure used at larger scale, the percentage of TIF was always > 70%. A complete cycle for processing 200 L at a flow rate of 1 L/min of clarified diluted fermentation broth with the S11K including equilibration, loading, washing, elution, and regeneration was completed in < 6 hours. Increasing the flow rate to 3 L/min reduced the complete process cycle to ~ 2 hr. As shown in Table 3, the performance of the S11K membrane adsorber was comparable at 1 L/min (lots 2 and 4) and 3 L/min (lot 5).

IV. CONCLUSIONS

Membrane chromatography for protein purification is a technology that has been used successfully in a wide range of applications with a variety of membrane geometries and interaction modes [50–52]. For this particular application, the apparent advantages of membrane adsorbers at this stage of development are still rather qualitative. The Sartorius S15 membrane adsorber was determined to be an attractive alternative to conventional adsorption column chromatography for the isolation of a recombinant immunofusion protein, TIF. The convenience of use, short setup time, low-pressure operability, high capacity, and high volumetric velocity capability are features of this separations technology that are likely to be transferable to the isolation of other proteins by ion exchange. Significant time savings result from the quick setup, which amounts to connecting tubing and loading a peristaltic pump head, and the high volumetric throughput of these membranes. In this work, the membrane adsorber performance was reproducible over multiple cycles with depyrogenation in NaOH solution between runs. Sufficient chemical and thermal stability for repeated sanitization is, of course, a critical requirement for economical scaleup in the biopharmaceuticals industry.

ACKNOWLEDGMENTS

The authors wish to thank Shau-Ping Lei for project support, George Baklayan for the spectrophotometric and HPLC results and HPLC assays, and Wilfredo Morales and Linda Lai for providing the TIF fermentation broth. The ion-exchange units used were kindly donated by Sartorius (Goettingen, Germany).

REFERENCES

1. Ding H, Yang MC, Schisla D, Cussler EL. Hollow-fiber liquid chromatography. AIChE J 35:814–820, 1989.
2. Briefs KG, Kula MR. Fast protein chromatography on analytical and preparative scale using modified microporous membranes. Chem Eng Sci 47:141–149, 1992.
3. Frey DD, Van de Water R, Zhang B. Dispersion in stacked-membrane chromatography. J Chromatogr 603:43–47, 1992.
4. Afeyan NB, Fulton SP, Gordon NF, Mazsaroff I, Várady L, Regnier FE. Perfusion chromatography: an approach to purifying biomolecules. Bio/Technol 8:203–206, 1990.
5. McGregor WC, Szesko DP, Mandaro RM, Rai VR. High performance isolation of a recombinant protein on composite ion exchange media. Bio/Technol 4:526–527, 1986.
6. Brandt S, Goffe RA, Kessler SB, O'Connor JL, Zale SE. Membrane-based affinity technology for commercial scale purifications. Bio/Technol 6:779–782, 1988.
7. Reif OW, Freitag R. Comparison of membrane adsorber (MA) based purification schemes for the down-stream processing of recombinant h-AT III. Bioseparation 4: 369–381, 1994.
8. Burke DJ, Duncan JK, Dunn LC, Cummings L, Siebert CJ, Ott GS. Rapid protein profiling with a novel anion-exchange material. J Chromatogr 353:425–437, 1986.
9. Church FC, Whinna HC. Rapid sulfopropyl-disk chromatographic purification of bovine and human thrombin. Anal Biochem 157:77–83, 1986.

10. Upshall A, Kumar AA, Bailey MC, Parker MD, Favreau MA, Lewison KP, Joseph ML, Maraganore JM, McKnight GL. Secretion of active human tissue plasminogen activator from the filamentous fungus *Apergillus nidulans*. Bio/Technol 5:1301–1304, 1987.

11. Unarska M, Davies PA, Esnouf MP, Bellhouse BJ. Comparative study of reaction kinetics in membrane and agarose bead affinity systems. J Chromatogr 519:53–67, 1990.

12. Tennikova TB, Belenkii BG, Svec F. High-performance membrane chromatography, a novel method of protein separation. J Liq Chromatogr 13:63–70, 1990.

13. Champluvier B, Kula MR. Microfiltration membranes as pseudo-affinity adsorbents: modification and comparison with gel beads. J Chromatogr 539:315–325, 1991.

14. Tennikova TB, Bleha M, Svec F, Almazova TV, Belenkii BG. High-performance membrane chromatography of proteins, a novel method of protein separation. J Chromatogr 555:97–107, 1991.

15. Gerstner JA, Hamilton R, Cramer SM. Membrane chromatographic systems for high-throughput protein separations. J Chromatogr 596:173–180, 1992.

16. Josic D, Reusch J, Löster K, Baum O, Reutter W. High-performance membrane chromatography of serum and plasma membrane proteins. J Chromatogr 590:59–76, 1992.

17. Prpic V,Uhing RJ, Gettys TW. Separation and assay of phosphodiesterase isoforms in murine peritoneal macrophages using membrane matrix DEAE chromatography and [^{32}P]cAMP. Anal Biochem 208:150–160, 1993.

18. Lütkemeyer D, Bretschneider M, Büntemeyer H, Lehmann J. Membrane chromatography for rapid purification of recombinant antithrombin III and monoclonal antibodies from cell culture supernatant. J Chromatogr 639:57–66, 1993.

19. Reif OW, Freitag R. Characterization and application of strong ion-exchange membrane adsorbers as stationary phases in high-performance liquid chromatography of proteins. J Chromatogr A 654:29–41, 1993.

20. Tennikova TB, Svec F. High-performance membrane chromatography: highly efficient separation method for proteins in ion-exchange, hydrophobic interaction and reversed-phase modes. J Chromatogr 646:279–288, 1993.

21. Luksa J, Menart V, Milicic S, Kus B, Gaberc-Porekar V, Josic D. Purification of human tumour necrosis factor by membrane chromatography. J Chromatogr A 661: 161–168, 1994.

22. Santarelli X, Domergue F, Clofent-Sanchez G, Dabadie M, Grissely R, Cassagne C. Characterization and application of new macroporous membrane ion exchangers. J Chromatogr B 706:13–22, 1998.

23. Agrawal A, Burns MA. Application of membrane-based preferential transport to whole broth processing. Biotechnol Bioeng 55:581–591, 1997.

24. Gebauer KH, Thommes J, Kula M-R. Membrane chromatography: performance and scale-up. Chimia 50:422–423, 1996.

25. Sloboda RD, Belfi LM. Purification of tubulin and microtubule-associated proteins by membrane ion-exchange chromatography. Protein Expr Purif 13:205–209, 1998.

26. Dancette OP, Taboureau J-L, Tournier E, Charcosset C, Blond P. Purification of immunoglobulins G by protein A/G affinity membrane chromatography. J Chromatogr B 723:61–68, 1999.

27. Gebauer KH, Thommes J, Kula MR. Breakthrough performance of high-capacity membrane adsorbers in protein chromatography. Chem Eng Sci 52:405–419, 1997.
28. Gebauer KH, Thommes J, Kula M-R. Plasma protein fractionation with advanced membrane adsorbents. Biotechnol Bioeng 54:181–189, 1997.
29. Hou KC, Mandaro RM. Bioseparation by ion exchange cartridge chromatography. BioTechniques 4:358–367, 1986.
30. Mandaro RM, Hou KC. Filtration supports for affinity separation. Bio/Technol 5: 928–932, 1987.
31. Menozzi FD, Vanderpoorten P, Dejaiffe C, Miller AOA. One-step purification of mouse monoclonal antibodies by mass ion exchange chromatography on Zetaprep. J Immun Methods 99:229–233, 1987.
32. Jungbauer A, Unterluggauer F, Uhl K, Buchacher A, Steindl F, Pettauer D, Wenisch E. Scaleup of monoclonal antibody purification using radial streaming ion exchange chromatography. Biotech Bioeng 32:326–333, 1988.
33. Kikumoto Y, Hong YM, Nishida T, Nakai S, Masui Y, Hirai Y. Purification and characterization of recombinant human interleukin-1β produced in *Escherichia coli*. Biochem Biophys Res Commun 147:315–321, 1987.
34. Kim M, Saito K, Furusaki S, Sato T, Sugo T, Ishigaki I. Adsorption and elution of bovine γ-globulin using an affinity membrane containing hydrophobic amino acids as ligands. J Chromatogr 585:45–51, 1991.
35. Nachman M, Azad ARM, Bailon P. Membrane-based receptor affinity chromatography. J Chromatogr 597:155–166, 1992.
36. Charcosset C, Su Z, Karoor S, Daun G, Colton CK. Protein A immunoaffinity hollow fiber membranes for immunoglobulin G purification: experimental characterization. Biotechnol Bioeng 48:415–427, 1995.
37. Camperi SA, Navarro del Canizo AA, Wolman FJ, Smolko EE, Cascone O, Grasselli M. Protein adsorption onto tentacle cation-exchange hollow-fiber membranes. Biotechnol Prog 15:500–505, 1999.
38. Kubota N, Muira S, Saito K, Sugita K, Watanabe K, Sugo T. Comparision of protein adsorption by anion-exchange interaction onto porous hollow-fiber membrane and gel bead-packed bed. J Membr Sci 117:135–142, 1996.
39. Demmer W, Nussbaumer D. Large-scale membrane adsorbers. J Chromatogr A 852: 73–81, 1999.
40. Nachman M. Kinetic aspects of membrane-based immunoaffinity chromatography. J Chromatogr 597:167–172, 1992.
41. Langlotz P, Kroner KH. Surface-modified membranes as a matrix for protein purification. J Chromatogr 591:107–113, 1992.
42. Huang SH, Roy S, Hou KC, Tsao GT. Scaling-up of affinity chromatography by radial-flow cartridges. Biotech Prog 4:159–165, 1988.
43. Josic D, Zeilinger K, Lim YP, Raps M, Hofmann W, Reutter W. Preparative isolation of glycoproteins from plasma membranes of different rat organs. J Chromatogr 484: 327–335, 1989.
44. Abou-Rebyeh H, Körber F, Schubert-Rehberg K, Reusch J, Josic D. Carrier membrane as a stationary phase for affinity chromatography and kinetic studies of membrane-bound enzymes. J Chromatogr 566:341–350, 1991.
45. Ritter K. Affinity purification of antibodies from sera using polyvinylidenedifluoride

(PVDF) membranes as coupling matrices for antigens presented by autoantibodies to triosephosphate isomerase. J Immunol Methods 137:209–215, 1991.

46. Woker R, Champluvier B, Kula MR. Purification of S-oxynitrilase from *Sorghum bicolor* by immobilized metal ion affinity chromatography on different carrier materials. J Chromatogr 584:85–92, 1992.
47. Champluvier B, Kula MR. Dye-ligand membranes as selective adsorbents for rapid purification of enzymes: a case study. Biotechnol Bioeng 40:33–40, 1992.
48. Reif OW, Nier V, Bahr U, Freitag R. Immobilized metal affinity membrane adsorbers as stationary phases for metal interaction protein separation. J Chromatogr A 664:13–25, 1994.
49. Serafica GC, Pimbley J, Belfort G. Protein fractionation using fast flow immobilized metal chelate affinity membranes. Biotechnol Bioeng 43:21–36, 1994.
50. Thommes J, Kula M-R. Membrane chromatography—an integrative concept in the downstream processing of proteins. Biotechnol Prog 11:357–367, 1995.
51. Charcosset C. Purification of proteins by membrane chromatography. J Chem Technol Biotechnol 71:95–110, 1998.
52. Zeng X, Ruckenstein E. Membrane chromatography: preparation and applications to protein separation. Biotechnol Prog 15:1003–1019, 1999.
53. Wang WK, Lei SP, Monbouquette HG, McGregor WC. Membrane adsorber process development for the isolation of a recombinant immunofusion protein. Biopharm 8: 52–59, 1995.
54. Laemmli UK. Cleavage of structural proteins during the assembly of the head of bacteriophage T4. Nature 227:680–685, 1970.

10

High-Performance Tangential-Flow Filtration

Andrew L. Zydney
University of Delaware, Newark, Delaware

Robert van Reis
Genentech, Inc., South San Francisco, California

I. INTRODUCTION

High-performance tangential-flow filtration (HPTFF) is a new membrane technology that can be used for the separation of protein mixtures without limit to their relative size [1,2]. HPTFF can potentially be used throughout the downstream purification process to remove specific impurities (e.g., proteins, DNA, or endotoxins), clear viruses, and/or eliminate protein oligomers or degradation products. HPTFF can also be used for the purification of natural protein products from whey [3], egg [4], or other protein sources. HPTFF is unique among available separation technologies in that it can effect simultaneous purification, concentration, and buffer exchange, allowing several different separation steps to be combined into a single scalable unit operation.

HPTFF is able to obtain the high selectivity required for effective protein purification by exploiting several recent developments. First, unlike traditional membrane processes, HPTFF is operated in the pressure-dependent regime [2], with the filtrate flux and device fluid mechanics chosen to minimize fouling and exploit the effects of concentration polarization [i.e., bulk mass transfer]. Since optimal separation in HPTFF is obtained at a specific filtrate flux, the membrane module should be designed to maintain a nearly uniform flux and transmembrane pressure throughout the module [2]. Second, several studies have shown that the selectivity in HPTFF can be significantly increased by controlling buffer pH and

277

ionic strength to maximize differences in the effective volume of the different species [1,5]. The effective volume of a charged protein (as determined by size exclusion chromatography) accounts for the presence of a diffuse electrical double layer surrounding the protein [6]. Increasing the net protein charge by adjusting the pH, or increasing the double-layer thickness by reducing the solution ionic strength, increases the effective protein volume, thereby reducing protein transmission through the membrane. Third, the electrical charge of the membrane can be modified to increase the electrostatic exclusion of all species with like charge [7,8]. Thus, a positively charged membrane will provide much greater rejection of a positively charged protein than will a negatively charged membrane of the same pore size. Fourth, protein separations in HPTFF are accomplished using a diafiltration mode in which the impurity (or product) is washed out of the retentate by simultaneously adding fresh buffer to the feed reservoir as filtrate is removed through the membrane. This buffer addition maintains an appropriate protein concentration in the retentate throughout the separation, minimizing membrane fouling and reducing protein aggregation/denaturation. Diafiltration also makes it possible to obtain purification factors for products collected in the retentate that are much greater than the membrane selectivity due to the continual removal of impurities in the filtrate [9]. Van Reis and Saksena have developed an optimization analysis that can be used to select the best operating conditions for HPTFF processes taking into account the practical constraints on membrane area and process time [9].

Although HPTFF is still an emerging technology, a number of recent studies have clearly demonstrated the potential of this separation technology. Several of these results are summarized in Table 1. The separation of bovine serum albumin (BSA) from an antigen-binding fragment (Fab) could be achieved with either protein collected in the retentate depending upon the choice of solution pH and membrane charge. In each case, purification factors were > 800-fold using a single-stage HPTFF device operated at the optimal pH and with the appropriate membrane. These results are discussed in more detail in Section V. The results

Table 1 Purification Factors and Yields for HPTFF Processes

Product (mw)	Impurity (mw)	Purification factor	Yield (%)	Reference
BSA (68,000)	Fab (45,000)	990	94	8
Fab (45,000)	BSA (68,000)	830	69	8
BSA (68,000)	Hb (67,000)	100	68	10
IgG (155,000)	BSA (69,000)	30	84	1
BSA (68,000)	BSA dimer (136,000)	9	86	1

for the BSA–hemoglobin system demonstrate that HPTFF can be used to separate proteins with essentially identical molecular weights by exploiting differences in protein charge. In this case, operation at pH 7 caused a strong electrostatic exclusion of the negatively charged BSA from the negatively charged membrane. The separation of BSA monomer and dimer occurs primarily because of the difference in protein size, with the more permeable monomer collected in the filtrate. However, electrostatic interactions are also important in this system due to the combined effects of size and charge on protein transmission and to possible differences in the charge-pH profiles for the monomer and dimer.

The next section of this chapter will examine the basic principles governing protein separations by HPTFF in more detail, with particular focus on the importance of the different phenomena in generating the high selectivities required for bioprocessing applications. We then review an optimization analysis that can be used to identify those operating conditions which give the best combination of yield and purification, taking into account the practical constraints on membrane area and process time. The final sections of this chapter examine process development in HPTFF, using data for the separation of BSA from an antigen-binding fragment as a model system. These results show how a fundamental understanding of the key principles of HPTFF technology can be used to develop highly selective protein separations with yield and purification factors that are potentially competitive with existing chromatographic systems.

II. PRINCIPLES OF HPTFF

A. Membrane Transport

The rate of protein transmission through a semipermeable membrane is conveniently described in terms of the intrinsic (also referred to in the literature as the actual) membrane sieving coefficient (S_i):

$$S_i = \frac{C_f}{C_w} \tag{1}$$

where C_f is the protein concentration in the filtrate solution and C_w is the protein concentration in the retentate solution at the upstream surface of the membrane. The sieving coefficient is equal to 1 minus the protein rejection coefficient. The sieving coefficient is determined by both thermodynamic (equilibrium partitioning) and hydrodynamic (frictional) interactions. However, the hydrodynamic interactions have a relatively minor effect on protein transport in HPTFF [11]. Thus, the membrane selectivity is determined almost entirely by the equilibrium partitioning of the solute between the solution adjacent to the membrane and the membrane pores.

In the absence of any long-range (e.g., electrostatic) interactions, the intrinsic protein sieving coefficient can be approximated as [11]:

$$S_i = \exp(-r_s/s) \tag{2}$$

where r_s is the effective hydrodynamic radius of the protein and s provides a measure of the mean pore size. Equation (2) implicitly accounts for the effects of a membrane pore size distribution, with S_i approaching zero asymptotically as r_s increases due to the presence of a small number of very large pores in the distribution. Decreasing the membrane pore size causes a reduction in the protein sieving coefficient; however, the ratio of the sieving coefficients for a small impurity to that of a larger protein increases with decreasing s. The net result is that there is an optimum pore size for HPTFF processes: membranes with a very small pore size will have good resolution but poor throughput, while membranes with very large pore size will have good throughput but poor resolution. The tradeoffs between throughput and selectivity can be quantified using the optimization analysis presented in Section III of this chapter.

One very powerful means for increasing protein resolution in HPTFF is to exploit electrostatic interactions between the protein and the membrane pores. The sieving coefficient of a charged protein in a charged membrane is determined by both steric and energetic interactions:

$$S_i = S_{i,\,steric} \exp\left(-\frac{E}{kT}\right) \tag{3}$$

where $S_{i,\,steric}$ is the sieving coefficient in the absence of any electrostatic interactions. The dimensionless electrostatic energy of interaction (E/kT) can be evaluated theoretically from solution of the linearized Poisson-Boltzmann equation as [12]:

$$\frac{E}{kT} = A_1\sigma_p^2 + A_2\sigma_p\sigma_m + A_3\sigma_m^2 \tag{4}$$

where σ_p and σ_m are the surface charge densities of the protein and membrane, respectively, and the A_i are positive coefficients which are complex functions of the protein and pore size and the solution ionic strength. The electrostatic energy of interaction has three distinct contributions. The first term in Eq. (4) accounts for the increase in free energy caused by the distortion or deformation of the electrical double layer surrounding the protein caused by the presence of the pore wall. This effect can be interpreted as an increase in the effective protein size (or hydrodynamic volume) associated with the diffuse ion cloud surrounding the charged protein. The middle term in Eq. (4) describes the direct charge-charge interactions between the protein and pore. This term is positive when the protein and pore have like charge, leading to a strong electrostatic exclusion under these conditions. This term becomes attractive (i.e., the energy is negative) when σ_p and σ_m are oppositely charged, which would be expected to cause an increase in protein transmission. However, membrane fouling by the protein is also enhanced when the protein and membrane are oppositely charged, and this fouling can cause a reduction in the effective pore size and in turn the protein sieving coefficient. The last term in Eq. (4) accounts for the distortion of the electrical double layer surrounding the pore wall caused by the presence of the protein. This effect can be interpreted in terms of a reduction in the effective pore size of the membrane, analogous to the increase in the effective protein radius associated with the electrical double layer surrounding the protein.

When the first term dominates the electrostatic energy of interaction, the protein sieving coefficient will attain its maximum value at the isoelectric point (i.e., at the pH where the protein has no net charge). Pujar and Zydney [6] developed an expression for the effective radius of a charged protein under these conditions:

$$r_{s,\,eff} = r_s + 0.0137\ z^2 r_s^{-1} I^{-1/2} \tag{5}$$

where z is the protein surface charge (in electronic charge units), r_s is the hard-sphere radius of the uncharged protein (in nm), and I is the solution ionic strength (in mol/L). For a protein like bovine serum albumin, which has a net charge of about -20 electronic units at pH 7, the effective radius increases from 3.6 nm at very high ionic strength to $r_{s,\,eff} = 18.8$ nm at $I = 10$ mM. This increase in radius would correspond to a reduction in BSA sieving coefficient of more than 2 orders of magnitude (for a membrane with $s = 3$ nm), a result which is in good agreement with experimental data for BSA sieving through a 100-kDa polyethersulfone membrane [13]. Note that the strong dependence of the effective protein radius on the protein charge makes it possible to reverse the selectivity in HPTFF, with either the product or the impurity being the more highly retained species depending upon the choice of membrane charge and buffer pH (i.e., protein charge). This phenomenon is discussed in more detail in Section V.

The effects of electrostatic interactions on the sieving coefficients for BSA, Pulmozyme®, and two monoclonal antibodies (E25 and anti-CD11a) are shown in Figure 1 [14]. All of the data were obtained using a Biomax 100-kDa polyethersulfone membrane at an ionic strength of 10 mM. The data are plotted as the normalized sieving coefficient, which is simply the ratio of the sieving coefficient at a given pH to the maximum value of the protein sieving coefficient, as a function of the difference between the solution pH and the protein isoelectric point. In each case, the maximum sieving coefficient was attained within 0.2 pH units of the protein isoelectric point. The sieving coefficients decrease at pH both above and below the protein isoelectric point due to the significant energetic penalty associated with the distortion of the electrical double layer around the protein (the first term in Eq. 4). However, the data for the normalized sieving

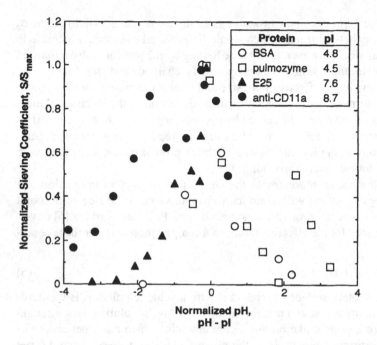

Figure 1 Normalized sieving coefficient (S/S_{max}) as a function of the difference between the solution pH and the protein isoelectric point. Data were obtained using Biomax 100-kDa polyethersulfone membranes at 10 mM ionic strength. (From Ref. 14.)

coefficient are not symmetric about the origin. This asymmetry is due to two distinct effects. First, the charge–pH profiles for the proteins are nonlinear due to the different number and pK_a of the various charged amino acid residues. Second, direct charge-charge interactions will be different at pH above and below the pI depending on the sign of the charge on the protein and membrane. The Biomax 100 membrane has an isoelectric pH after protein adsorption of ~ pH 5. Thus, the E25 and anti-CD11a are oppositely charged from the membrane for pH > 5, which causes an "attractive" contribution to the energy of interaction (as described by the second term in Eq. 4). This attraction is absent for BSA and pulmozyme since the protein and membrane have similar isoelectric points and thus have like charge at pH both above and below the pI.

B. Fluid Dynamics

HPTFF processes are operated using tangential-flow filtration [2], in which the feed flow is parallel to the membrane surface, with a fraction of the feed driven through the membrane by the applied transmembrane pressure drop to form the filtrate solution. This tangential flow "sweeps" the membrane surface, reducing the extent of fouling and increasing the filtrate flux (the volumetric filtrate flow rate per unit membrane area) compared to that obtained in dead-end systems.

One of the critical factors governing the performance of HPTFF systems is the accumulation of retained protein at the upstream surface of the membrane, a phenomenon typically referred to as concentration polarization [15]. The extent of concentration polarization can be estimated using a stagnant film model as [11,15]:

$$\frac{C_w}{C_b} = \frac{\exp(J/k)}{1 - S_i + S_i \exp(J/k)} \tag{6}$$

where J is the filtrate flux and C_b is the protein concentration in the bulk solution. The mass transfer coefficient (k) characterizes the rate of protein transport away from the membrane and back into the bulk solution. It is a function of the module geometry, the feed flow rate, and the solution properties (e.g., solution viscosity and protein diffusivity). The observed sieving coefficient S, which is equal to the ratio of the protein concentration in the filtrate (C_f) to that in the bulk solution (C_b), is a function of the intrinsic sieving coefficient (S_i) and the extent of polarization:

$$S = S_i\left(\frac{C_w}{C_b}\right) \tag{7}$$

Concentration polarization has often been cited as an inherent limitation in using membrane systems for high-resolution separations. However, proper choice of filtrate flux and mass transfer coefficient can be used to improve the performance of HPTFF systems by ensuring operation at the optimal C_w. At low filtrate flux, increasing J causes an increase in protein transmission due to the increase in C_w. However, at very high values of C_w protein fouling can become significant, leading to a reduction in the intrinsic sieving coefficient. The implications of concentration polarization for HPTFF are shown in Figure 2 for the separation of BSA monomer and oligomer using a Pellicon 2 cassette with a Biomax 150-kDa membrane [1]. The results are plotted as the selectivity, defined as the ratio of the observed sieving coefficient for the monomer (S_1) to that of the dimer (S_2):

$$\psi = S_1/S_2 \tag{8}$$

There is a clear maximum in the selectivity at an intermediate filtrate flux, with the value of this critical flux increasing with increasing feed flow rate due to the

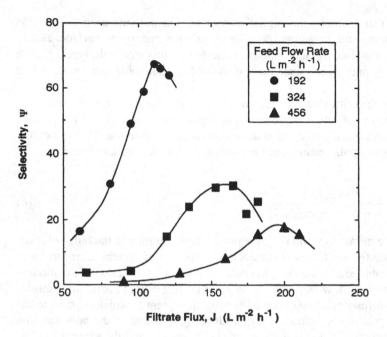

Figure 2   Selectivity as a function of filtrate flux for three values of the feed flux (feed flow rate per unit membrane area) for the separation of BSA monomer and dimer using a Pellicon 2 cassette. (From Ref. 1.)

corresponding increase in protein mass transfer coefficient. It is important to note that concentration polarization can potentially be used to reverse the selectivity in HPTFF since larger proteins have smaller mass-transfer coefficients and thus polarize to a greater extent [5].

III.   OPTIMIZATION DIAGRAM
A.   Governing Equations

The goal of HPTFF process optimization is to achieve the highest yield and purity taking into account the constraints on membrane area and process time. Yield and purification factor are functions of both the process selectivity and throughput [9]. The selectivity (ψ) is equal to the ratio of the sieving coefficients for the less and more highly retained proteins (S_1 and S_2) as defined by Eq. (8). The process throughput is defined as:

$$N\Delta S = (S_1 - S_2)\frac{JAt}{V} \qquad (9)$$

where N is the number of diavolumes (equal to the total collected filtrate volume divided by the retentate volume), J is the filtrate flux (L m^{-2}h^{-1}), A is the membrane area (m^2), t is the process time (h), and V is the retentate volume (L).

HPTFF is performed in a diafiltration mode to achieve the desired purification. The most common approach is to maintain constant retentate volume with the diafiltration buffer added to the feed at the same rate as the filtrate is removed. Van Reis and Saksena [9] have developed a set of design equations for HPTFF diafiltration which provide a rational method for process optimization. For a filtrate product, the purification factor (P_1) can be expressed in terms of the fractional yield (Y_1) and selectivity as:

$$P_1 = \frac{Y_1}{1 - (1 - Y_1)^{1/\psi}} \qquad (10)$$

The purification factor and yield for a filtrate product can also be related to the throughput ($N\Delta S$) as:

$$P_1 = \frac{Y_1}{1 + (Y_1 - 1)e^{N\Delta S}} \qquad (11)$$

For a retentate product, the purification factor (P_2) can be expressed as a function of the yield (Y_2) and selectivity as:

$$P_2 = Y_2^{1-\psi} \qquad (12)$$

The purification factor for a retained product varies with throughput as:

$$P_2 = e^{N\Delta S} \qquad (13)$$

Experimental optimization of membrane, buffer, and fluid dynamics can be performed by evaluating ψ and $N\Delta S$ from the experimentally determined values of J, S_1, and S_2. Membrane area and process time are independent engineering parameters chosen based on process and economic constraints. The retentate volume can be adjusted by an initial ultrafiltration based on the optimum bulk concentration and the total mass of product. The optimum bulk concentration can be determined experimentally by evaluating ψ and $N\Delta S$ as a function of concentration. For a process designed to purify a retentate product (with $S_2 \ll 1$) in which the impurity has a sieving coefficient $S_1 \approx 1$, the minimum process time is achieved at a bulk concentration directly related to the wall concentration of the retained solute [16]:

$$C_b^* = C_w/e \qquad (14)$$

where C_w is the concentration of the retained product at the membrane wall and e is the natural logarithm base. The optimum flux under these conditions is simply [17]:

$$J^* = k \tag{15}$$

where k [L m^{-2}h^{-1}] is the mass-transfer coefficient based on the stagnant film model.

In addition to using these optimization equations, it is convenient to use a set of optimization diagrams that express the process yield and purification factor as functions of the selectivity and throughput. These diagrams consist of a family of ψ and $N\Delta S$ curves on yield and purification factor coordinates.

B. Filtrate Product Diagram

The optimization diagram for a filtrate product is shown in Figure 3. The process starts at a purification factor limited to a value equal to the selectivity [9]. The process proceeds along a line of constant selectivity (darker curves in Figure 3) toward increasing yield at the expense of a lower purification factor. The endpoint on the yield–purification factor curve is determined by the magnitude of the process throughput parameter ($N\Delta S$). Processes with large ΔS can achieve high yield with only a small number of diavolumes. The diagram shows that it is relatively

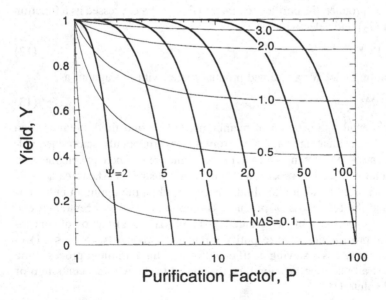

Figure 3 Optimization diagram for a filtrate product showing the tradeoff between yield and purification factor. (From Ref. 9.)

easy to obtain high yields by increasing the number of diavolumes, but high purification factors are dependent on achieving high selectivity values.

C. Retentate Product Diagram

The optimization diagram for a retentate product is shown in Figure 4. The process starts at 100% yield and a purification factor of 1 and follows a line of constant selectivity (darker curves in Figure 4) toward increasing purification factor at the expense of reduced yield [9]. Performance is highly dependent on selectivity. The final purification factor obtainable with a given process is determined by the value of $N\Delta S$ as shown by the vertical lines in Figure 4. The purification factor can thus be increased by increasing membrane area, process time, or both (Eq. 9). With high selectivity values ($\psi \geq 100$), enhanced purification factors can be achieved with only moderate yield losses. The choice of having the product in either the retentate or filtrate is not solely determined by the relative molecular weights of the product and impurities. Complete reversal of selectivity can be obtained by choosing the appropriate pH and membrane charge (see Sections IIA, IIB, and VA). It is hence possible to explore both retentate and filtrate product options for overall optimization of the process.

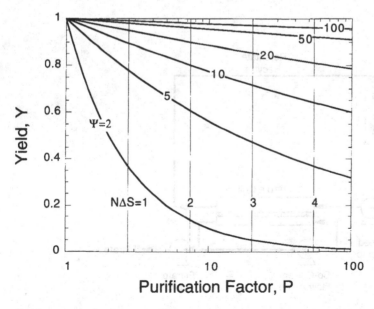

Figure 4 Optimization diagram for a retentate product showing the tradeoff between yield and purification factor. (From Ref. 9.)

IV. PROCESS CONFIGURATIONS

A. Filtrate Flow Configurations

HPTFF is typically performed using flat-sheet cassettes. These cassettes employ a support plate (which also defines the filtrate flow path), membrane, and channel spacer in a sandwich arrangement. The channel spacer has a meshlike structure to promote mixing and increase mass transport. A more detailed description of the flat-sheet cassette and other available modules for ultrafiltration and HPTFF is provided by Van Reis and Zydney [18].

    As discussed in Section IIA, the optimal combination of selectivity and $N\Delta S$ occurs at a specific value of the filtrate flux. To obtain optimal system performance, the filtrate flux, and thus the transmembrane pressure, must be maintained essentially uniform over the full filtration region. This goal can be accomplished by placing a recirculation pump on the filtrate line to generate a pressure drop in the filtrate channel that balances the pressure drop in the feed channel [2]. A schematic of this cocurrent flow operation is shown in Figure 5 for a system in which the filtrate flow rate is set using a filtrate pump. The recirculation flow rate on the filtrate side is adjusted so as to provide equal transmembrane pressures at the device inlet and outlet. Alternatively, an "open"-channel configuration, e.g., a flat-sheet cassette with a suspended screen or a hollow-fiber module, can

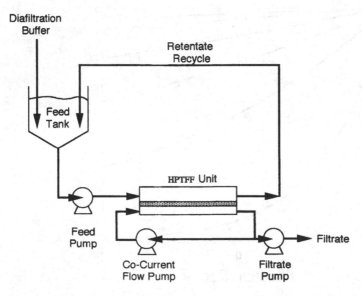

Figure 5 Schematic of cocurrent filtrate flow operation used to maintain nearly uniform transmembrane pressures and filtrate flux.

be used to reduce the pressure drop associated with feed flow through the module. The overall performance of an open-channel module will likely be somewhat less than the theoretical optimum due to the small pressure gradient from device inlet to outlet, but the elimination of the cocurrent filtrate flow reduces the operational complexity of the system.

B. Cascade Diafiltration

Although HPTFF is typically performed as a standalone operation, it can also be used as part of a cascaded system in which the filtrate from the first stage is directly used as a feed to a second stage (typically an ultrafiltration device employing a fully retentive membrane). In the open-loop design, the filtrate from

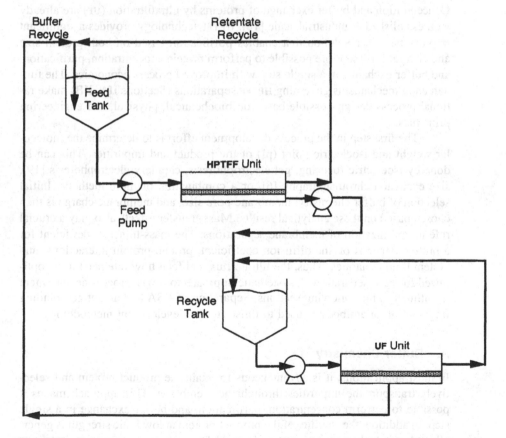

Figure 6 Two-stage closed-loop cascade configuration with the second UF stage used to regenerate diafiltration buffer for the HPTFF.

the second stage is directed to drain, with the retentate recycled to a holding tank to concentrate the filtrate product collected from the HPTFF module. In the closed-loop design (Figure 6), the ultrafiltrate from the UF module is used as diafiltration buffer for the first stage. This design allows HPTFF processes to be run with a very large number of diavolumes without excessive buffer consumption. Cascade configurations with two (or more) HPTFF units can be used to separate a feed stream into multiple product fractions. In this case, the initial HPTFF module would be designed to retain the least permeable component, with all of the other species passing into the filtrate for separation in the subsequent HPTFF units.

V. PROCESS DEVELOPMENT

Concentration and buffer exchange of proteins by ultrafiltration (UF) are already well established at industrial scale [18]. HPTFF technology provides a significant improvement in resolution that enables purification based on both protein size and charge. It is therefore possible to perform protein concentration, purification, and buffer exchange in a single step with improved process economics. The fundamental mechanisms governing HPTFF separations (Sections IIA, IIB) make rational process design possible based on biochemical, physical, and engineering principles.

The first step in the process development effort is to determine the molecular weight and isoelectric point (pI) of the product and impurities. This can be done by isoelectric focusing, gel electrophoresis, capillary electrophoresis [19], size exclusion chromatography [6], or a combination of these methods. Initial selection of buffer chemistry, membrane pore size, and membrane charge is then chosen based on these analytical results. Mass-transfer effects also play a crucial role in optimization of membrane separations. The mass-transfer coefficient for a protein depends on the diffusion coefficient, protein-protein interactions, and system fluid dynamics. Thus, the filtrate flux and feed flow rate need to be optimized for each separation. A systematic approach to HPTFF process development is outlined in the following sections. Separation of BSA and an antigen-binding fragment of an antibody is used to illustrate the development methodology.

A. Buffer Chemistry

In most applications, it is advantageous to retain the product protein and selectively transport the impurities through the membrane. This approach makes it possible to perform concentration, purification, and buffer exchange in a single step. In addition, the stability of the product protein at low ionic strength is generally enhanced when operating at a pH away from the isoelectric point. Retentate products can also be purified to values much higher than the selectivity while maintaining high yields using constant retentate volume diafiltration (Figure 4). In contrast, the purification factor for a filtrate product is mathematically limited to a value that is less than the selectivity [9]. It is also possible to design processes that remove multiple impurities from a retentate product by performing the diafiltration at several pH values or by establishing a pH gradient during the diafiltration using systems developed for chromatography. A filtrate product process should be considered when concentration and buffer exchange are not required and when the selectivity and throughput values for such a process result in higher yields and purification factors. This would typically occur when the molecular weight of the impurity is so much larger than the product that charge effects are insufficient to enable reversal of selectivity.

Initial process development is typically performed with the pH of the process buffer close to the pI of the impurity. Experiments can hence be limited to a few pH values around the pI of the impurity. It is not clear that there is a preference for operating either immediately above or below the pI. Although the net charge of the protein near the pI changes sign, charge patches on the protein could play a more important role by binding to either the membrane or other proteins. Significant transmission of a protein is typically achieved within 1 pH unit of the pI, with an optimum value close to the pI (Figure 1). Conversely, proteins will tend to have high retention values when the pH is 2 units above or below the pI. Experiments should therefore be performed within 1 pH unit of the pI of one of the proteins and at least 2 pH units away from the pI of the retained species. Note that HPTFF can also be used to separate proteins that have the same or similar pI, as illustrated by the separation of BSA monomer and dimer [1]. pH optimization is still important since the monomer and dimer may have different charge-pH profiles. In addition, one can use pH and ionic strength to optimize the hydrodynamic volumes to best match the pore size characteristics of the available membranes.

A single ionic strength can be chosen for initial experiments. The ionic strength should in principle be as low as possible to enhance the difference in hydrodynamic volumes of the solutes to be separated. In practice, the concentration of buffer species also needs to be sufficiently high to maintain buffering capacity at the optimum process pH. Furthermore, increasing ionic strength can prevent aggregation and denaturation of solutes with pI values close to the buffer pH, thereby preventing membrane fouling and enhancing process selectivity and throughput parameters. The quality (denaturation, aggregation) of a filtrate product may also be affected by operating at low ionic strength near the pI. It may therefore be necessary to investigate the addition of nonionic solutes to stabilize the protein. Such nonionic additives can also be used to reduce membrane fouling and subsequent deterioration of selectivity and throughput values. A 10-mM buffer is a good starting point, since this concentration should maintain adequate buffer capacity while providing significant electrostatic interactions. Size exclusion chromatography can complement initial buffer screening experiments since it rapidly identifies conditions which maximize the difference in effective hydro-dynamic volume of the product and impurities [6]. Final buffer optimization should be performed in small-scale HPTFF systems [8].

After the optimum pH has been determined, one can then proceed to opti-mize the ionic strength of the buffer. Increasing ionic strength will increase the sieving coefficients of both product and impurity, which will typically reduce selectivity while increasing the $N\Delta S$ value. Optimization of ionic strength can also be used to adjust the hydrodynamic volume of both product and impurities to best match one of a finite number of different membrane pore sizes. This optimization may be especially important for separation of proteins with the same or similar pI. In addition to optimizing buffer pH and ionic strength, it should be recognized that the selection of buffer species can also play an important role in HPTFF processes because of the potential for ion binding to proteins which can change the net protein charge [20]. This effect can be quite significant for some proteins. For example, the net charge on BSA at pH 3.5 in 10 mM ionic strength Na_2SO_4 is only half that in NaCl, leading to more than a six-fold difference in BSA sieving coefficient [20].

When using HPTFF, it is generally possible to select a single membrane pore size (Section VB) for the initial buffer screening experiments. One can also start with a single feed rate of approximately 300 L $m^{-2}h^{-1}$ (volumetric feed flow rate divided by the membrane area). Filtrate flux values should typically be examined in the range of 10 to 100 L $m^{-2}h^{-1}$ (Section VC). These experiments can be conducted using total recycle of both retentate and filtrate. The optimum buffer species, pH, ionic strength, and nonionic additives should be evaluated by measuring the sieving coefficients versus flux followed by data analysis using the optimization diagrams (Section III). In cases where it is not readily feasible to measure the sieving coefficient of impurities, it may be necessary to carry out complete HPTFF processes with diafiltration and measure the overall yield and purification factor. Examples include virus removal applications and processes with multiple impurities that lack specific assays.

The development of an HPTFF process is illustrated by the separation of BSA (68 kDa, pI = 4.8) and Fab (45 kDa, pI = 8.4) [8]. Initial experiments were performed around the pI of both BSA and Fab in order to examine the forward and reverse separations (i.e., processes with either BSA or Fab as the retentate product). Data were obtained using Biomax 100-kDa nominal molecu-lar-weight cutoff membranes in a linear scale-down system at a feed flow rate of 323 L $m^{-2}h^{-1}$. Protein sieving coefficients at each pH were evaluated as a function of filtrate flux, with the optimum flux determined by analyzing the selec-tivity and $N\Delta S$ values using the optimization diagrams (Section III). The best combinations of selectivity and throughput for the forward and reverse separa-tions are shown in Table 2. High selectivity and throughput could be achieved

Table 2 Forward (BSA/Fab) and Reverse (Fab/BSA) Selectivity (ψ) and Throughput ($J\Delta S$) for the Separation of BSA and a Fab Using Positively and Negatively Charged Membranes[a]

Retentate product	Membrane	Buffer	pH	Selectivity (ψ)	$J\Delta S$ [Lm^{-2}h^{-1}]
BSA	Biomax 100−	Tris	8.4	200	44
Fab	Biomax 100+	Acetate	5.0	110	44

[a]Feedstock contained 3 mg/mL of each protein.
Source: Ref. 8.

with either protein in the retentate by proper choice of buffer pH and membrane charge. HPTFF using the negatively charged Biomax 100 membrane at pH 8.4 (close to the Fab pI of 8.5) led to very high rejection of the negatively charged BSA with nearly complete transmission ($S = 0.8$) of the Fab. HPTFF with the positively charged membrane at pH 5.0 (close to the pI of BSA) resulted in the reverse separation; the positively charged Fab was strongly rejected while the BSA sieving coefficient was $S = 0.7$. Note that use of a membrane with opposite charge (i.e., using a positively charged membrane for the separation at pH 8.4 or the negatively charged membrane at pH 5.0) caused more than an order of magnitude reduction in both selectivity and $J\Delta S$ [8]. Thus, the high selectivity for the BSA-Fab separation was accomplished by simultaneous optimization of the buffer pH and membrane surface charge. Note that in this particular study, no optimization was done of the membrane pore size, solution ionic strength, or feed flow rate. The 100-kDa membrane was chosen based on past experience suggesting that this membrane would provide high throughput (with the possibility for high resolution), while the ionic strength (10 mM) and feed flow rate (323 L m^{-2}h^{-1}) were chosen based on the rules of thumb discussed previously.

B. Membranes

Membrane selection should start with the choice of a high-quality vendor since robustness and reliability are of paramount importance in any bioprocessing application. Several attributes are important in selecting a membrane for HPTFF processes. Selectivity, throughput, fouling characteristics, cleanability, extractables, and economics must all be taken into consideration. Selectivity and throughput are influenced by membrane charge, pore size, and pore size distribution. In order to obtain the high selectivity needed for HPTFF processes, membranes must be completely free of large defects. The highest selectivity and throughput are obtained by selecting a membrane that has the same charge as the solutes that are to be retained by the membrane. Pore size and pore size distributions have a significant impact on selectivity, throughput, and permeability. Minimal lot-to-lot variation in membrane pore size and pore size distribution is critical in HPTFF technology. A liquid-liquid intrusion method has been developed to ensure consistency in the manufacture of HPTFF membranes [21].

Membrane chemistry and charge can also have a significant affect on the extent of fouling. Cellulosic and polysulfone (including polyethersulfone) are the most commonly used membrane polymers, although these can be surface modified to produce more desirable fouling and electrostatic characteristics. It should be noted that membrane fouling can, under some circumstances, enhance the overall separation by reducing the breadth of the pore size distribution and/or altering the surface charge. Chemical compatibility of the membranes needs to be considered with respect to the feed solution, cleaning cycle, and storage chemicals. Long-term stability of the membranes in bacteriostatic storage solutions should be evaluated by studying product retention and process flux after appropriate storage times. Suitable conditions for membrane cleaning often need to be worked out for specific membranes and process feed streams, although some general guidelines are available [18]. The level of membrane extractables must also be determined under worst-case conditions for the specific membrane and chemical exposure conditions in compliance with regulatory guidelines. Membrane economics is based on purchase price, number of reuses, and optimization of membrane area (Section VC).

High-permeability membranes (with equal selectivity) permit operation at lower transmembrane pressures. This can be important when using small-pore-size membranes to attain maximum $J\Delta S$ values within system pressure limits. However, using a membrane with a lower permeability may enable operation without a cocurrent filtrate loop (Section IVA). The purpose of the cocurrent filtrate loop is to maintain an approximately constant transmembrane pressure along the length of the HPTFF module since the optimum combination of selectivity and throughput occurs at a single filtrate flux. By decreasing membrane permeability, the percentage difference between the inlet and outlet transmembrane pressures and filtrate flux is minimized, so the system performance will approach the theoretical optimum. Cocurrent filtrate flow can also be eliminated by decreasing the pressure drop through the HPTFF module, which can be accomplished by reducing the feed flow rate (Section VC) or by using an open-channel configuration (Section IVA).

Initial experiments can be performed with a single molecular-weight cutoff membrane, with data obtained at several pH values and ionic strengths over a range of filtrate flux. If higher selectivity is required, experiments should proceed with a smaller-pore-size membrane and/or at lower ionic strength. Higher throughput ($J\Delta S$) can be obtained by choosing a membrane with larger pore size or by operating at higher ionic strength to reduce the electrostatic interactions.

C. Fluid Dynamics

HPTFF fluid dynamics depend on module mass-transfer characteristics and operating conditions. The mass transfer characteristics of a module are primarily governed by channel dimensions and the presence or absence of turbulence promoters. Turbulence promoters are usually in the form of polymeric screens that have been inserted into the feed channels. Mass transfer can also be enhanced by induction of Taylor [22] or Dean vortices [23]. The mass-transfer coefficient also depends on the diffusion coefficient of the protein, which in turn is influenced by protein size, charge, concentration, and buffer chemistry. For a conventional module and a given buffer chemistry, the mass-transfer coefficient can be varied by altering the feed flow rate. The optimum selectivity and throughput will be obtained at a specific wall concentration [17]. The wall concentration in turn will depend on the mass transfer coefficient and the filtrate flux as given by Eq. (6). It should hence be possible to obtain optimum operating conditions at several combinations of feed rate and filtrate flux with various filter modules. A good starting point with the Millipore Pellicon modules is a feed rate of 300 L m^{-2}h^{-1} and filtrate fluxes in the range of 10 to 100 L m^{-2}h^{-1}. Fluid dynamic studies can be done in a total recycle mode with both retentate and filtrate recycled to the feed solution. The results are evaluated the same way as with buffer and membrane selection by measuring sieving coefficients and determining the optimum combination of selectivity and throughput using the optimization diagrams (Section III).

D. Process Design

Once the optimum buffer conditions, membrane, and fluid dynamics have been determined from total recycle experiments, it is important to perform a test of the actual HPTFF process. Using optimum ψ and $J\Delta S$ values, it is possible to select a membrane area and the number of diavolumes that will result in the desired yield and purification factor within a given process time (Section III). The selectivity and throughput values obtained in the initial screening experiments may, however, not be maintained in the full HPTFF process. For example, significant reductions in both ψ and $J\Delta S$ were found in the BSA-Fab system during process runs compared to the values obtained in total recycle experiments. However, the selectivity and throughput in these processes were still sufficient to generate > 800-fold purification of either the BSA or Fab, as summarized in Table I. Changes in ψ and $J\Delta S$ may be due to membrane fouling or to changes in the concentration of the permeable solutes during diafiltration. Membrane fouling can be addressed by changes in buffer pH, ionic strength, and/or the addition of nonionic additives. The goal is to enhance protein solubility and reduce protein adsorption and denaturation while maintaining pH and ionic conditions that maximize protein resolution. The membrane matrix and ligand chemistry must also be considered. The overall process design should also include optimization of the bulk protein concentration as previously described (Section III.A). A complete process may include concentration to the optimum bulk value, diafiltration for protein purification, and additional diafiltration for formulation or in preparation for a subsequent process step.

VI. SCALEUP

Observed sieving coefficients are related to the intrinsic sieving characteristics of the membrane, the mass-transfer coefficient, and the filtrate flux. The mass-transfer coefficient in turn depends on the system geometry and the operating parameters. The mass-transfer coefficient is also dependent on the solute diffusivities which are functions of solution pH, ionic strength, solute concentration, and solute-solute interactions. Reproducing the ψ and $J\Delta S$ values obtained at small scale in an industrial scale system requires that the membrane, system mass-transfer coefficient, and operating parameters are all the same. It is hence critical that small-scale process development studies be performed with equipment that has been validated to accurately represent an industrial-scale system using linear-scale principles [24]. Linear scaleup is achieved by designing systems that have equal channel path length and channel height, and the same turbulence promoter (if applicable). Maintaining constant channel path length keeps the flux versus path length profiles independent of scale. The channel height and turbulence promoter influence the mass transfer coefficient.

Membrane area is enlarged by increasing channel width and the number of channels in parallel. Increasing the number of channels in parallel inherently leads to the addition of static and dynamic pressure losses in the system. It is therefore important to use a system with carefully controlled and well-established flow distribution profiles [24]. Many systems with constant channel path length are now commercially available. Most of these systems do not, however, meet linear-scale criteria because manufacturing tolerances have not been adequate to ensure reproducible mass transfer coefficients and flow distribution. The mass-transfer coefficients are highly dependent on channel height. The effective channel height is determined by cassette-manufacturing tolerances, cassette compression in the system, and channel deflection from process fluid pressure. Significant variability in pressure drops caused by entrance and exit effects associated with poor manufacturing control have been observed in some systems. Cassette-to-cassette variations in pressure drop will affect the flow distribution and hence the mass-transfer coefficient throughout the system. A simple test of manufacturing control consists of comparing the pressure drop (feed-retentate pressure) of various cassettes at the same feed rate and filtrate flux.

Controlled tolerances on membrane and cassette manufacturing and validated system designs are essential to successful scale up of both UF and HPTFF processes. These criteria must be validated by the vendor and verified by the end user. The end user can then perform linear scale up by simply maintaining the same feed, filtrate, and cocurrent filtrate flow rates normalized by the membrane surface area. This methodology is already well established in UF, with 400-fold scaleup of processes for human pharmaceutical proteins produced with recombinant DNA methods [17,24].

REFERENCES

1. van Reis R, Gadam S, Frautschy LN, Orlando S, Goodrich EM, Saksena S, Kuriyel R, Simpson CM, Pearl S, Zydney AL. High performance tangential flow filtration. Biotechnol Bioeng 56:71–82, 1997.
2. van Reis R. Tangential flow filtration process and apparatus. U.S. patents 5,256,294 (1993); 5,490,937 (1996); and 6,054,051 (2000).
3. Zydney AL. Protein separations using membrane filtration: new opportunities for whey fractionation. Int Dairy J 8:243–250, 1998.
4. Ehsani N, Pakinnen S, Nystrom M. Fractionation of natural and model egg-white protein solutions with modified and unmodified polysulfone membranes. J Membr Sci 123:105–119, 1997.
5. Saksena S, Zydney AL. Effect of solution pH and ionic strength on the separation of albumin from immunoglobulins (IgG) by selective membrane filtration. Biotechnol Bioeng 43:960–968, 1994.
6. Pujar NS, Zydney AL. Electrostatic effects on protein partitioning in size exclusion chromatography and membrane ultrafiltration. J Chromatogr A 796:229–238, 1998.
7. Nakao S, Osada H, Kurata H, Tsuru T, Kimura S. Separation of proteins by charged ultrafiltration membranes. Desalination 70:191–205, 1988.
8. van Reis R, Brake JM, Charkoudian J, Burns DB, Zydney AL. High performance tangential flow filtration using charged membranes. J Membr Sci 159:133–142, 1999.
9. van Reis R, Saksena S. Optimization diagram for membrane separations. J Membr Sci 129:19–29, 1997.
10. van Eijndhoven HCM, Saksena S, Zydney AL. Protein fractionation using membrane filtration: role of electrostatic interactions. Biotechnol Bioeng 48:406–414, 1995.
11. Zeman L, Zydney AL. Microfiltration and Ultrafiltration: Principles and Applications. New York: Marcel Dekker, 1996.
12. Smith FG, Deen WM. Electrostatic double-layer interactions for spherical colloids in cylindrical pores. J Colloid Interf Sci 78:444–465, 1980.
13. Pujar NS, Zydney AL. Electrostatic and electrokinetic interactions during protein transport through narrow pore membranes. Ind Eng Chem Res 33:2473–2482, 1994.
14. Burns DB, Zydney AL. Effect of solution pH on protein transport through semipermeable ultrafiltration membranes. Biotechnol Bioeng 64:27–37, 1999.

15. Blatt WF, Dravid A, Michaels AS, Nelsen L. Solute polarization and cake formation in membrane ultrafiltration. Causes, consequences, and control techniques. In: JE Flinn, ed. Membrane Science and Technology. New York: Plenum Press, 1970:47–97.

16. Ng P, Lundblad J, Mitra G. Optimization of solute separation by diafiltration. Sep Sci 2:499–502, 1976.

17. van Reis R, Goodrich EM, Yson CL, Frautschy LN, Whiteley R, Zydney AL. Constant Cwall ultrafiltration process control. J Membr Sci 130:123–140, 1997.

18. van Reis R, Zydney AL. Protein Ultrafiltration. In: MC Flickinger, SW Drew, eds. Encyclopedia of Bioprocess Technology: Fermentation, Biocatalysis, and Bioseparation. New York: John Wiley & Sons, 1999:2197–2214.

19. Menon MK, Zydney AL. Effect of ion binding on protein transport through ultrafiltration membranes. Biotechnol Bioeng 63:298–307, 1999.

20. Menon MK, Zydney AL. Measurement of protein charge and ion binding using capillary electrophoresis. Anal Chem 70:1581–1584, 1998.

21. Gadam S, Phillips M, Orlando S, Kuriyel R, Pearl S, Zydney AL. A liquid porosimetry technique for correlating intrinsic protein sieving: applications in ultrafiltration processes. J Membr Sci 133:111–125, 1997.

22. Hallstrom B, Lopez-Leiva M. Description of a rotating ultrafiltration module. Desalination 24:273–279, 1978.

23. Chung KY, Bates R, Belfort G. Dean vortices with wall flux in a curved channel membrane system. 4. Effect of vortices on separation fluxes of suspensions in microporous membrane. J Membr Sci 81:139–150, 1993.

24. van Reis R, Goodrich EM, Yson CL, Frautschy LN, Dzengeleski S, Lutz H. Linear scale ultrafiltration. Biotechnol Bioeng 55:737–746, 1997.

11

Virus Removal by Ultrafiltration

A Case Study with Diaspirin Crosslinked Hemoglobin (DCLHb)

Kristen F. Ogle
Baxter Healthcare Corporation, McGaw Park, Illinois

Mahmood R. Azari
Baxter Healthcare Corporation, Thousand Oaks, California

I. INTRODUCTION

A. Brief History of Virus Removal by Membrane Ultrafiltration

Membrane ultrafiltration has long been recognized as a powerful and reliable technique for bioseparation, concentration, or reformulation of proteinaceous solutions. Many examples can be found in the literature for concentrating viruses from very dilute solutions using dead-end or tangential flow filtration [1–3]. At some point along the way, scientists realized that ultrafiltration could not only be used to concentrate viruses but also that it could remove them from the final product by retaining virus while allowing product to pass through the filter. Recent uses of ultrafiltration for removing virus particles from solutions are cited below. This is not intended to be a thorough review of the literature, but rather a representative guide of virus removal by filtration.

Flat-sheet membrane or pleated-sheet membrane ultrafiltration using an Ultipor VF grade DV50 membrane was demonstrated to reduce the titer of viruses > 50 nm by > 6 log and smaller viruses to a lesser degree [4,5]. However, the smallest viruses tested (Bacteriophage PP7 and poliovirus) were retained very little when suspended in a complex medium. Burnouf-Radosevich et al. demonstrated the removal of virus from Factor IX and Factor XI concentrates using

Planova 15N and Planova 35N filters operated in the dead-end mode [6]. These filters were constructed of cuprammonium-regenerated cellulose hollow fibers and had mean pore sizes of 15 and 35 nm, respectively. They reduced the titer of enveloped viruses (human immunodeficiency virus-1, bovine viral diarrhea virus, and pseudorabies virus) and nonenveloped viruses (bovine parvovirus, poliovirus, simian virus-40, and reovirus type 3) ranging from ~ 20 to 200 nm in size by > 5.7 log for both types of filter.

A 1988 European patent application [7] claimed virus removal using a Millipore polysulfone membrane in a tangential flow cassette system (Minitan or Pellicon). A 100,000-Dalton cutoff membrane was claimed to process biological fluids under pressure to remove contaminants such as human immunodeficiency virus and simian virus 40. It also claimed efficacy in reduction of prions from biological samples.

The Millipore Viresolve/70 membrane filters were evaluated for their ability to remove viruses from solutions [8,9]. These systems used a composite membrane having a "pre-formed microporous membrane plus a thin asymmetric, finely porous retentive layer" [8]. The membrane sheets had protein sieving characteristics corresponding to a slightly less than 100,000-Dalton cutoff and were used in fabricated plate-and-frame tangential-flow modules. Viruses with larger diameters (e.g., > 32.6 nm) were removed by an average of > 4.6 logs. Viruses < 32.6 nm were removed by 2.8 to 4.2 logs, depending upon the conditions.

In 1988 a patent was issued [10] claiming the use of hollow-fiber membrane filtration for removal of virus from an aqueous protein solution. This same approach has been demonstrated using the Asahi Planova 15N filter, which has a mean pore size of 15 nm, to remove sindbis virus, semliki forest virus, herpes simplex virus, polio-1 virus, encephalomyocarditis virus, bovine parvovirus, and hepatitis A virus [11]. The Planova 35 N filter, which has a larger pore size (35 nm), did not remove the small viruses (e.g., polio) to the same degree, confirming that virus removal was primarily dependent upon size. Hollow-fiber filters with mean pore sizes with molecular-weight cutoffs (MWCO) of 50,000, 13,000, or 6,000 Da were also evaluated for virus removal from process solutions [12]. In this case, poliovirus and phage PP7 were removed by more than 6 log from spiked growth medium using 13,000- and 6,000-Da MWCO membranes. The 50,000-Da MWCO filters were not effective in removing significant amounts of these viruses from the solutions.

Generally, several types of ultrafilters have been demonstrated to remove enveloped and some nonenveloped viruses to varying degrees. The virus removal tends to be based mostly on size, with a smaller contribution by adherence. Because there is no industrywide standard for classifying membranes, each type of membrane filter must be evaluated for virus removal in a case-specific situation.

The filters listed above are not the only ones acceptable for virus removal. Other ultrafiltration devices are commercially available which may be used to reduce virus titers in therapeutic products. A process for purifying human hemoglobin from red blood cell lysate for use in the production of the acellular oxygen transport solution diaspirin crosslinked hemoglobin (DCLHb) was designed using a hollow fiber filter (A/G Technology Corporation, average pore size distribution of 500,000 Da MWCO); the procedure was also used simultaneously for the removal of bloodborne viruses. The validation of this process is described in the remainder of this chapter.

B. Filtration in DCLHb Production

1. Process

DCLHb solutions have been the subject of clinical studies since 1992. The production process for DCLHb is well defined and uses established, well-known technology. The process was developed through extensive production experience and was designed to facilitate future scaleup. Although the red blood cell units used for production of DCLHb are screened for the presence of pathogens, virus inactivation/removal were included in the process to eliminate the potential for undetected contaminants from red blood cells being carried over into the final product; this also served to meet the virus removal requirements of regulatory agencies.

The process summary for production of DCLHb is shown in Figure 1. During production, red blood cells are pooled, the macroscopic debris is removed, and the cell suspension is diafiltered to reduce the extracellular proteins. The cells are then lysed to release the hemoglobin and the hemolysate is filtered to remove the cell membrane fragments (stroma) and viruses. The stroma-free hemoglobin is concentrated and deoxygenated in preparation for chemical crosslinking. After the crosslinking process, which stabilizes the hemoglobin tetramer, the solution is heat treated. The heat-treatment step inactivates viruses, precipitates residual nonhemoglobin proteins, and precipitates uncrosslinked hemoglobin. The crosslinked hemoglobin is then oxygenated and filtered to remove the protein precipitate. The recovered product (DCLHb) is ultimately concentrated and diafiltered into an electrolyte solution to assure physiological compatibility. After adjustment of the hemoglobin concentration and pH, the final product is sterile filtered to reduce the bioburden, filled into the final containers, and packaged for distribution.

2. Choice of Filters

The choice of membrane filter for use in virus removal from DCLHb was made based upon efficacy and economics. To ultimately be successful in manufacturing, a virus removal filter must be able to process the material properly and efficiently, be robust enough to be used many times, and pass post-use, virus-correlated integrity testing. The ideal virus removal filter would also be part of the

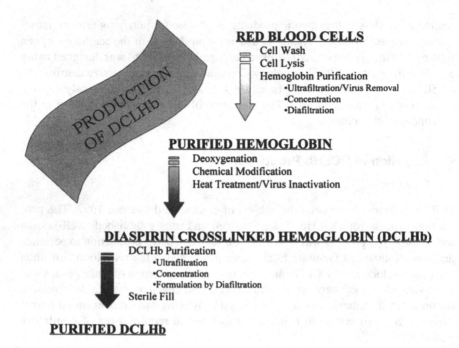

Figure 1   DCLHb production process summary.

bioprocess operation, although adding an additional virus removal filtration step would be an acceptable, more costly alternative.

Several commercially available filter types were evaluated for application to virus removal from the process stream: stacked, dead-end, and hollow-fiber filters. Various Asahi Planova (75 nm, 35 nm, and 15 nm pore size; dead-end operation), Pall Ultipor VF-UDV50 (dead-end operation), Millipore Viresolve/70 (stacked membrane module), Romicon (Koch CTG hollow-fiber), and A/G Technology (500,000 MWCO, 100,000 MWCO and 0.1-μm pore size) filters were subjected to evaluation under controlled conditions to determine the effect on processing and/or virus removal.

In particular, hollow-fiber modules operated in the tangential-flow filtration mode processed hemoglobin very efficiently. The result of the preliminary processing evaluation was that A/G Technology hollow-fiber filters were chosen for use in the stroma removal step because the processing rate was acceptable (60 to 100 mL/min/ft^2), the filter was robust enough to be reused and subjected to nondestructive integrity testing, and the filter removed significant amounts of virus. The ultimate goal for an economically feasible filter was to be reusable. However, with each use and cleaning, filters risk damage of the membrane. If the membrane becomes torn or damaged and the integrity is compromised, then permeation of larger particles or viruses may occur. Therefore, demonstration of filter integrity after each use was required to assure adequate virus removal properties. A test method for this purpose was designed so that damage to the membrane during the test procedure was avoided. In addition, the integrity test was correlated to virus removal to ensure adequate virus retention during manufacture.

C. Need for Virus Removal/Inactivation Steps

1. Regulation

This chapter is intended to be a case study on virus validation and therefore will not be an in-depth study of regulations for virus removal. Several excellent discussions of the topic have been written [13–18].

Regulations of particular importance to our decision of how to validate the filtration process for virus removal were the 1991 Committee for Proprietary Medicinal Products (CPMP) guidelines on virus removal/inactivation validation [19], the Paul Ehrlich Institute (PEI) announcement concerning virus safety of drugs derived from human blood [20], the CPMP guidance on virus validation studies [21], the CPMP guidance on plasma-derived products [22], and the International Conference on Harmonisation of Technical Requirements (ICH) guidelines for viral safety evaluations [23]. It was determined that two distinct steps which complement each other would be used to remove and/or inactivate virus from the process stream, at least one of which is effective against nonenveloped viruses. Process requirements were defined such that enveloped viruses must be reduced by at least 10 logs using two or more steps which each remove at least 4 logs of virus and nonenveloped viruses must be reduced by 6 logs with at least one step reducing a minimum of 4 logs of virus. It was recognized that the interpretation of virus removal is dependent upon the test model and the log reductions and that each step must be considered carefully.

2. Virus Removal/Inactivation in DCLHb Production

Because DCLHb is produced from human packed red blood cells, robust methods for virus removal were required. The process for producing DCLHb from red blood cells was designed to produce high quality material with high standards of safety. The two process steps which were implemented and validated with respect to virus removal or inactivation were stroma removal from hemolysate by filtration and crosslinked hemoglobin heat-treatment. The use of two very different techniques provided higher probability of inactivating/removing virus than two steps of similar types, allowing the overall log removal to be determined as the sum of the individual log removal values.

(a). Heat. The heat treatment step used to precipitate uncrosslinked hemo-globin and contaminating proteins from DCLHb reaction mixture proved to be very robust with respect to virus removal. The validation procedure was carefully designed to represent a scaled-down model of the manufacturing process step and to accurately reflect virus inactivation. Validation studies for this process step [24–27] were performed in conjunction with contract validation laborato-ries (Microbiological Associates, Gaithersburg, MD and Quality Biotech, Inc., Camden, NJ). The use of a scaled-down test system in these studies enabled attainment of high titers of virus, which would be impossible to achieve at large scale, and also eliminated the possibility of contamination of actual processing facilities.

The heat treatment system was validated using model viruses and viruses which were considered relevant to blood derived products (see Section IIB.). Greater than 7.9 log removal of enveloped viruses and > 7.2-log removal of nonenveloped virus was achieved [24–28].

Even though the heat treatment step was shown to be robust and highly effective, it was somewhat selective in that some nonenveloped virus remained following the process. To demonstrate adequate virus removal in the finished product, a second virus removal step (filtration) was validated.

(b). Filtration. The processing scheme for DCLHb contains several fil-tration steps that were possible options for validation of virus removal. The pro-cess step in which stroma fragments are removed from hemolysate was selected for this purpose. This step was relatively early in the production process and used A/G Technology 500,000 MWCO pore size filters (500K filters). Preliminary experiments using simple recirculation systems indicated that the A/G Technol-ogy 500K filters did indeed remove virus from the process stream [29].

The validation system used to assess virus removal during hemolysate fil-tration was designed to represent a realistic model of the filtration system and also provide a measurable evaluation of virus removal capacity. In addition, a nondestructive integrity test was developed which was correlated with virus re-moval. Finally, the scaled-down process system and integrity testing system were applied to the validation of virus removal by the 500K filter.

II. VALIDATION DESIGN

Scaled-down model systems are used in validation of virus removal for three very important reasons:

1. *Virus titer.* Limitations on the yield of model viruses used for introduc-tion into the systems make large-scale inoculations difficult, if not impossible.

2. *Statistical analysis.* Virus titer determination from biological samples

is typically tedious and time-consuming. In addition, relatively large samples are required to accurately determine very low titers of virus, as was typical for the virus removal step. Therefore, it is easier and more accurate to analyze some or all of the test material from a small model system than from a large system.

3. *Contamination control.* Use of model systems eliminates contamination of the manufacturing system with potentially pathogenic viruses.

The design of the model system is very important to the validation of the virus removal process. Care must be exercised to ensure that all variable parameters of the relevant process step represent the worst, practical case. To do this, operating parameters must be identified which may potentially adversely or favorably affect virus removal by the process step (i.e., filtration). In this validation study, variable filtration parameters such as transmembrane pressure (TMP) and crossflow rates were evaluated. Filtration volume load and protein concentration were also considered as variables, as was filter pore size distribution. Other factors such as temperature and pH, which could potentially affect filtration, were not varied, as extremes in these factors cause irreversible degradation of the product and were therefore beyond the allowable manufacturing criteria.

The scale of the models was 1:300, relative to the manufacturing process. The degree of scaling was selected as the smallest practical scale of the system that could be accurately instrumented and operated, relative to manufacturing. Special considerations were given to the construction of the filter so as to reflect the diameter and length of the manufacturing-scale filters, yet have reduced filter surface area. That is, the fibers were the same diameter, length, pore size and manufacturing lot as those used in manufacturing; the surface area was reduced, relative to manufacturing, by using fewer fibers in the filter unit.

A. Design of Filtration Models

1. Recirculation Model

Preliminary studies to define the effect of filter pore size and test solution concentration on virus removal were performed using a simplified, closed-loop filtration system (recirculation model; Figure 2). This system consisted of a retentate vessel and hollow-fiber filter. The filter retentate and filtrate were returned to the retentate vessel. Monitoring sensors for filter inlet, outlet, and permeate pressures; filter inlet and filtrate fluid flow rates; and solution temperature were installed in the system. A filter feed pump and filter retentate line back-pressure clamp were used to maintain the system parameters at the desired settings. Aseptic sample ports were inserted in the retentate vessel, filter retentate line, and filter permeate (filtrate) line.

To operate, the system was stabilized and then inoculated with viable virus. Samples were removed for virus titration at defined intervals to determine the

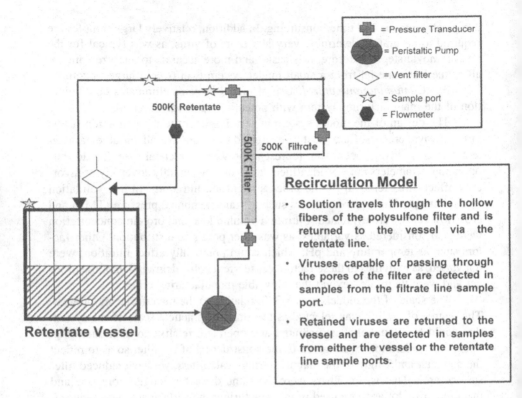

Figure 2   Schematic diagram of recirculation model.

degree of virus removal over time. Care was given to the calculation of remaining solution volume to account for materials balance.

2.   Scaled-Down Process Model

For some experiments, a scaled-down version of the process filtration and concentration steps was used to more accurately represent the specific process step being validated (scaled-down model, Figure 3). This model system consisted of two circuits: the stroma removal circuit, and the stroma-free hemoglobin (SFHb) concentration circuit. The stroma removal circuit consisted of a hemolysate reservoir that was connected to the feed and retentate ends of a 500,000 nominal MWCO filter (500K). The permeate ports of the 500K filter were fed to the SFHb retentate vessel, which was connected to the feed and retentate ports of a 10,000 nominal MWCO filter (10K). The permeate port of the 10K filter was fed either to waste or to the hemolysate vessel (diafiltration step). The vessels were both equipped

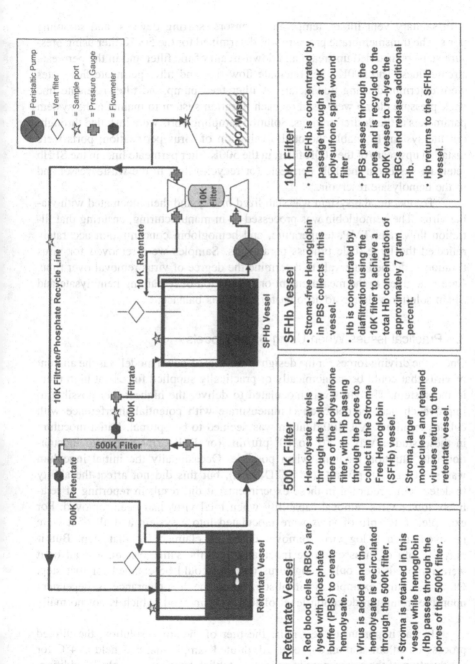

Figure 3 Schematic diagram of scaled-down model.

= Peristaltic Pump
= Vent filter
= Sample port
= Pressure Gauge
= Flowmeter

PO, Waste

10K Filter
10K Retentate

SFHb Vessel

500K Filtrate

10K Filtrate/Phosphate Recycle Line

500K Filter

500K Retentate

500K Filtrate

Retentate Vessel

Retentate Vessel

· Red blood cells (RBCs) are lysed with phosphate buffer (PBS) to create hemolysate.

· Virus is added and the hemolysate is recirculated through the 500K filter.

· Stroma is retained in this vessel while hemoglobin (Hb) passes through the pores of the 500K filter.

500 K Filter

· Hemolysate travels through the hollow fibers of the polysulfone filter, with Hb passing through the pores to collect in the Stroma Free Hemoglobin (SFHb) vessel.

· Stroma, larger molecules, and retained viruses return to the retentate vessel.

SFHb Vessel

· Stroma-free Hemoglobin in PBS collects in this vessel.

· Hb is concentrated by diafiltration using the 10K filter to achieve a total Hb concentration of approximately 7 gram percent.

10K Filter

· The SFHb is concentrated by passage through a 10K polysulfone, spiral wound filter.

· PBS passes through the pores and is recycled to the 500K vessel to re-lyse the RBCs and release additional Hb.

· Hb returns to the SFHb vessel.

with sanitary vent filters, temperature sensors, stirring devices, and sampling ports. The transmembrane pressure was determined for the 500K filter using pressure sensors installed upstream and downstream of the filter and in the permeate stream lines. The 500K filter retentate flow rate and filter permeate flow rates were determined using flowmeters. A filter feed pump and filter retentate line back-pressure clamp were used on each filtration system to maintain the system parameters at the desired settings. Solution sampling ports were installed throughout the system to enable thorough evaluation of virus permeation; ports were installed upstream of the 500K filter, in the 500K filter permeate line, in the SFHb retentate vessel, in the 10K permeate (or recycle) line, in the waste vessel and in the hemolysate reservoir.

To operate, the system was stabilized briefly and then inoculated with viable virus. The hemoglobin was processed as in manufacturing, ensuring that filtration flow rate, TMP, temperature, and hemoglobin concentration accurately reflected the large-scale process parameters. Samples were removed for virus titration at defined intervals to determine the degree of virus removal over time. Care was given to the measurement or calculation of remaining hemolysate and SFHb solution volumes to account for materials balance.

3. Practical Issues When Using Small Models

One of the driving forces for the design of the scaled-down model was the amount of virus that could be economically or practically supplied for use at high titers in the system. The virus was inoculated to deliver the highest titer possible to improve chances of detection commensurate with potential interference with culture solutions. A 1:100 dilution was deemed to be optimal. Initial inoculum in these studies was $\sim 10^8$ to 10^9 pfu/mL (or $TCID_{50}$/mL) prior to inoculation into filtration medium, when possible. Occasionally the initial inoculum was $< 10^8$ to 10^9 pfu/mL (or $TCID_{50}$/mL), but this did not affect the ability to detect virus removal in these experiments; it did result in reporting of relatively lower virus removal capability when total virus had been removed. For example, if 10^9 pfu of virus were inoculated into a system and all of it were removed, then 9 log virus removal could be claimed for that step. But if only 10^5 pfu of virus could be inoculated into the same system, and all of it were removed, then only 5 log virus removal could be claimed for that step. Of course, this example is simplistic, since the log clearance is dependent upon the minimum detection level of the assay method, which is not normally zero.

It was important to determine the titer of the stock solution, the diluted process inoculum, and an unprocessed, diluted sample that was held at 4°C for the duration of the experiment in order to establish baseline controls. In addition, preliminary studies were performed with the challenge virus in the presence of process materials to determine if the latter possessed properties that were inhibitory or lethal to the virus. If materials had been found to be lethal or inhibitory to the virus, then interpretation of the virus titers would have been compromised. Samples that were not expected to contain virus, or contain low levels of virus, were difficult to interpret because absence of virus does not indicate sterility. Rather, these samples were titered in larger volumes to account statistically for relevant reductions in titer. Likewise, reporting of results in which no virus particles were detected could only be reported as "below a threshold level of detection"—not as zero infectious particles.

For safety reasons, all sample ports were designed for needleless sampling (Baxter, Fenwal InterLink sample ports and cannulae). As an additional measure of safety, all test systems were operated in biological safety cabinets with HEPA-filtered air, and operators were required to wear protective clothing, eyewear, and gloves. This was especially important because some challenge viruses were bloodborne, human pathogens such as HIV and HAV.

Another driving force for the design of the model was the size limitation of equipment and instrumentation. Solution transfer lines were miniaturized as much as possible and still fit standard, off-the-shelf equipment, such as pressure transducers, pumps, sanitary connectors, and flowmeters. However, because the solution volumes were relatively low for standard process-monitoring equipment, small turbine flowmeters (EG & G Instruments, Phoenix, AZ) used with 1/4'' sanitary connectors were purchased and calibrated for use in the low flow range (20 to 1000 mL/min). The permeate solutions were more difficult to measure, as they flowed in the range of 10 to 50 mL/min. In this case, ultrasonic flowmeters (Transonic Systems, Inc., Ithaca, NY) were purchased which measured the flow through the flexible tubing. Pressure transducers or gauges (0 to 60 psi) were purchased with 3/4'' sanitary connector adapters. The equipment used in the virus evaluation/validation studies was designed for pharmaceutical sanitary operation and none was reused in large-scale production.

The test filter, an A/G Technology 500K hollow-fiber filter (model UFP-500-E-3X2, custom made, now sold as VAG-500-E-3X2), was custom made for the system. The length of the filter was important for evaluation of fluid flow dynamics within a typical filter and therefore was manufactured at the same length as the large-scale system filters. To scale the filter down to model size, fewer fibers of the same diameter as manufacturing-scale filters were used in the model to generate a filter with ~ 1:300 the area of the large-scale process filters.

Solution vessels (Labglass or Applicon) were purchased in glass to allow visual evaluation of the process solutions and were jacketed to enable circulation of coolant for maintenance of temperature at ≤ 10°C. These vessels were fitted with custom-made stainless steel headplates with sanitary fittings. Stirring assemblies (Applicon) were purchased and the shafts were modified to fit the vessels to ensure adequate mixing.

Finally, to establish that the filters were intact at the conclusion of the study and to show a correlation with virus removal, a nondestructive filter integrity test had to be designed and validated. The test was initially designed for use with the custom-designed filter and was later scaled-up for manufacturing use (see Section III).

B. Model Viruses

The viruses used in the studies were chosen to reflect a variety of sizes, composition, and robustness (Table 1). The chemical and genetic composition of model viruses is less important for removal by filtration than for inactivation by other means (e.g., heat, detergent, etc.) because removal is primarily due to sieving in the case of filtration. The inclusion of a variety of viruses in these studies was to confirm size exclusion as the primary removal mechanism, for compliance with regulatory guidelines, and to facilitate the calculation of overall virus removal factors for the manufacturing process as a whole.

Feasibility studies were performed with non-pathogenic, model viruses: Bacteriophage ϕX174 (PHG), encephalomyocarditis virus (EMC), pseudorabies virus (PRV), and bovine viral diarrhea virus (BVDV) to determine the best model viruses to use in determination of worst-case filtration parameters. The filters in these recirculation model studies quantitatively retained EMC, PRV, and BVDV. PHG was partially retained by representative 500K filters. EMC and PHG were the two smallest viruses used in the studies, and were used in subsequent virus removal evaluation studies.

Validation studies were performed using the scaled-down filtration model and relevant human blood-borne viruses or their models: BVDV, PRV, hepatitis

Table 1 Physical Properties of Viruses Used in Filtration Removal Studies

Virus	Size (nm)	Envelope	Genome	Chemical/ heat resistance	Model for
PHG	22	no	DNA	NA*	(industry standard)
EMC	25–30	no	RNA	NA	polio
PPV	18–24	no	DNA	very high	parvo B-19
HAV	28	no	RNA	high	hepatitis A
BVDV	40–70	yes	RNA	medium	hepatitis C
HIV	80–110	yes	RNA	low	HIV
PRV	120–200	yes	DNA	low–medium	herpes hepatitis B

* NA, not available. *Source*: Refs. 13, 23, 30.

A virus (HAV), human immunodeficiency virus (HIV), and porcine parvo virus (PPV).

C. Determination of Worst, Practical Case Conditions for Virus Removal

A series of experiments was performed to determine the worst-case parameters to be used in the virus validation studies. These experiments evaluated the pore size distribution of the filter membrane, the effect of red cell hemolysate concentration, and the effect of the filtration crossflow rate and the filter transmembrane pressure. Log virus reduction (Eqs. 1 and 2) was used to evaluate relative effectiveness of the variations in the recirculation and scale-down procedures, respectively.

$$\log \left(\frac{Average\ retentate\ solution\ virus\ titer,\ \text{pfu/mL}}{Average\ permeate\ solution\ virus\ titer,\ \text{pfu/mL}} \right) \tag{1}$$

$$\log \left(\frac{Initial\ hemolysate\ titer,\ \text{pfu/mL} \times solution\ volume,\ \text{mL}}{Final\ SFHb\ titer,\ \text{pfu/mL},\ \times solution\ volume,\ \text{mL}} \right) \tag{2}$$

Unless otherwise stated, temperatures, solution concentrations, TMPs and crossflow rates were operated within the current manufacturing specifications. The solution processing volume per filter area was increased by $> 50\%$ above that used in manufacturing to represent worst case, relative to processing capabilities. This increased the total amount of protein, cellular debris, and virus particles contacting the filter over time as compared to typical filter operation in manufacturing. However, the particulate density in the study solution was the same as the most concentrated allowable in manufacturing. This resulted in a realistic representation of the solution composition while providing worst-case processing load conditions.

1. Filter Pore Size/Pore Distribution

The virus retention characteristics of filter membranes with various pore sizes were determined using the recirculation model with PHG or EMC inoculated 1:100 into phosphate saline buffer. The filters tested had nominal MWCO pore sizes of 100,000 (100K) Da, 500,000 (500K) Da, and 1,000,000 (1000K) Da. Additionally, a filter with a membrane having a 0.1-µm pore size was also evaluated. EMC was completely retained by 100K and 500K membrane filters, and partially retained by 1000K and 0.1-µm filters (Figure 4). PHG was highly retained by 100K and 500K filters, but not retained by 1000K and 0.1-µm filters in phosphate buffer (Figure 4). Therefore, 500K filters with more open-pore dis-

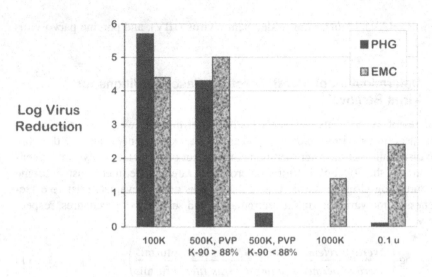

Figure 4   Virus retention by hollow-fiber filters differing in pore sizes. Pore size cutoffs: 100K, 100,000 Da; 500K, 500,000 Da; 1000K, 1,000,000 Da; 0.1 u, 0.1 µm. PHG, bacteriophage ΦX174 (dark bars); EMC, encephalomyocarditis virus (light bars).

tribution as determined by the manufacturer's integrity test (PVP K-90; Section III) represented the worst-case situation for virus removal. 1000K and 0.1-µm filters were not used in validation studies because they did not retain virus and were not representative of the filter type used in manufacturing.

2. Test Solution

The relationship between the concentration of hemolysate in the filtration medium and virus retention by 500K filters was examined using the recirculation model. Three test solutions were tested at 77%, 100%, and 133% of the hemolysate dilution for the manufacturing process step (Figure 5). The models used the same volume of test solutions, which varied only in protein concentration due to dilution. The least retention of PHG, and therefore worst-case condition, was with the hemolysate diluted by only 77% of the typical process dilution. This solution (77% of process dilution) contained higher numbers of intact RBCs, fewer small stromal fragments, and relatively more protein per volume than typical diluted hemolysate (100% of process dilution). As the hemolysate dilution increased, the protein content per volume decreased.

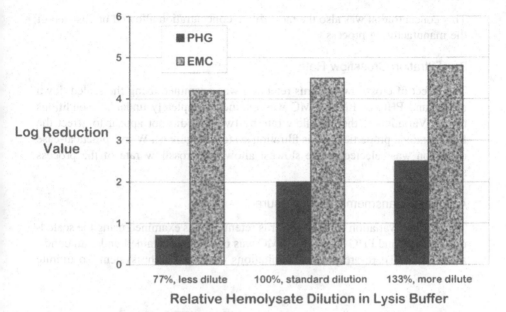

Figure 5 Virus retention by 500K filters in presence of different hemolysate concentrations. 100% refers to typical lysed red blood cell solution during production. PHG, Bacteriophage ΦX174 (dark bars); EMC, encephalomyocarditis virus (light bars).

Virus retention was inversely proportional to hemolysate dilution; that is, the virus retention was greater in hemolysate which was diluted more than the typical process solution (> 100% of process dilution), and fewer viruses were retained by filtration in hemolysate which was diluted less than the typical process solution (< 100% of process dilution). The nature of this phenomenon is poorly understood, but is thought to be related to competition with stromal fragments. For example, as dilution of red blood cells with lysis buffer increases, so does the efficiency of lysing the cells. Consequently, more efficient lysis results in higher quantities of smaller stromal fragments. These small fragments may form a barrier by blocking portions of the pores from access by intact viruses, thereby increasing virus retention.

Because hemolysate made at 77% of the standard dilution was relatively more viscous and the filtration/diafiltration of hemolysate this concentrated was difficult, it was not practical to use this dilution in filtration studies. Therefore, hemolysate generated at 90% of the standard dilution was used in the remaining prevalidation and validation filtration studies as the worst, practical-case solution.

This concentration was also the most dilute concentration allowed in this step of the manufacturing process.

3. Filtration Crossflow Rate

The effect of crossflow on virus retention was examined using the scaled-down model and PHG or EMC. EMC was retained completely under all conditions tested. Variation of the crossflow rate by two-fold did not appear to affect the PHG removal properties of this filtration system (Figure 6). Worst, practical-case operation was selected as the slowest allowable crossflow rate of the process step.

4. Filter Transmembrane Pressure

The effect of variation of TMP on virus retention was examined using the scaled-down model and PHG and EMC. EMC was completely retained under all conditions tested. There are physical limitations of TMP on the system, so infinite

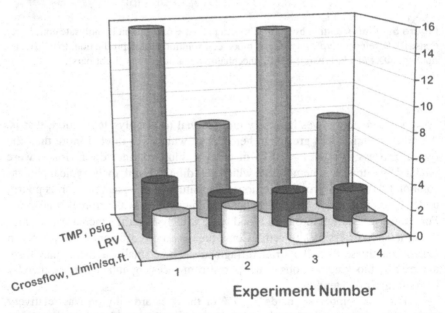

Figure 6   Bacteriophage ΦX174 log reduction by 500K filtration when varying TMP and crossflow rate. Four experiments were performed by varying crossflow and TMP: (1) high crossflow and high TMP; (2) high crossflow and low TMP; (3) low crossflow and high TMP; and (4) low crossflow and low TMP. (Crossflow, light columns; LRV, dark columns; TMP, medium-shade columns.)

adjustments to TMP could not be made. However, within the tested, feasible range of variation for this filter, doubling the average TMP approximately doubled the PHG retention (Figure 6). Therefore, to operate the model under worst-case conditions, the lowest practical TMP was selected which would still allow permeation of SFHb.

III. FILTER INTEGRITY TESTING

The ultimate selection of the 500K filter for virus removal properties depended on the ability to demonstrate that the filter remained intact during operation. Since virus challenge subsequent to every process use is impractical, a physical or chemical test that can be correlated with virus retention needed to be identified. Such a postuse test can assure that filter integrity was maintained. Ideally, the test will be nondestructive, permitting filter reuse in processing.

Two different nondestructive methods were evaluated for use as correlatable integrity tests: air diffusion rate testing, and PVP K-90 retention.

A. PVP K-90 Retention

A/G Technology, the manufacturer of the 500,000 nominal MWCO membranes used in the production of DCLHb, historically characterized filters using a solute retention/rejection test. This test measures the amount of polymer retained by the membrane under defined conditions. The polymer, poly(N-vinyl-2-pyrrolidone), also called PVP, is a hygroscopic powder that is colorless in liquid and dissolves in both water and organic solvents. PVP solution consists of a series of different-chain-length polymers; the molecular weight is expressed as an average of the various molecular weights of the different chain length units that comprise the polymer. The PVP molecules have been historically classified based on the K value, which represents a function of the average molecular weight, the degree of polymerization, and the intrinsic viscosity. The solution used to characterize the filters was PVP K-90, which has an average molecular weight of 630,000.

In the retention test, an aqueous solution of known PVP K-90 concentration is recirculated through a filter for a specified period of time under defined TMP, cross flow rate, and temperature conditions. The percentage of the solution permeating the filter is determined after a period of equilibration and recorded as the relative retention value for this filter. For example, if 85% of PVP K-90 is retained by the filter, then the value is reported as 85% PVP K-90 retention. This type of filter will have a more open pore size distribution than will a filter with a PVP K-90 retention rate > 85%.

B. Air Diffusion Rate Test

Another nondestructive test used to characterize the pore size distribution of 500K filters was also developed, based on the measurement of gas diffusion through appropriately wetted filters. The air diffusion rate (ADR) test exploits the flow rate of a gas passing through a prewetted membrane as a function of applied pressure. The theory of the test is summarized below:

> The shape of a typical flowrate-versus-pressure curve may be explained in terms of the relative contributions of diffusive and convective flow through a completely wetted membrane. At applied pressures sufficiently below the bubble point of the wetted membrane, the flowrate increases linearly with pressure due to simple diffusion of dissolved gas through the liquid in the pores. For ideal membranes, which contain uniform cylindrical pores, when the pressure reaches the bubble point of the membrane, the gas pressure will overcome the surface tension, which holds the liquid in the pores. At this point the liquid is forced out of the pores and gas flows convectively through the open pores, resulting in a sharp increase in flowrate above the linear dependence of diffusive flow on pressure (i.e., the bubble point is reached).
>
> For nonideal membranes which contain randomly shaped pores whose cross-sectional area may vary along the length of the pore, as the applied pressure nears the bubble point of the membrane, the diffusive pathlength through the membrane may decrease as liquid is forced partway out of a pore. This results in a more gradual departure from the linear dependence of diffusive flow on pressure. When the pressure reaches the bubble point of a membrane with nonuniform pore size distribution, the gas pressure will overcome the surface tension in the largest pores, forcing the liquid out of those pores. This allows convective gas flow through the largest pores, and gives rise to a more marked increase in the gas flow rate above and beyond the linear diffusive flow. At higher pressures, liquid is forced out of successively smaller and smaller pores, leading to further nonlinear increases of gas flow rate [28].

The ADR test was developed for filters with irregular pores and exploited the traditional low-pressure ADR test. The volume of gas diffusing through the filter per unit time was determined at certain pressures. As the pressure increased, so did the gas diffusion rate until the liquid barriers in the largest pores were disrupted, causing a significant increase in gas flow (bubble point). The characteristic curve generated by plotting gas flow versus pressure was evaluated for a variety of filters in the 100,000-Da, 500,000-Da, and 1,000,000-Da MWCO pore size ranges. The curve was similar for each class of membrane, but the smaller pore size membranes had a left-shifted curve and the larger pore size membranes had a right-shifted curve, relative to the 500K membranes, as expected.

Because the curves are reproducible for each filter, they can be used to compare filters to each other. By evaluating the gas flow rate at pressures within the linear portion of the curve and below the bubble point, the ADR was reliably used as a relative measure of the filter pore size distribution for a particular filter. To simplify operation of the test, the ADR for each filter was determined at 6.1 bar after an initial recirculation period. The ADR at 6.1 bar was determined for each filter and used as an indicator of filter pore size distribution. The ADR value was reproducible and able to identify filters that had been compromised in filter pore size or integrity.

This ADR test was validated for use with the small, virus validation filters as well as with the larger, manufacturing-size filters.

C. Correlation of Virus Retention with Integrity Test

Representative 500K filters were selected and sorted by the manufacturer's PVP K-90 value or the experimentally derived ADR value. The degree of virus removal was determined for these filters to generate a correlation between virus removal and the integrity test. To provide a relative comparator for the study, filters with larger pores (0.1 μm or 1,000,000 MWCO) were also tested.

The filters were tested using the recirculation model system. Three viruses differing in size—encephalomyocarditis (EMC, 25–30 nm), porcine parvovirus (PPV, 18–24 nm), and bacteriophage φX174 (PHG, 25–28 nm)—were analyzed in the system for their degree of removal by the filter.

The amount of virus removed during a defined period of time was determined for the filters in each PVP K-90 retention category (Figure 7). The filters that retained PVP K-90 by 80% or more were able to retain at least 2.3 log of PPV or EMC suspended in saline buffer. Bacteriophage φX174 was not retained to a significant degree by any filter with a PVP K-90 retention of < 99%. As a reference, the large-pore filter, 0.1 μm, did not retain any virus to a significant degree and the PVP K-90 retention was < 73%.

The amount of virus removed during a defined period of time was also determined for the filters in each ADR category (Figure 8). Therefore, the typical 500K filters (PVP K-90 retention of 80% to 97%) and 100K filters (PVP K-90 retention of ≥ 99%) correlated with retention of mammalian viruses, while larger-pore-size filters did not retain virus.

Some variability was seen in the LRV values in these studies. This was most likely due to the variability in pore-size distribution and pore shapes of the 500K filter membranes. The pore-size distributions for A/G Technology 500K filters is fairly tight, relative to filters in general. However, some variability does exist in the filters; some filters in this MWCO range may have a greater number of larger pores, which would produce a low or average LRV and a correspondingly high ADR value. In contrast, some filters may have a greater number of smaller pores, producing the opposite effect of a higher LRV and a lower ADR value. Filters with a greater number of large pores would be excluded from use

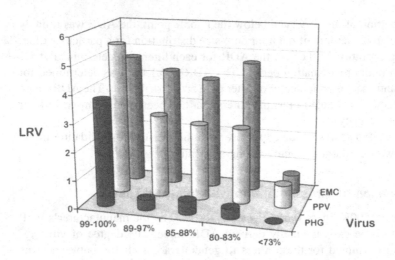

Figure 7 Virus retention by hollow-fiber filters differing in PVP K-90 retention. PHG, Bacteriophage ΦX174 (dark columns); PPV, porcine parvovirus (light columns); EMC, encephalomyocarditis virus (medium-shade columns).

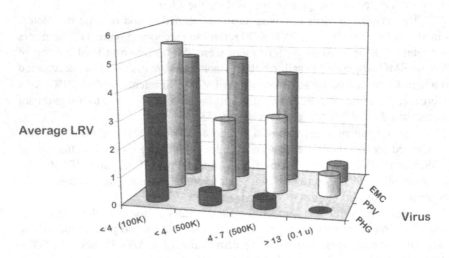

Figure 8 Virus retention by hollow fiber filters differing in ADR values. ADR was measured in standard cubic centimeters (SCCM). Filter pore sizes: 100K, 100,000 Da; 500K, 500,000 Da; 0.1 u, 0.1 μm. PHG, bacteriophage ΦX174 (dark columns); PPV, porcine parvovirus (light columns); EMC, encephalomyocarditis virus (medium-shade columns).

because the ADR value would lie outside the range determined to be acceptable for this application. Filters with ADR values outside the acceptable range were used in other processing steps that were not designated for virus removal.

The acceptable criteria for new A/G Technology 500K filters which retain viruses to a significant degree were conservatively defined as those with PVP K-90 retention $\geq 89\%$ and ADR value of ≤ 27.5 SCCM/ft^2 at 6.1 bar. The filters were demonstrated to retain viruses as long as the ADR was < 27.5 SCCM/ft^2.

D. Practical Considerations of Integrity Testing

Nondestructive integrity tests can be very powerful for qualifying filters as virus removal steps in the production of pharmaceuticals. However, performance of these tests can be difficult and correlation to virus removal can be problematic.

Several parameters must be considered when developing an integrity test for hollow-fiber filters. TMP, cross flow rate, solution composition, and temperature must be optimized when developing the integrity test because small variations in these parameters may make big differences in the test outcome. When correlating the parameters with virus removal, the integrity test should be performed under conditions that are reproducible; in particular, airflow should be stable or changing only slowly in a predictable fashion.

Once the operational parameters of the test are optimized, the solution lot-to-lot variability, instrument-to-instrument variability, and analyst-to-analyst variability must be determined to set limits for acceptance criteria. It is then prudent to adjust the acceptable limits to accommodate the test variability. Strict adherence to the developed test method and rigorous analysis of the test materials will allow a high degree of confidence in the application of the method.

IV. FORMAL VALIDATION OF VIRUS RETENTION BY 500K FILTERS

The establishment of worst, practical-case operating parameters and a robust integrity test enabled the 500K filtration system to be validated with regard to virus removal. Due to the infectious nature of some of the viruses used, the study was performed at a contract site.

A. 500K/10K Filtration System

1. Validation Model Design

The system used in the virus validation studies was the scaled-down 500K/10K filtration system described above. The system was operated with $\sim 60\%$ greater processing volume per filter area and a 10% higher protein concentration per volume than used in manufacturing. The crossflow rate, filter permeate flow rate, TMP, and filter pore-size distribution were worst, practical case. The processing and product concentration endpoints were the same as in manufacturing to assure that the model performed properly.

The experiments were performed in duplicate with five model viruses: HIV, HAV, PPV, BVDV, and PRV. The virus titer analysis was performed by Microbiological Associates, and results were evaluated using statistical analysis.

2. Results

The model system used to validate virus removal by 500K filters was evaluated with respect to two criteria: Suitability of the scaled model as a representative of full-scale manufacturing and efficacy of virus removal.

The model was evaluated with regard to crossflow rate, TMP, process temperature, solution volume filtered, and hemoglobin and methemoglobin concentrations of the pre- and postprocessing solutions. The results obtained from the validation experiments were compared with the corresponding values from full-scale manufacturing. Results from these experiments showed consistency between the model system and the manufacturing process. This confirmed that the scaled-down test system used to validate the A/G Technology 500K filter in the stroma removal step was representative of the full-scale manufacturing process and can reliably predict the minimum viral retention for that step.

Some portions of model systems may not scale proportionately and must be considered for the model. In this case, the filtration flow rate was slower than anticipated for a direct scale. The decision was made to maintain the length of the filter and scale the area proportionately. The same fibers were used for the model filter that were typically used in the large-scale filters. Therefore, the only physical differences between the model filter and full-scale process filters were the number of fibers and filter housing diameter. The important consideration when parameters do not scale directly is that the model must not bias the results of the study. In this case, since the same fibers were used in the two filters, and the process parameters were proportionate to the full-scale filters based upon area, a reasonable scaledown was achieved. The slower filtration by the model filters was most likely due to the problem of filter soiling during processing. A few fibers routinely become clogged with debris during this stage in processing. This is not a problem with large filters, which contain thousands of fibers. However, the smaller filters only contain around a dozen fibers where clogging of a single fiber would represent 100 times the clogging of a large filter. However, clogging of the filter fibers still represented worst case because the filter in these conditions would have experienced more virus exposure per fiber.

Virus removal by the filters was validated in these experiments. The viruses were selected to possess a variety of properties for nucleic acid, encapsulation,

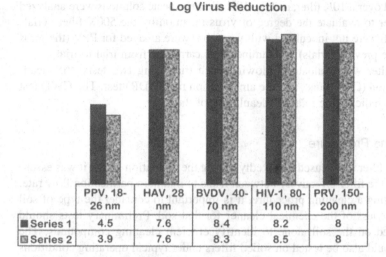

Log Virus Reduction

	PPV, 18-26 nm	HAV, 28 nm	BVDV, 40-70 nm	HIV-1, 80-110 nm	PRV, 150-200 nm
■ Series 1	4.5	7.6	8.4	8.2	8
⊠ Series 2	3.9	7.6	8.3	8.5	8

Figure 9 Log virus reduction by 500K filters in formal validation study. Virus sizes indicated in table. PPV, porcine parvovirus; HAV, hepatitis A virus; BVDV, bovine viral diarrhea virus; HIV-1, human immunodeficiency virus 1; PRV, pseudo-rabies virus.

particle size, and chemical/heat resistance. The filters retained all encapsulated viruses with no detectable filter permeation (Figure 9). Only the smallest virus, PPV, was able to permeate the filter membrane. However, the retention of even the smallest virus tested was ~ 4 logs, making this a nonselective, robust virus removal step in the production of DCLHb.

B. Repeated Use of 500K Filter

In the initial validation studies, clean, virgin filters were used for each run to ensure robustness of the experiment design and to eliminate the potential of filters to vary in performance upon use. A second study was performed to validate virus removal by the filters during multiple reuse.

1. Validation Model Design

The system used the same design as the previous validation study (i.e., the 500K/10K scaled-down model). The evaluation of virus removal during multiple reuse of an A/G Technology 500K hollow-fiber filter was accomplished by performing 16 stroma removal filtration/SFHb concentration cycles. The hemolysate solution was spiked with a nonenveloped virus (PPV) or an enveloped virus (BVDV) at the beginning of selected trials. The 500K filter retentate, 500K filter permeate concentrated over a 10K filter, and the 10K filter permeate solutions were analyzed for virus titer to evaluate the degree of virus retention by the 500K filter. Trials 5, 11, and 16 were not inoculated with virus, but were assayed for PPV (the virus spiked in the previous trials) to examine virus carryover from trial to trial.

The filter was evaluated following each trial using two tests: the clean-water diffusion (CWD) test, and the air diffusion rate (ADR) test. The CWD test was used to indicate the relative cleanliness of the filter.

2. Cleaning Procedure

Because the filter was reused repeatedly during the validation study, it was essential to clean the filter effectively between trials to maintain permeation flow rate. When selecting a cleaning procedure, it is important to consider the type of soil and effectiveness of the chemical cleaner for that soil. Preliminary tests should be performed on the soil alone to identify efficient cleaning compounds. The cleaners should also be tested on soiled filters under typical operating conditions to determine effectiveness of the cleaners *in situ*. This is especially important when using filters such as polysulfone, which bind proteins. Chemical properties of the cleaning agent must be considered. The cleaning agent must be safe for use with the membrane material and should not damage process equipment. The components and their breakdown products should not be toxic to humans, should there be an inadvertent carryover problem. In addition, if clean-in-place systems are used, then any detergents should be evaluated for excessive foam generation.

The cleaner used may be preformulated, or be generated in house from one or many ingredients. The use of preformulated cleaners is recommended, when available, because of the vast amount of compliance data available from the vendors. These data may include such items as toxicity testing, formulation documentation, and residual testing methods. It should be noted that the composition of formulated cleaners is extremely variable, and certain commonly used detergents, such as nonionic detergents, are difficult to remove from polysulfone membranes. Therefore, it may be desirable to formulate cleaners in-house to control component composition or to use simple cleaning agents such as caustic. If compound cleaners are formulated in-house, considerable resources may be required to comply with regulatory requirements for toxicity and validation testing.

Cleaning agent must be thoroughly removed from filters and equipment following the procedure to prevent leaching into final product. Residual testing should be performed on the equipment surfaces, as rinse water may not accurately reflect rinsability. However, most filters are contained in permanent housings, which prevents surface testing. In these cases, an extended soak (3 to 5 days) of the test filter in dilute caustic or other appropriate extraction solution which is compatible with the filter will likely indicate the relative cleaning as well as detergent removal from the membrane. It is recommended that testing of the rinse water for proteins or cleaning agents using such techniques as total organic carbon (TOC) testing may not accurately reflect adequate cleaning/rinsing of the filters. Rather, the caustic storage solution generally gives more reliable TOC and other analytical test results for the amount of leached protein and/or cleaning components. Adjustments to the standard analytical techniques may be necessary because the pH of the storage material may interfere with the test method. The data from these evaluations may then be used to determine the maximum allowable residual protein or cleaning component levels for the cleaning steps.

The cleaning procedure used in the validation of 500K filter reuse was the same as that used in large scale manufacturing, consisting of a large-volume water flush, a hot caustic detergent flush and recirculation, another water flush, and a hot caustic flush and recirculation. The cleanliness was determined using a clean-water diffusion test which measured the amount of flux at a defined temperature, crossflow, and TMP per area membrane; water flux is slower through the more soiled filters than through cleaned filters.

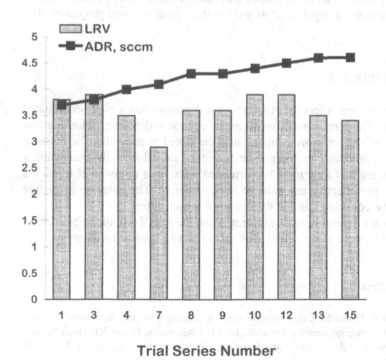

Figure 10 PPV reduction and ADR values for 500K filter reused repeatedly (16 times). Virus removal reported as LRV (log reduction value). ADR reported in sccm (standard cubic centimeters) for 0.24 ft^2 filter at 6.1 bar. PPV, porcine parvovirus.

3. Results

The virus removal capability of the filter was determined after each trial. BVDV was completely retained during trials in which it was used (data not shown; > 7.2 log reduction). The smaller, nonenveloped virus, PPV, was retained by approximately 3.6 ± 0.3 log in trials in which it was used (Figure 10). No carryover was detected from previous trials into unspiked trials. Therefore, the filter was found to retain virus to a significant and consistent degree over the course of sixteen uses.

To ensure filter integrity, the air diffusion rate (ADR) was determined for the filter following each run. A slight increase in the ADR over the course of experiments was observed up to 21% above the original value (Figure 10). The increase in ADR, however, was not associated with a change in viral log reduction values. Therefore, even though the pore size distribution may have been modified throughout the course of experimentation, the change was not significant enough to change the virus retention properties.

The net result from these studies was that the filters were demonstrated to retain their processing capabilities as well as their virus removal properties after repeated use.

V. CONCLUSIONS

Validation of process filters for virus removal depends upon a well-orchestrated plan of action. Preliminary experiments must be made to determine which parameters of the step affect virus removal, and the decision must then be made to define worst-case conditions. Integrity tests must be available for the acceptability of the filter's use and they must be correlated with virus removal. If filters are to be reused, the integrity test must be nondestructive. The cleaning procedure should also be considered in the validation of reused filters.

These experiments have demonstrated that the A/G Technology 500K filter is robust and reliable in the removal of virus from biological solutions.

REFERENCES

1. Payment P, Trudel M. Concentration and purification of viruses by molecular filtration and ultracentrifugation methods. In: PN Cheremisinoff and RP Ouelette, eds. Biotechnology: Applications and Research. Technomic Pub., Lancaster, PA, 1985: 436–450.
2. Crooks AJ, Lee JM, Stephenson JR. The purification of alphavirus virions and subviral particles using ultrafiltration and gel exclusion chromatography. Anal Biochem 152:295–303; 1986.

3. Divizia M, Santi AL, and Pana A. Ultrafiltration: an efficient second step for hepatitis A virus and poliovirus concentration. J Virol Methods 23:55–62, 1989.

4. Aranha-Creado H, Oshima K, Jafari S, Howard G Jr, Brandwein H. Virus Retention by Ultipor VF™ Grade DV50 Membrane Filters. Scientific and Technical Report, Pall Corporation, Port Washington, NY, 1995.

5. Aranha-Creado H, GJ Fennington Jr. Cumulative viral titer reduction demonstrated by sequential challenge of a tangential flow membrane filtration system and a direct flow pleated filter cartridge. PDA J of Pharm Sci Technol 51:208–212, 1997.

6. Burnouf-Radosevich M, Appourchaux P, Huart JJ, Burnouf T. Nanofiltration, a new specific virus elimination method applied to high-purity factor IX and factor XI concentrates. Vox Sang 67:132–138, 1994.

7. Eshkol A, Maillard F, Stiles G. Purification of biological fluid by filtration through ultrafiltration membrane. European patent No. EP0307373A2 (1988).

8. DiLeo AJ, AE Allegrezza Jr, Builder SE. High resolution removal of virus from protein solutions using a membrane of unique structure. Bio/Technology 10:182–188, 1992.

9. DiLeo AJ, Vacante DA, Deane EF. Size exclusion removal of model mammalian viruses using a unique membrane system Part I. Membrane qualification. Biologicals 21:275–286, 1993.

10. Manabe S-I Satani M. A porous hollow fiber membrane and a method for the removal of a virus by using the same. Patent No. 87/5900 (12-2-1988).

11. Roberts P Feldman P. Removal of viruses from factor IX by filtration: process validation. Paper presented at IBC Symposium on Viral Clearance: Novel Viral Removal Techniques and Detection Methodologies for the Biopharmaceutical Industry, Philadelphia, PA, Oct 27–28, 1997.

12. Oshima KH, Evans-Strickfaden TT, Highsmith AK, Ades EW. The removal of phages T1 and PP7, and poliovirus from fluids with hollow-fiber ultrafilters with molecular weight cut-offs of 50 000, 13 000, and 6000. Can J Microbiol 41:316–322, 1995.

13. Darling AJ Spaltro JJ. Process validation for virus removal: considerations for design of process studies and viral assays. BioPharm 9:42–50, 1996.

14. Grun JB, White EM, Sito AF. Viral removal/inactivation by purification of biopharmaceuticals. BioPharm 9:22–30, 1992.

15. White EM, Grun JB, Sun C-S, Sito AF. Process validation for virus removal and inactivation. BioPharm 4:34–39, 1991.

16. Levy RV, Phillips MW, Lutz H. Filtration and the removal of viruses from biopharmaceuticals. In: TH Meltzer and MW Jornitz, eds. Filtration in the Biopharmaceutical Industry. Marcel Dekker, 1998:619–646.

17. MD Stern. Viral safety of biotech products. Gen Eng News, May 15:26, 35, 1998.

18. White EM and Woodward RS. Recent issues raised in the evaluation of virus removal and inactivation. Gen Eng News 15:6, 1995.

19. CPMP (Committee for Proprietary Medicinal Products). Note for guidance: validation of virus removal and inactivation procedures, CPMP/III/8115/89. Biologicals 19:247–251, 1991.

20. Federal Health Office and Paul Erlich Institute Federal Office for Sera and Vaccines. Notice on the registration of drugs: Requirements for validation studies to demon-

strate the virus safety of drugs derived from human blood or plasma. Bundesanzeiger 84:4742–4744, 1994.

21. CPMP Biotechnology Working Party. Note for Guidance on Virus Validation Studies: The Design, Contribution, and Interpretation of Studies Validating the Inactivation and Removal of Viruses, CPMP/BWP/268/95 Final Version 2. London: Canary Wharf 1996.

22. CPMP Biotechnology Working Party. Note for Guidance on Plasma-Derived Medicinal Products, CPMP/BWP/269/95. Canary Wharf, London 1996.

23. International Conference on Harmonisation (ICH) Tripartite Guideline. Viral Safety Evaluation of Biotechnology Products Derived from Cell Lines of Human or Animal Origin. 1997.

24. Azari M, Catarello J, Burhop K, Camacho T, Ebeling A, Estep T, Guzder S, Krause K, Marshall T, Rohn K, Sarajari R. Validation of the heat treatment step used in the production of diaspirin crosslinked hemoglobin (DCLHb) for viral inactivation–effect of crosslinking. Art Cells Blood Subs Immobil Biotech 24:303, 1996.

25. Azari M, Ebeling A, Baker R, Burhop K, Camacho T, Catarello J, Estep T, Guzder S, Marshall T, Rohn K, Sarajari R. Validation of the heat treatment step used in the production of diaspirin crosslinked hemoglobin (DCLHb) for viral inactivation. Art Cells Blood Subs Immobil Biotech 24:304, 1996.

26. Azari M, Catarello J, Burhop K, Camacho T, Ebeling A, Estep T, Guzder S, Krause K, Marshall T, Rohn K, Sarajari R. Validation of the heat treatment step used in the production of diaspirin crosslinked hemoglobin (DCLHb) for viral inactivation—effect of crosslinking. Art Cells Blood Subs Immob Biotech 25:521–526, 1997.

27. Azari M, Ebeling A, Baker R, Burhop K, Camacho T, Catarello J, Estep T, Suzder S, Marshall T, Rohn K, Sarajari R. Validation of the heat treatment step used in the production of diaspirin crosslinked hemoglobin (DCLHb) for viral inactivation. Art Cells Blood Subs Immob Biotech. In press.

28. Farmer M, Ebeling A, Marshall T, Hauck W, Sun C-S, White E, Long Z. Validation of virus inactivation by heat treatment in the manufacture of diaspirin crosslinked hemoglobin. Biomat Art Cells Immobil Biotech 20:429–433, 1992.

29. Azari M, Ebeling A, Sun C-S. New blood product: diaspirin crosslinked hemoglobin as a blood substitute, virus inactivation and removal. Transfusion 32:85, 1992.

30. Lewin B. Gene Expression 3: Plasmids and Phages. New York: Wiley, 1977:724–727.

31. Ofsthun NJ. Internal Research Report No. TP06ME9638. Round Lake, II: Baxter Healthcare Corporation 1997.

12

Scaleup and Virus Clearance Studies on Virus Filtration in Monoclonal Antibody Manufacture

Ping Yu Huang and John Peterson
Protein Design Labs, Inc., Fremont, California

I. INTRODUCTION

In the manufacture of biological pharmaceuticals, the elimination of potentially infectious agents, such as viruses, is a critical component in the purification process. Without sufficient virus clearance, regulatory agencies will not allow the purified pharmaceutical to be used in a clinical trial. In the production of monoclonal antibodies for therapeutic uses, a purification process must be capable of eliminating substantially more virus (3 to 5 logs more) than that is estimated in a single-dose-equivalent of unprocessed bulk [1].

Unprocessed bulk from a normal stirred tank fermentation process in antibody production typically contains no infectious viruses using sensitive infectivity assays. However, retroviruslike particles (RVLP) are detected in the unprocessed bulk by transmission electron microscopy (TEM). The quantitation of these particles is used to estimate the amount of RVLP in a single-dose-equivalent of unprocessed bulk. Depending on the antibody titer and the amount of antibody in a dose, the viruses in the single-dose-equivalent are estimated in the ranges from 10^{10} to 10^{15} particles/mL. Therefore, a purification process must be capable of eliminating more than 15 to 20 logs of viruses. To achieve such a high clearance factor, several independent, orthogonal operation units are incorporated in a purification process for viral removal and inactivation. They include virus filtration, chromatography, extreme pH condition, detergent, heat treatment, etc. [2–6].

327

Virus filtration is the removal of viruses from the product stream using dead-end or tangential-flow filters. In principle, any filter that can remove virus is a virus filter. In the early stage of development, ultrafiltration (UF) membranes were tested for virus removal [5,6]. From <1 to >7.2 logs of virus clearances have been reported on different viruses. Many of such filters are still in use. However, theoretical analysis and scanning electron microscopic image revealed the problem of using UF membrane for virus filtration. The construct of UF membranes contains mainly small pores, but also some large pores which do not retain viruses [5]. To achieve a 6 log virus clearance factor, the large pore number density must be 1 in 10^9 small pores.

A new generation of virus filters specially designed for virus retention were produced. They are DV50 (Ultipor VF DV50) virus filters from Pall, Viresolve virus filters from Millipore, and Planova virus filters from Asahi Chemical Industry Co. [5]. In this chapter, we will describe the factors affecting the selection of virus filters and where to incorporate the filter into a purification process. The scaleup process development on DV50 virus filter, and virus clearance studies on the scaled-down version of DV50 virus filter will be presented.

II. SELECTION OF VIRUS FILTERS

There are two situations for which one might select a virus filter. The first situation is where the purification process was developed without using virus filters. Virus filtration is relatively new in comparison to other conventional purification methods, such as chromatography and microfiltration. However, when virus clearance studies on the overall purification process reveal insufficient virus clearance, virus filters can be selected and incorporated into the process.

The second situation is the design of a new purification process with one or two virus filters incorporated. If a virus filter was used for previous products, that virus filter would be incorporated into a new purification process for the new product. This has the obvious advantage of reducing the amount of development work, equipment cost, procedure development, and documentation times.

In general, the selection of an appropriate virus filter should first address the concerns of product safety, then consider the operational procedures, flexibility of scaleup and scaledown, cost, and the overall purification process.

A. Product Safety Concern

In the production of monoclonal antibodies using cell lines such as Chinese hamster ovary (CHO) or murine hybridoma, the presence of type A and type C retroviruslike particles detected by TEM measurement indicate the potential risk of enveloped retrovirus contamination [7]. It is understood by the biopharmaceutical industry and regulatory agencies around the world that substantial safety factors should be obtained from a purification process using model viruses similar to the particles observed by TEM [8,9]. Type A and type C particles are viruslike particles between 60 and 130 nm. Murine leukemia viruses (size 80 to 130 nm) have been used as model viruses for these types of RVLP. Therefore, the fundamental requirement for a virus filter used to remove RVLP is a pore size <60 nm.

In addition to the presence of type A and type C particles in harvest fermentation broth, there exist the possibilities of adventitious contamination with different kinds of viruses in every step of the manufacturing process [10]. Such adventitious contaminants can be enveloped or nonenveloped viruses. Some nonenveloped viruses are very stable and as small as 25 nm in size [5]. From this point of view, the smaller the pore size of the virus filters, the better, as long as it does not significantly affect product recovery and product quality.

The pore size rating of the three virus filters (Pall's DV50, Millipore's Viresolve, and Asahi Chemical's Planova) is shown in Table 1. Pall's DV50 membrane has pore size rating of approximately 40 nm, which means that it can

Table 1 Comparison of Three Virus Filters

	Pall's Ultipor VF DV50	Millipore's Viresolve	Asahi Chemical's Planova
Membrane and construction	3 Layers of PVDF (polyvinylidene) pleated and packed into a cartridge	PVDF 10-μm UF skin and 120-μm open support	Hollow fiber, each made of about 150 layers of cuprammonium regenerated cellulose
Operation	Dead-end filtration	Tangential flow	Dead-end or tangential flow
Test before use	30% IPA (isopropyl alcohol) in WFI at 60 psi	Ammonium sulfate solution and PEG (polyethylene glycol)	Pressure hold
Test postuse	WFI at 80 psi	Ammonium sulfate solution and PEG	Gold particle slurry
Pore size rating	~40 nm	70K and 180K MWCO	15 nm, 35 nm, 75 nm
Available surface areas (m²)	0.0013, 0.075, 1.63, 3.26, 4.89, 6.52	0.001, 0.03, 0.1, 0.7, 1.4	0.001, 0.003, 0.01, 0.03, 0.06, 0.12, 0.3, 1.0

effectively retain virus particle >40 nm. The size distribution of monoclonal antibody is between 10 and 15 nm. Therefore, antibody should flow freely through the DV50 membrane.

Millipore defines the pore size of Viresolve as molecular-weight cutoff (MWCO). Two available sizes are 70K and 180K MWCO. Protein molecules <180 kDa will flow through 180K membrane freely. Such size definition has the advantage of simplicity in membrane selection for protein purification. For example, the molecular weight of an antibody is ~150 kDa, so the 180K membrane will be used for antibody production.

Planovavirus filters have more pore size range selections than the others. They are available in sizes 15, 35, and 75 nm. Monoclonal antibody IgG can flow freely through any of the three filters.

In comparison, the pore sizes of Planova 15 nm and Viresolve 180 are certainly tighter than the DV50 filters. Both Planova 15 nm and Viresolve 180 have demonstrated 3 to 5 logs virus clearance on poliovirus (about 28 nm) in virus spiking experiments [3,5,11]. They can also remove more than 6 logs of large enveloped viruses, such as murine leukemia virus. For Pall's DV50 filter, no clearance of poliovirus was obtained [12]. Therefore, from the point of removing both nonenveloped and enveloped viruses, Planova 15 or Viresolve 180 should be used.

B. Operational Procedures

Table 1 also shows that the three virus filters are made of different materials and in different configurations. Pall's DV50 filter is a dead-end filter cartridge. It can be easily installed into existing filter housings and incorporated between any two steps in the purification process. The filter integrity test uses 30% IPA in WFI (water for injection) as forward airflow test [2]. It is simple and clean. The forward airflow diffusion rates are correlated to retention of viral particles [2].

The nondestructive liquid porosimetric integrity test from Millipore is not difficult, but the tangential flow for product filtration is unnecessary. Tangential flow is an effective operation to prevent the fouling of membrane by the retentate, the viruses, in the case of virus filtration. However, virus filter is often placed after initial concentration and purification, where there is no detectable amount of viruses in product stream to foul the membrane. In contrary, the shear stress during the tangential flow operation can cause antibody aggregation and product loss.

The Planova filters can be operated as a dead-end filter or a tangential-flow filter. The gold particle integrity test on the Planova membrane is the most complicated among the three filters. Therefore, from the point of simple operation, DV50 filter has the most advantages.

C. Flexibility of Scaleup and Scaledown

The flexibility of scaleup and scaledown is very important in selecting a virus filter. Process development requires small and pilot-scale filters for process optimization and troubleshooting. Process validation requires small-scale filters for virus clearance validation. Clinical manufacturing requires appropriate surface area and capacity for production. Commercial-scale manufacturing requires larger-capacity filters for large-scale manufacturing.

All filter manufacturers provide filters of various surface areas. Pall's DV50 filters include many different surface areas. A 47-mm membrane disk (surface area 0.0013 m^2) is optimal for process development and virus clearance validation. A 2″ virus filter (surface area 0.075 m^2) can be used for process development, scaleup, and small-scale clinical manufacturing. A 10″ virus filter (surface area 1.63 m^2) can be used for clinical manufacturing. Larger filters such as 20″, 30″, and 40″ filters (up to 6.5 m^2 surface area) will be sufficient for commercial scale manufacturing.

As shown in Table 1, Millipore's Viresolve filters are supplied with surface areas at 0.001, 0.03, 0.1, 0.7, and 1.4 m^2. Planova filters are also supplied with 0.001, 0.003, 0.01, 0.03, 0.06, 0.12, 0.3, and 1.0 m^2 surface areas. In comparison, DV50 filters have more room for expansion and scaleup.

D. Cost

Cost of virus filters is not a significant factor in selecting a virus filter. In general, the cost of virus filters is in the order of Planova > Viresolve > DV50. However, virus filter accounts for less than 5% of the raw material cost in the purification process. In the manufacture of clinical material, the cost of virus filter is even less significant.

E. Overall Purification Process

Virus filtration is not the only unit operation in the purification process that can remove viruses. Many chromatography steps are equally effective. Since the adventitious contamination exists only in statistics, a very small pore size virus filter is not mandatory in the manufacture of monoclonal antibodies from cell culture [9]. If there is an effective step in the overall purification process that can eliminate small nonenveloped viruses, product safety is not sacrificed. An effective step means that it can consistently eliminate 4 logs, or more of viruses under the process conditions [9].

In our purification process, protein A affinity chromatography is used to capture and purify monoclonal antibodies. It is considered an effective step in virus removal/inactivation [9], because it can consistently remove and inactivate more than 4 logs of nonenveloped and enveloped viruses from our virus clearance studies [12]. In this case, the Pall's Ultipor DV50 virus filter is the best filter for us because it is easy to operate and it has wide range of surface area for scaleup and scaledown. In our purification process discussed below, we have incorporated the Pall DV50 virus filter. The process scaleup and validation of DV50 virus filtration are described in the following sections.

III. SCALEUP OF VIRUS FILTRATION IN MONOCLONAL ANTIBODY MANUFACTURE

A. Design of Scaleup Process and Experimental Setup

Scaleup process includes three major steps: (1) small-scale (bench scale) experiment for feasibility studies and process optimization; (2) pilot-scale experiment to determine if the process is scalable, and to provide a more reliable estimate of the filtration parameters, such as filtration capacity, flux, etc.; and (3) manufacturing-scale operation.

The design of a scaleup process must start with the end process in mind; that is, to assume a suitable filter with sufficient surface area at manufacturing scale, then scale down to define experiment at bench and pilot scales. The estimate of filter surface area for manufacturing scale can come from the water flux data provided by filter manufacturer. The final decision on the surface area of a filter must be supported by the small- and pilot-scale data from experiments performed on real materials.

The real materials can have up to 100-fold different volumes depending on what material is selected for virus filtration. In our purification process, the volume of harvest or microfiltrate pool is 30- to 100-fold larger than any product pool post protein A affinity chromatography. The selection of materials post protein A will reduce filtration time at least 30-fold using the same surface area of virus filter. At our current scale, the product pool volume at every step after protein A affinity chromatography is <30 L.

Table 2 shows the scaledown and scaleup considerations for the virus filtration of <30 L product pools. According to Pall, the water flux on the DV50 filters is ~0.061 mL/min/cm^2. If same flux can be obtained with our product stream, it will take <30 min for a 10″ filter (1.63 m^2 surface area) to process <30 L product pool. If a 2″ filter (750 cm^2 surface area) is used, it will take up to 500 min. The possibility of clogging the 2″ element is much greater than a 10″ filter because of much larger volume-to-surface area ratio. Therefore, the 10″ element filter is an appropriate surface area to be used in our manufacturing scale.

If the 10″ filter is to be used for manufacturing, and the virus filtration process can be scaled up linearly, the small-scale virus filter must demonstrate

Table 2 Scaledown and Scaleup Considerations for Virus Filtration of a Monoclonal Antibody Product

Filter	Load volume	Filtration capacity	Water flux (mL/min/cm^2)	Process time (min)
AB1UDV507PH4 10″ filter, surface area 1.63 m^2	<30 L	1.8 mL/cm^2	~0.06	<30
SBF1DV50PH4 2″ filter, surface area 750 cm^2	30 L	39 mL/cm^2	~0.06	500
47-mm disk, surface area 12.5 cm^2	>23 mL	>1.8 mL/cm^2	~0.06	>30

the filtration capacity of >1.8 mL/cm^2 volume-to-surface ratio on our antibody solution.

The experimental setup of small-scale virus filtration is shown in Figure 1. A 47-mm disk is installed in a membrane housing purchased from Fisher Scientific (Pittsburgh, PA). The effective filtration area of an installed membrane is 12.5 cm^2. The membrane is wetted by passing 5 to 10 mL water for injection (WFI) through the membrane at 30 psi. Then a 5-min pressure hold test is performed at 60 to 70 psig.

The holder with the membrane is autoclaved at 121°C for 60 min. Protein solutions, with or without spiked viruses, are loaded on the upstream of the membrane holder. The holder is pressurized to 30 psi and the filtrate is collected from

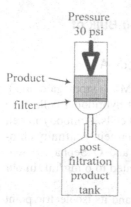

Pressure
30 psi

Product

filter

post
filtration
product
tank

Figure 1 Virus filtration setup at small scale using 47-mm disk.

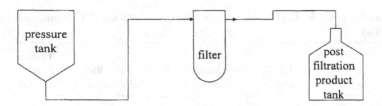

Figure 2 Virus filtration setup at pilot and manufacturing scales.

the outlet. The flux is calculated by dividing the collected filtrate volume by filtration time and filter surface area.

After filtration, the pressure is increased to 60 to 80 psi and the outlet of the filter is immersed in a pool of water for postfiltration integrity test. If no air bubble is observed from the outlet in 5 min, the test is passed.

The setup of pilot- and manufacturing-scale filtration using 2″ or 10″ virus filters is shown in Figure 2. The filter is installed into a stainless-steel filter holder (from Pall) and forward tested according to the procedure from Pall [2]. The product pool is filtered through the membrane at a pressure of 30 to 35 psig. After the product pool flows through, the filter is washed with 0.27 mL buffer/cm² surface area to recover product in the filter housing.

IgG concentrations in the sample load and filtrate were measured by absorbance at 280 nm. The percentage of IgG monomer before and after filtration was measured by size exclusion chromatography using Tosohaas columns (Montgomeryville, PA). HPLC systems were purchased from Waters (Milford, MA). Spectrophotometer was model UV2101PC from Shimazu (Columbia, MD). Excipient T was obtained from Amresco, national formulary (NF) grade. All solutions were made using WFI and chemicals of biological grade or NF grade.

B. Scaleup of Virus Filtration on the Formulated Bulk of Antibody A

1. Characteristics of the Formulated Bulk of Antibody A

Antibody A is a human antibody IgG subclass 1 against CMV (cytomegalovirus). The genes for the antibody A were inserted into Sp2/0 hybridoma cells. The antibody is produced by a fed batch culture of the Sp2/0 cells. Antibody in cell culture supernatant was purified through microfiltration, protein A affinity chromatography, low-pH viral inactivation, size exclusion, and Q Sepharose ion-exchange chromatography. Purified antibody was formulated to 6 mg/mL in our formulation buffer containing 0.01% excipient T [12].

The molecular weight of antibody A is ∼150 kDa and its isoelectric point is around 8.

2. Effect of Filtration Volume on the Flux at Small Scale Using 47-mm Disk Membrane

Using the scaled-down version of filtration process, the effect of loading volume on the flux was determined by challenging the membrane with a large volume of formulated bulk of antibody A. Figure 3 shows the filtrate flux as a function of cumulative volume. The flux remained at ~80% to 90% of initial value after 90 mL filtrate was collected. This was equivalent to a 7.2 mL/cm^2 volume-to-surface ratio, which was much higher than the required 1.8 mL/cm^2. The variation of the two flux decaying lines was most likely due to the effect of temperature. In the filtration of the first sample of formulated bulk, temperature remained at 20°C during the entire course of filtration. The flux decay was minimum. The filtration of the second sample of formulated bulk was carried out to simulate our manufacturing process in which the solution at 4°C would be directly filtered through the filter. Lower temperature might have caused a quick drop in flux at the beginning of the filtration. The two lines became almost parallel at the end of filtration, indicating similar plugging rates.

In process scaleup consideration, Figure 3 shows that the 47-mm filter disk is able to filter >23 mL of the formulated bulk of antibody A, and the decrease

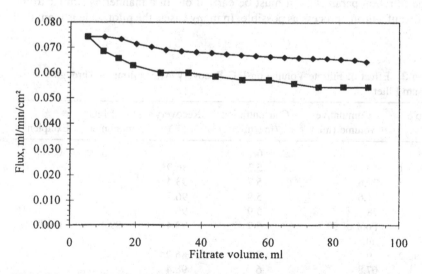

Figure 3 Effect of loading volume on the flux for the filtration of formulated bulk of antibody A. The effective surface area of membrane is 12.5 cm^2. The data points ◆ were from first set of experiment where temperature was maintained at 20°C. The data points ■ were from second set of experiment where antibody A at 4°C was applied to the filter holder at 20°C.

in flux is minimum. This implies that we should be able to filter 30 L of the formulated bulk through a 10-inch element filter, if the scaleup process is linear.

3. Effect of Filtration Volume on the Recovery of IgG and Excipient T at Small Scale

The effect of filtrate volume on the antibody A recovery is shown in Table 3. Antibody A formulated bulk was filtered immediately after the membrane was wetted with 5 mL WFI (without autoclave). Due to some water left in the holder, IgG recovery in the first fraction of 10 mL was only 86.9%. The other fractions had IgG recovery of 93.4% to 100%. The overall recovery in the 90-mL filtrate pool was 93.4%. The percentage of IgG monomer for the two filtrate pools remained unchanged—99.5% before and after filtration.

The excipient T in the load was 0.01%. It was reduced to 0.0083% in filtrate pool. Although the excipient concentration was reduced, it met our specification set at 0.01 ± 0.002%. The filtration process was recommended for scaled-up to pilot scale.

4. Pilot-Scale Development Using a 10″ Virus Filter

Since the goal of pilot-scale experiment is to provide a more reliable estimate of the filtration parameters, it must be carried out in a manner as similar to the real manufacturing process as possible. In many cases, the pilot scale is manyfold

Table 3 Effect of Filtrate Volume on IgG Recovery After Filtration Through a 40-nm Filter

Sample name	Cumulative volume (mL)	Concentration (mg/mL)	Recovery (%)	% of IgG monomer	Excipient T
Load		6.1		99.5	0.01%
Filtrate 1	10	5.3	86.9*		
	24.6	5.7	93.4		
	51.6	5.9	96.7		
	78	5.9	96.7		
	Pool of 90 mL	5.7	93.4	99.4	0.0083%
Filtrate 2	10	5.3	86.9*		
	67.8	6	98.4		
	87.6	6.1	100.0		
	Pool of 90 mL			99.5	

* Low IgG recovery was due to dilution from water used to wet the membrane.

smaller than the manufacturing scale in order to save money and conserve material. During our scaleup process, we were in a unique situation where we could use an entire lot of formulated bulk and a 10″ filter for our experiment. In this case, the scaleup experiment was performed exactly the way it would be done at manufacturing scale.

A 10″ filter cartridge was autoclaved and used for the filtration of 8.8 L of the formulated bulk. At a pressure of 30 psi, the formulated bulk was filtered through the cartridge in about 8 min. The filtered bulk solution was analyzed by our quality control department. The results are summarized in Table 4. It shows that all the specifications are met except the concentration of excipient T. Approximately 50% of excipient T was lost during the filtration.

5. Determine the Loss of Excipient T at Small Scale

The pilot-scale data revealed a mistake we made during the small-scale development. At small scale, we challenged the filter up to 7.2 mL/cm^2 to determine the filtration capacity. Only 0.54 mL/cm^2 was processed at pilot scale. The results of large volume-to-surface ratio at small scale provided us with confidence that

Table 4 Specifications and Measured Results for DV50 Filtered Formulated Bulk

	Specification	Measured results
Identity		
Anti-ID ELISA	Identified as antibody A	Identified as antibody A
Isoelectric focusing	SSPCR*	SSPCR
Purity		
HPLC-SEC	≥98% monomer	99.5%
SDS-PAGE (silver stain)	SSPCR	SSPCR
Reduced and nonreduced endotoxin (LAL)	<6 EU/mL	<0.32 EU/mL
Sterility	Pass	Pass
Strength		
IgG concentration	6.0 ± 0.6 mg/mL	6.01 mg/mL
Bioassay		
Potency by ELISA	70 ~ 130% of reference	100%
Excipients		
Excipient T	0.01 ± 0.002%	0.0051%**
pH	6.0 ± 0.2 at 25°C	6.0

* Sample staining pattern consistent with reference.
** Out of specification.

the 10″ virus filter would not be plugged, but it minimized the loss of excipient T in the filtrate pool.

To understand the loss of the excipient T, we performed more small-scale experiments. Using the 47-mm filter disk, we collected fractions during filtration and measured the excipient T concentration in all fractions. Figure 4 shows the excipient T concentration in the filtrate as function of filtrate volume. It started with only 0.03% in the first fraction, and increased with volume until an equilibrium was reached. The breakthrough curve looked very much like an adsorption isotherm, so we speculated that the excipient T was adsorbed by the filter membrane during the first 25 mL of filtration. Once the membrane was saturated, the excipient T concentration was the same pre- and postfiltration. (Two years later, we obtained more experimental results showing that the excipient T might have formed micelles in solution and the micelles were retained by the membrane [data not shown].)

6. Strategy in Solving Problem of Excipient T Loss

The problem of excipient T loss could be solved either by adding concentrated excipient T solution into the filtered formulated bulk or by preequilibrating the filter with formulation buffer (placebo buffer containing 0.01% excipient T). The

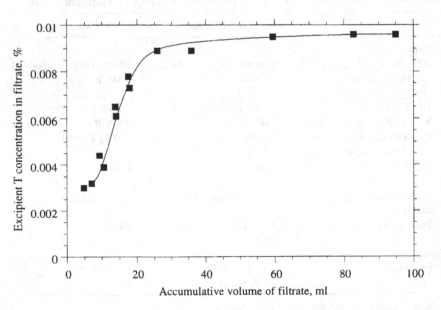

Figure 4   Adsorption/retention of excipient T by DV50 filter (47-mm disk membrane).

second option seems to be better because there is no need to add any solution to the final product pool.

According to the breakthrough curve of excipient T in Figure 4, a 30-mL placebo buffer should be sufficient to equilibrate the 47-mm disk membrane. Considering that the membrane of a 10-inch filter cartridge was pleated and potentially more difficult to be completely saturated with excipient T, we doubled the saturation volume. To demonstrate this process in a small-scale experiment, we equilibrated a 47-mm disk filter with 60 mL of placebo buffer, followed by 5.2 mL of antibody A formulated bulk. This represented the same ratio of solution volume to surface area as in pilot scale (0.54 mL/cm^2). The excipient T concentration in the 5.2-mL formulated bulk after filtration was 0.0084%.

The new process was tested at pilot scale again. Excipient T concentration in the filtered formulated bulk was adjusted from 0.0052% to 0.01% by adding 1% standard excipient T solution. A 10-inch filter cartridge was equilibrated with 100 L of placebo buffer, followed by filtration of the formulated bulk. After filtration, the excipient T concentration in the formulated bulk was measured as 0.010%. Therefore, we had a process that could be used to filter 8.8 to 20 L formulated bulk.

7. Results from Production Scale Using 10″ Virus Filter

The filtration of another three batches of formulated bulk showed consistent results. All three filtered formulated bulks met the specifications as listed in Table 5. The product recovery rates were ~95% for all three batches, as listed in Table 5.

8. Summary of the Virus Filtration Scaleup for Antibody A

In summary, we have successfully developed and scaled up a virus filtration process for antibody A formulated bulk. Since there were excipients in the formulated bulk, we had to consider both filtration capacity and excipient loss at small-scale development. To prevent the excipient loss, the virus filter was equilibrated with excess amount of placebo buffer prior to the filtration of the formulated bulk.

After our successful filtration of several lots of antibody A formulated bulk, we reexamined our strategies. First, if we move the virus filter upstream in our purification process prior to formulation, we don't even need to consider the excipient loss. The filtration process will be simpler because the filter will not need buffer equilibration. Second, if the virus filter is placed after protein A affinity chromatography, we can physically separate our purification process into two major groups. The first group includes microfiltration, protein A chromatography, low-pH virus inactivation, and virus filtration. They are designed for product recovery and major virus reduction steps. These steps are carried out in the fermentation suite. The second group includes size exclusion chromatography,

Table 5 Specifications and Test Results for DV50 Filtered Formulated Bulks at Manufacturing Scale

	Measured results		
	Lot 1	Lot 2	Lot 3
Identity			
Anti-ID ELISA	Antibody A	Antibody A	Antibody A
Isoelectric focusing	SSPCR	SSPCR	SSPCR
Purity			
HPLC-SEC	99.6%	99.5%	99.5%
SDS-PAGE (silver stain) reduced and nonreduced	SSPCR	SSPCR	SSPCR
Endotoxin (LAL)	<0.16 EU/mL	<0.16 EU/mL	<0.08 EU/mL
Sterility	Pass	Pass	Pass
Strength			
IgG concentration	5.92 mg/mL	6.03 mg/mL	5.91 mg/mL
Bioassay			
Potency by ELISA	92%	109%	109%
Excipients			
Excipient T	0.0099%	0.0087%	0.0092%
pH	6.0	6.0	6.0
Product recovery	94.3%	96.7%	95.1%

Q Sepharose ion-exchange chromatography, and formulation. They are polishing steps and are performed in a separate suite, which is designed to be cleaner than the fermentation suite. Such consideration can be regarded as building product safety into our process.

C. Scaleup of Viral Filtration on the Post Protein A Affinity Chromatography Product Pool of Antibody B

1. Characteristics of Antibody B Solution

Antibody B is a humanized antibody. It was constructed by combining the complementarity-determining regions (CDR) of a murine antibody with human framework and constant regions. Its molecular weight is ~150 kDa and its isoelectric point is ~8.8.

The antibody is produced by a fed-batch culture of engineered Sp2/0 hybridoma cells. After cell removal by microfiltration (MF), the antibody in the

MF filtrate is captured and purified by protein A affinity chromatography. The antibody concentration in the post protein A product pool is between 7 and 22 mg/mL.

2. Effect of Filtration Volume on Flux and Product Recovery at Small Scale Using a 47-mm Disk Membrane

The effect of filtration volume on the flux for the filtration of post protein A product pool of antibody B is shown in Figure 5. The two flux lines were obtained from duplicate experiment and they showed consistent decay in the flux with increasing filtration volume. In comparison to Figure 3, the decrease in the flux for filtration of post protein A product pools was faster than that for the formulated bulk, because the post protein A pool is not as pure as the formulated bulk. However, after 100 mL of protein A eluate pool passed through the membrane, the flux remained ~55% of the initial measurement.

The recovery of antibody B in the filtrate is shown in Table 6. The filtration of post protein A product pools was performed using autoclaved membrane holder assemblies. The IgG recovery was between 95.2% and 99%. The percentage of IgG monomer was 98.2% before and after filtration. These results suggested that the scaleup should be straightforward.

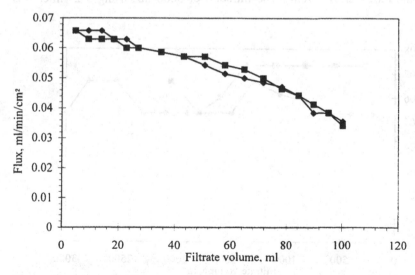

Figure 5 Effect of filtration volume on the flux for the filtration protein A pool of antibody B. The effective surface area of membrane is 12.5 cm². Two data lines are from duplicate experiments.

Table 6 Effect of Filtrate Volume on Recovery of Antibody B After Filtration Through a DV50 Virus Filter

	Cumulative volume (mL)	Concentration (mg/mL)	Recovery (%)	% of IgG monomer
Load	100	8.11		98.2
Filtrate 1	9	7.98	98.4	
	46.2	7.98	98.4	
	92.1	7.72	95.2	
	Pool of 101.1 mL	7.94	99.0	98.1
Filtrate 2	18.2	7.94	98.5	
	50.6	7.96	98.8	
	95.2	7.66	95.0	
	Pool of 101.2 mL	7.95	99.2	98.2

3. Effect of Filtration Volume on Flux and Product Recovery at Pilot Scale Using a 2″ Virus Filter

The virus filtration of protein A product pools of antibody B was scaled up using a 2″ virus filter. As shown in Figure 6, the filtration flux at pilot scale was between 0.05 and 0.074 mL/min/cm². The filtration of 3000 mL using a 2″ filter was

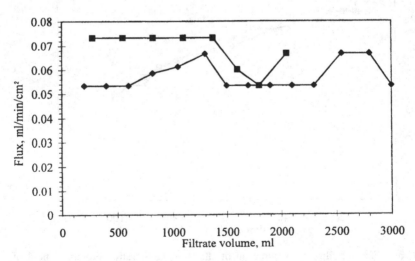

Figure 6 Effect of filtration volume on the flux for the filtration protein A pool of antibody B at pilot scale using a 2″ virus filter. The effective surface area of filter is 750 cm². The data points ◆ and ■ were from the filtration of two different lots.

Table 7 Antibody B Recovery After Virus Filtration at Pilot Scale

Lot no.	Filtration volume (L)	Product recovery (%)	% of IgG monomer	
			Before filtration	After filtration
D150F2	1.06	97.2	97.6	97.4
D150F4	3.00	97	99.0	99.1
D150F5	2.25	92	99.0	99.1
D150F6	2.35	98	97.2	97.2

equivalent to 50 mL using a 47-mm disk filter. During the filtration of the first 50 mL using a 47-mm disk filter, the flux was also in the range of 0.05 to 0.07 mL/min/cm^2, as shown in Figure 5. This indicated a linear scaleup for the filtration process.

The recovery of antibody B after filtration through the 2″ virus filter is shown in Table 7. In the four lots summarized, the filtration volume varied from 1 L to 3 L, and the product recovery varied from 92% to 98%. There were minimal changes in the percent IgG monomer before and after virus filtration.

4. Results from Large-Scale Manufacturing Using 10″ Viral Filter

Table 8 shows product recovery and percent monomer for antibody B after filtration at manufacturing scale. The 10″ virus filter was successfully used to filter 12 to 18.5 L of antibody B post protein A product pools. The recovery in the three lots of manufacturing was 91% to 100%. The percent monomer of antibody B was not changed significantly before and after filtration.

Table 8 Antibody B Recovery from DV50 Virus Filtration of Post Protein A Affinity Chromatography Product Pool at Manufacturing Scale

Lot no.	Filtration volume (L)	Product recovery (%)	% of IgG monomer	
			Before filtration	After filtration
No. a	14.0	91	97.0	97.0
No. b	18.5	99	97.2	97.3
No. c	12.3	100	98.1	98.0

5. Summary of Virus Filtration Process Scaleup for Antibody B

In summary, scaleup of the virus filtration on process intermediates, such as post protein A product pool, was more straightforward than on the final formulated bulk. The scaleup of DV50 virus filtration process was linear from 12.5 cm² to 1.63 m² surface areas.

IV. VIRUS CLEARANCE STUDIES ON THE VIRUS FILTER

A. Selection of Model Viruses

Three model viruses—XMuLV (Xenotropic murine leukemiavirus), PRV (pseudorabiesvirus), and PV (poliovirus)—were used in virus spiking studies on DV50 virus filters. They were selected to represent a wide variety of sizes and chemical and physical properties. These viruses were also used in virus spiking studies on several other purification steps. Table 9 shows the characteristics of these viruses.

B. Design of Virus Spiking Experiment

The virus spiking studies were performed at BioReliance (formerly Microbiological Associates, Rockville, MD) with PDL personnel performing the scaled-down manufacturing process. Three model viruses were spiked into our formulated bulk solution or post protein A product pool at a ratio of approximately 1 to 20. The spiked solutions were filtered through the DV50 filters. After filtration, the viral concentration in the filtrate was assayed by BioReliance. The overall design of virus spiking, filtration, and sampling are illustrated in Figures 7 through 9 for XMuLV, PRV, and PV, respectively.

In Figure 7, the virus filtration of XMuLV spiked solution was performed in duplicate to test the reproducibility of the process. In Figures 7 and 8, large sample assays (12.5 mL for XMuLV and 13 mL for PRV) were designed to

Table 9 Characteristics of Model Viruses

	Xenotropic murine leukemiavirus (XMuLV)	Pseudorabiesvirus (PRV)	Poliovirus (PV)
Family	Retro	Herpes	Picorna
Genome	RNA	DNA	RNA
Enveloped	Yes	Yes	No
Size (nm)	80–110	120–200	~30
Resistance to physico-chemical reagents	Low	Medium	High

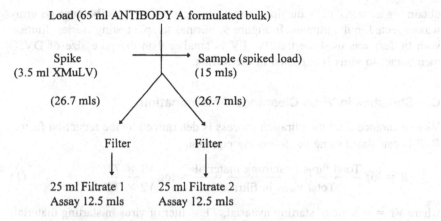

Figure 7 Schematic drawing of XMuLV spiking, sampling, viral filtration in duplicate, and 12.5-mL sample assay in both filtrates 1 and 2.

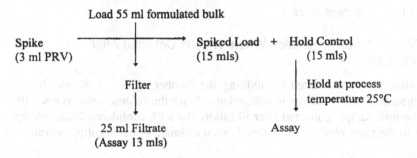

Figure 8 Schematic drawing of PRV spiking, sampling, viral filtration, and 13-mL sample assay in the filtrate.

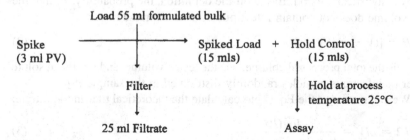

Figure 9 Schematic drawing of PV spiking, sampling, viral filtration, and sample assay in the filtrate.

obtain higher statistical reduction factors for these two viruses, because no virus was expected in the filtrates. In Figure 9, normal sample testing (series dilution with buffer) was used for the PV. PV is smaller than the pore size of DV50 membrane, so virus is expected in the filtrate.

C. Statistics in Virus Clearance Determination

Virus clearance from the filtration process is determined by the reduction factor, R. R is calculated using the following equation:

$$R = \log \frac{\text{Total virus in starting material}}{\text{Total virus in filtrate}} = \log \frac{V1 \times T1}{V2 \times T2} \tag{1}$$

where $V1$ = volume of starting material, $T1$ = titer of virus in starting material, $V2$ = volume of material after process, and $T2$ = titer of virus after process. The $V1$, $T1$, and $V2$ can be measured accurately. However, $T2$ may not be accurately measured when there is no detectable virus in the filtrate. The estimate of the $T2$ can significantly affect the final reduction factor R. These are demonstrated in the following three cases.

1. Virus Titer Determination When Virus Is Detected After Filtration

The virus titer is calculated by dividing the number of foci or plaques by the volume added to Petri dishes or microplates. Since the volume tested is normally not the total sample collected after filtration, the 95% confidence limits are applied to the titers obtained. In general, such estimate is statistically accurate.

2. Virus Titer Determination When Virus Is Not Detected and Large Sample Volume Is Assayed

If there is no virus detected in the sample tested, a theoretical virus titer is calculated by a statistical method. Based on the definition, the probability, P, that the tested volume does not contain infectious virus is expressed by:

$$P = [(V - v)/V]^n \tag{2}$$

where V is the total processed volume, v is the tested volume, and n is the absolute number of infectious particles randomly distributed in the sample [9].

We can rearrange the Eq. (2) to calculate the theoretical titer in the filtrate:

$$T2 = n/V = \frac{\log(P)}{V \times \log[(V - v)/V]} \tag{3}$$

Given a probability of 0.05 and 25 mL of total processed volume (V), $T2$ can be calculated as function of the tested volume, v. The calculated results are shown

Table 10 Effects of Tested Volume* on
Titers of Filtered Sample

Tested volume v(mL)†	$T2$ from Eq. (3)	$\log(T2)$
0.0444	67.397	1.83
0.444	6.672	0.82
1	2.920	0.47
5	0.521	−0.28
10	0.216	−0.67
15	0.108	−0.97
24.999	0.015	−1.82

* No virus is detected in the tested volume.
† Total processed volume is 25 mL and the probability
  is 0.05.

in Table 10. When the tested volume (v) is 0.0444 mL, the $T2$ is estimated as 67.397 particles/mL, or 1.83 logs. If 24.999 mL out of 25 mL total sample is tested and no virus is found in the tested volume, the $T2$ is as little as 0.015 particles/mL, or −1.82 logs. Based on Eq. (1), the overall log reduction R increases by 3.65 logs simply by increasing the volume tested. Even when the tested volume is 10 mL, the reduction factor R is 2.5 logs more than that when 0.0444 mL is tested. The 0.0444 and 0.444 mL are typical test volumes in a microtiter plate with serial 1:10 dilutions.

The understanding of the statistics in the determination of $T2$ helped us design our virus filtration studies. Large samples were assayed to maximize the reduction factors in the virus filtration of XMuLV and PRV.

3. Virus Titer Determination When Virus Is Not Detected and a Small Sample Volume Is Assayed

If the volume tested is a small portion of a filtrate pool (<10% v/v), a theoretical titer, $T2$, can be obtained from Poisson distribution:

$$T2 = -\ln(P)/v \tag{4}$$

where P is probability (equal to 0.05 for 95% confidential limit) and v is the sample volume tested. Equation (4) merely simplifies the calculation of theoretical titers.

D. Virus Reduction Factors from XMuLV, PRV, and Poliovirus Clearance Studies

The virus reduction factors for three different viruses are listed in Table 11. For the filtration of XMuLV and PRV, no virus was detected in 12.5- and 13-mL

Table 11 Viral Clearance from Spiking Experiment on Antibody A Formulated Bulk

	XMuLV	PRV	PV
Size (nm)	80–110	120–200	~30
Total virus in start- ing material ($\log_{10}$)	7.30 ± 0.41 FFU*	8.15 ± 0.02 PFU†	8.34 ± 0.30 PFU
Total virus in fil- trate ($\log_{10}$)	(1) ≤0.63 FFU‡ (2) ≤0.63 FFU‡	≤0.60 PFU§	7.62 ± 0.09 PFU
Reduction factor R	(1) ≥6.67 logs (2) ≥6.67 logs	≥7.55 logs	<1 log

* Focus formation unit.
† Plaque formation unit.
‡ No virus was detected. The theoretical titer is based on a 95% confidence of sampling 12.5 mL from a 25-mL filtrate pool.
§ No virus was detected. The theoretical titer is based on a 95% confidence of sampling 13 mL from a 25-mL filtrate pool.

samples of 25-mL filtrate pools. Theoretical virus reduction factors are ≥6.67 logs for XMuLV, and ≥7.55 logs for PRV. The filtration of XMuLV was performed in duplicate and consistent results were obtained. Poliovirus was not significantly removed because its size was smaller than the pore size of the DV50 filter (~40 nm). This negative result demonstrates that the mechanism of virus removal by the membrane is based on size.

Virus filtration studies on the protein A eluate for antibody B were designed and carried out similar to antibody A. The reduction factors are summarized in Table 12. It shows reduction factors of ≥7.17 for XMuLV, ≥6.92 for PRV, and <1 for PV. These results are very similar to those obtained from the filtration

Table 12 Viral Clearance from Spiking Experiment on Antibody B in Post Protein A Affinity Chromatography Pool

	XMuLV	PRV	PV
Size (nm)	80–110	120–200	~30
Total virus in starting material ($\log_{10}$)	(1) 8.24 (2) 7.94	(1) 8.58 (2) 7.60	8.82
Total virus in filtrate ($\log_{10}$)	(1) ≤0.64 (2) ≤0.77	(1) ≤0.65 (2) ≤0.68	8.60
Reduction factor R	(1) ≥7.60 logs (2) ≥7.17 logs	(1) ≥7.93 logs (2) ≥6.92 logs	<1 log

of formulated bulk of antibody A. Based on these consistent results from duplicate runs, different products, and different process steps, we believe that this virus filtration step using Pall's DV50 membrane is robust, reproducible, and reliable.

V. PROSPECTS

Pall's Ultipor VF DV50 virus removal filter (40-nm filter) was selected and successfully incorporated into our downstream process to increase the virus clearance capacity. In selecting this virus filter, we emphasized the simplicity of operation, and ease of integrity testing. In the future, as regulatory agencies raise the standard for virus clearance, pore size <40 nm may be required. By that time, we hope the membrane manufacturers have developed smaller-pore-size filters and simpler operations to meet the challenges.

In the scaleup of the filtration of antibody A formulated bulk, we found that the Pall's DV50 membrane was not significantly plugged by our product. The retention of excipient T was a problem discovered during our scaleup process. This problem was resolved by preequilibration of the membrane with sufficient amount of placebo buffer to saturate the filter with excipient T.

The virus filtration procedure was further simplified by placing the virus filter further upstream after the protein A affinity chromatography step. Such design also physically segregated the major virus reduction operations in the fermentation suite from the polishing operations in the cleaner purification suite. The possibility of cross contamination is minimized.

The scale up of the filtration of antibody B at post protein A affinity chromatography step was straightforward. The flux at small scale was almost the same at pilot and manufacturing scales. The product recovery was excellent (92% to 100%) and the product quality was not affected after filtration.

Virus clearance studies on antibody A and antibody B showed >6.6 logs reduction of XMuLV and PRV by the virus filter. The clearance factors obtained were similar for the two antibody products and at different purification steps. This provides assurance that the filtration is a reliable process. Less than 1 log reduction on PV indicates that the filter removes viruses based on the size of the viruses.

ACKNOWLEDGMENTS

Special thanks to the downstream purification group for performing the large-scale filtration test. We are grateful to the Quality Control department for the excipient T and size exclusion HPLC assays.

REFERENCES

1. ICH Topic Q 5 A. Quality of biotechnological products: viral safety evaluation of biotechnology products derived from cell lines of human or animal origin. 3/4/1997.
2. Aranha-Creado H, Oshima K, Jafari S, Howard G Jr, Brandwein H. Virus Retention by Ultipor® VF™ Grade DV50 Membrane Filters, Scientific and Technical Report, Pall Inc. 1995.
3. Hoffer L, Schwinn H, Biesert L, Josic DJ. Improved virus safety and purity of a chromatographically produced factor IX concentrate by nanofiltration. J Chromatogr B 669:187–196, 1995.
4. Grun JB, White EM, Sito AF. Viral removal/inactivation by purification of biopharmaceuticals. BioPharm Nov.:22–27, 1992.
5. Levy RV, Phillips MW, Lutz H. Filtration and the removal of viruses from biopharmaceuticals. In: Meltzer TH, Jornitz MW, eds. Filtration in the Biopharmaceutical Industry. New York: Marcel Dekker, 1998:619–646.
6. Walter JK, Nthelfer D, Werz W. Virus removal and inactivation—a decade of validation studies: critical evaluation of the data set. In: Kelley BD, Ramelmeier RA, eds. Validation of Biopharmaceutical Manufacturing Processes. ACS Symposium Series 698, 1998:114–124.
7. Bierley ST, Monticello TM, Morgan EM, Leininger JR. Morphometric estimation of viral burden in cell culture material by electron microscopy (TEM). In: Bailey GW, Bentley J, Small JA, eds. Proceedings, Electron Microscopy Society of America Annual Meeting. San Francisco: San Francisco Press, 1992:732–733.
8. Lubiniecki AS. Dev Biol Stand 70:187–191, 1989.
9. CPMP Biotechnology Working Party. Note for Guidance on Virus Validation Studies: The Design, Contribution and Interpretation of Studies Validating the Inactivation and Removal of Viruses, Final Version 2, Feb. 29, 1996.
10. Garnick RL. Experience with viral contamination in cell culture. Dev Biol Stand 88:49–56, 1996.
11. Bartlett RT. Virus clearance validation for biopharmaceutical manufacturing processes. A case study of monoclonal antibody manufacturing process. Paper presented at BioPharm Conference Proceedings, San Francisco, May 1998.
12. Reindel K. Comparability protocols and the design of viral clearance studies for monoclonal antibody purification processes. Paper presented at the Fifth Annual Antibody Production and Downstream Processing, San Diego, March 1999.

13

Virus Removal by Tangential-Flow Filtration for Protein Therapeutics

Brian D. Kelley and Jon T. Petrone
Genetics Institute, Andover, Massachusetts

I. INTRODUCTION

The assurance of the safety for therapeutic agents derived from human plasma must include an evaluation of the clearance of known or unknown viruses by the production process. Similarly, for products produced by cell culture, the potential for the propagation of viruses during the cell culture process requires an assessment of the risk incurred by the manufacturing process. Critical elements of the safety programs for products derived from both plasma and cell culture must include control and quality assurance of the starting materials. For plasma products, this will involve control over the plasma source, tracking of the donors, look-back procedures, and routine testing of the pooled plasma. For cell culture processes, the characterization of the cell bank, elimination or characterization of proteins or serum used in media, and postproduction testing for adventitious viral contamination all assure a consistent and safe product source.

In addition to assurance of the virological safety derived from the control of the starting or source materials for these products, manufacturers must demonstrate that the purification process has the ability to remove or inactivate viruses, which may have escaped the controls set for the plasma or cell culture sources. Documents from various regulatory agencies provide guidance as to the expectations for the manufacturing process [1–4] and the validation studies, which will establish the ability of the purification process to clear and/or inactivate virus.

The purification processes will almost invariably employ column chromatography for product purification, combined with other purification or concentration steps such as precipitation, ultrafiltration, or partitioning/extraction. Some of these steps have the ability to clear viruses from the product-containing pools, through either physical removal of virus particles or inactivation of viruses under conditions encountered during the unit operation. Technologies have been developed solely to remove or inactivate viruses. If these technologies are shown to provide a large reduction in viral titer and are robust to typical variations in process operations, they can provide a high degree of assurance that the product is devoid of viruses.

Examples of process steps that have been developed solely for the inactivation of viruses include pasteurization [5], solvent-detergent treatment [6,7], microwave heating [8], pH inactivation [9], and chemical treatment [10]. These methods have varying success in inactivating viruses [11]. Certain families of virus, such as the Parvoviridae, are characterized by their resistance to inactivation by heat or chemical methods [12]. While all enveloped viruses are prone to inactivation by solvent-detergent treatment, these steps do not affect nonenveloped viruses.

Physical removal of viruses can be achieved by recently developed membrane filters that separate viruses and the product based on their size differences [13–15]. These filtration processes provide a simple and robust means of removing viruses, often with a predictable performance based only on the size of the virus particle to be removed.

This chapter will focus on the use of tangential-flow filtration techniques for virus removal, and outline typical development experiments that can be used to establish operating conditions for successful implementation. While none of the products being processed are identified in this chapter, the chapter includes data generated during the development of a purification process for recombinant blood coagulation Factor IX which includes a virus filtration step [16–18], in addition to data from processes developed for other recombinant proteins.

II. STEP OBJECTIVES

Various objectives should be defined for a virus removal filtration (VRF) step early in the process development cycle, to ensure that the appropriate VRF module, operating conditions, and validation package are selected. Some objectives for consideration are described below.

A. Assurance of Virus Removal

A virus-removing filtration step should ensure that a high degree of virus removal would consistently be achieved during the processing step. For plasma-derived products, there is always the potential that a virus that is not detected by the battery of tests for starting material could be present. In the case of a protein derived from Chinese hamster ovary (CHO) cell culture, there should be no infective virus present in the material entering the virus removal filtration step. However, CHO cell cultures do contain noninfective type A and C retroviruslike particles (RVLPs) [19]. For proteins derived from murine hybridoma cell culture, infective retroviruses may be present in the cell culture–conditioned media [20]. No reliable test method is available which can demonstrate the performance of the virus filtration step, especially when there are no viruses detected in the feedstream. For this reason, the performance of the virus removal filtration step is measured by spiking and clearance experiments conducted as part of the process validation.

The assurance of viral safety is provided by validation studies that directly measure the reduction of virus by the processing of the protein either at scale (for low-volume feedstreams) or using an appropriately scaled-down prospective scale system. Model viruses are introduced into a product-containing feedstream, and the reduction in viral load is measured by comparing the amount of virus in the feedstream and the product pool. The design and execution of such studies are described in regulatory documents [1] and literature reviews [21], and provide a consistent and rational basis for comparison of various viral removal and inactivation steps. Typically, the removal of viruses is expressed as a base 10 log reduction value (LRV), which is the log of the quotient of the quantity of virus in the spiked feedstream during the validation study and the quantity of virus measured in the product pool. Clearly, a processing step that provides a large LRV for all possible viruses is ideal. Unfortunately, no processing step for virus inactivation or removal is ideal, and tradeoffs must be struck between maximal assurance of virus removal and the various selection criteria described below.

B. Product Recovery

The intention of a virus removal filtration step is to separate any virus particles that may be present in a feedstream containing the protein of interest. The virus particles are retained by the membrane and remain behind in the retentate stream. The protein passes through the membrane and is recovered in the permeate stream. The process step should be designed to maximize the recovery of the product; in some cases, with the appropriate filtration module and operating conditions, quantitative recovery can be achieved. The presence of the product in the permeate pool allows additional latitude in the selection of the module and fluid flow path equipment such as pumps, valves, and tubing [22–25]. In most ultrafiltration operations, the product remains in the retentate, and the choice of membrane area and the design of the retentate fluid flow path may be restricted so that the system can be adequately flushed to maximize product recovery, or to ensure that a sufficiently high product concentration can be achieved. This restriction is relaxed with virus removal filtration systems, where additional holdup volume in the retentate fluid flow path or large membrane area modules may result in increased dilution of the product in the permeate, but need not necessarily result in a decrease in product recovery due to holdup losses. Frequently, the VRF step is placed before an ultrafiltration step in the purification process where the product is concentrated prior to being diafiltered into the formulation buffer. Any dilution as a result of the VRF step can be accommodated by appropriately sizing the ultrafiltration membrane area of the subsequent step.

C. Integrity Testing

The integrity test of a virus removal filter is critical to the assurance of virus removal, by demonstrating that the membrane is integral postuse. In the typical operation of an ultrafiltration membrane for product concentration or buffer exchange, a failure of the integrity of the membrane may result only in product loss without a change in product quality. However, an integrity failure of a virus removal filter results in the failure of the process step to achieve its intended purpose. This may either render the product unsuitable for use, or require reprocessing of the product through a second VRF device. The VRF user relies fully on the integrity test to verify that the VRF module performed as intended and as demonstrated in validation studies. The correlation between integrity test values and the performance of the VRF membranes for retention of actual virus or viral marker (such as an appropriate-size particle or a bacteriophage) must also be experimentally sound. This correlation of retention and the integrity test value is generally provided by and validated by the VRF vendor. The data used to set specifications for the integrity test values should be examined thoroughly and the specification justified in light of the expected LRV of the test virus.

The method by which a virus removal filter is deemed integral is an important attribute of the modules available for use. Several different test methods exist including the retention of virus-size test particles, air diffusion, bubble point, and liquid porosimetry. These are described below.

Monodispersed gold particles have been used to challenge a hollow-fiber VRF device with a mean pore size of 30 nm [26,27]. The 30-nm particle size is similar to the smallest viruses used to challenge VRF devices (Minute Virus of Mice and other Parvoviridae are 18 to 24 nm and poliovirus is ~28 nm). Suspensions of 20-, 30-, 50-, and 60-nm gold particles were used to challenge the VRF membrane. The concentration of gold particles passing through the membrane was determined by measuring the absorbance of light at 530 nm. The two mechanisms of particle removal observed were plugging of particles within the capillary pores and the entrapment of particles in the larger voids of the membrane structure. The passage of particles was dependent on the operating conditions of the test such as particle size, particle concentration, and the volume processed. Asahi uses a similar procedure to correlate the retention of gold particles to the removal rate of viruses for their PLANOVA 15 and 35 VRF devices. This is a test of effective pore size. However, it is a destructive test that must be performed post-use. A less stringent leakage test using compressed air or nitrogen may be used to give additional assurance of membrane integrity preuse.

Air diffusion–based integrity tests may be correlated to bacteriophage LRV values. The air diffusion procedure involves the wetting of the membrane with an appropriate fluid. The filter housing is drained of fluid, and feed side of the VRF filter is pressurized with nitrogen or air to a specified pressure. The rate of air diffusion through the membrane (Eq. 1) can be described by the integration of Fick's law [28].

$$N = \frac{D \cdot H \cdot (P_1 - P_2) \cdot \varepsilon}{L \cdot \tau} \tag{1}$$

where N = the diffusion rate, D = the diffusivity of the gas in the liquid, H = the solubility coefficient of the gas, $(P_1 - P_2)$ = the differential pressure, ε = the porosity or void volume of the membrane, L = the thickness of the membrane, and τ = the tortuosity of diffusion path through non-straight pores.

The membrane properties that affect the air diffusion rate are the membrane porosity and thickness, not pore size. The pore size of the membrane is not relevant to the diffusion rate as long as the membrane's bubble point has not been exceeded. The air diffusion integrity test is a relatively quick test to ensure that no major damage has occurred to the membrane, but it will not confirm that the VRF device has the appropriate pore size to retain virus. The VRF user must rely on the manufacturer's internal testing (porosimetry or retention studies) of the VRF membrane and device and validation package when using an integrity test that does not measure or correlate to pore size.

The bubble point based integrity test of VRF filters is described by Eq. (2) [29–31]. The bubble point pressure (P_{bp}) is defined as the point at which air completely evacuates a wetted pore of the membrane. The bubble point is inversely related to the membrane pore diameter. However, the water bubble point of Viresolve membrane with a 3-LRV capability of a 30-nm virus would be in the order of 500 psig using water [32]. The use of an alcohol-in-water solution to lower the surface tension still results in a bubble point in excess of 200 psig. This would require the use of high-pressure systems, possibly exceeding the pressure rating of the VRF device and irreversibly compressing the VRF membrane.

$$P_{bp} = (P_g - P_1) = \frac{K \cdot 4 \cdot \sigma \cdot \cos \Theta}{d} \tag{2}$$

where P_{bp} = bubble point pressure, P_g = pressure in the gas phase, P_1 = pressure in the liquid phase, σ = surface tension of the wetting liquid, Θ = the contact angle against the solid, usually assumed to be ~0 for hydrophilic membranes, K = correction factor to account for nonideality, and d = diameter of the pore.

The use of a liquid-liquid porosimetry based integrity test (CorrTest) was described by Phillips and DiLeo [33] to overcome the high-pressure limitation of the bubble point test. Figure 1 shows an example of a correlation between the retention of the bacteriophage and the result of a liquid porosimetry test value for a 70-kDa membrane. The use of two immiscible fluids allows the integrity test to be performed without measurement of a diffusional flow rate. The fluids

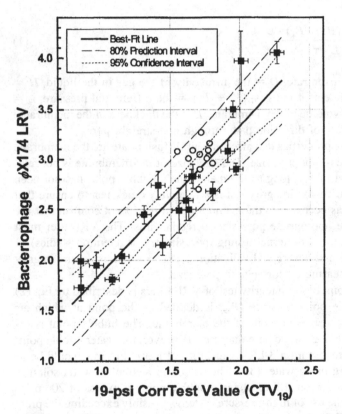

Figure 1 Experimentally determined correlation between the bacteriophage φX174 log reduction value (LRV) and the 19-psi CorrTest value. The ■ and ○ symbols represent flatstock membrane and module data, respectively. The solid line represents a least-squares linear regression of the membrane flatstock data, the dotted lines represent upper and lower 95% confidence limits on the mean, and the dashed lines represent upper and lower 80% prediction intervals. (By permission of Millipore Corporation.)

were chosen to reduce the surface tension between the phases and thus allow the porosimetry test to be conducted at more reasonable test pressures of 15 to 19 psig. The CorrTest values have been correlated to the retention of a 28-nm bacteriophage, ϕX174. This type of integrity test is nondestructive and therefore may be conducted preuse (provided the complete removal of the test solutions can be achieved and validated) as well as postuse. Another advantage of a nondestructive test is that it may be repeated in the event that a clerical or operation deviation causes an integrity test value failure or invalid result.

Performing a preuse integrity test provides added assurance that the VRF module is integral prior to use and that the VRF step will perform as intended. Most vendors pretest their VRF modules prior to shipment. However, this test may not be an integrity test that is correlated to viral retention. The user must perform a postuse integrity test that is correlated to a viral or viral-marker retention to assure that the step performed as expected. The VRF module and system must be cleaned prior to conducting the postuse integrity test. Cleaning removes residual protein and colloidal material on the VRF membrane that could affect the integrity test. The cleaning operation should also inactivate any viruses trapped in the VRF module and system. In many cases in the biopharmaceutical industry, there are no viruses or RVLPs present in the product by the VRF step in the purification process. However, the ability of this viricidal cleaning solution is of critical importance when conducting validation studies with live virus. It is important to prevent virus carryover between experiments if equipment is not dedicated to individual virus validation studies. The decontamination of the apparatus is also necessary to limit operator exposure when the system is dismantled or the module is removed for integrity testing.

If a VRF device fails the postuse integrity test, the integrity test may be repeated if the original test was nondestructive. If the postuse test has been deemed as failed (or cannot be repeated), it must be assumed that the VRF step did not provide viral clearance and the virus removal step must be repeated. It is important to consider this potential situation when developing the VRF step and develop an appropriate reprocessing procedure.

D. Validation

Validation studies employing process streams that have been spiked with model viruses are used to measure the removal of virus by the virus removal filtration process. Information from these studies as well as a description of the postuse integrity test are then reported to the regulatory agencies in the form of investigational new drug applications (INDs), biological license applications (BLAs), and other filings. It is on the basis of these validation studies that a virus removal filtration step is judged to be effective in removing virus particles. The design and execution of these studies is critical. Strict attention should be paid to the scaledown of the process system to allow testing in a laboratory. Details of validation studies will be described below.

III. TYPES OF VRF FILTERS

A. Filtration Options (TFF vs. NFF)

The removal of virus by VRF membranes is achieved through a combination of adsorption to the membrane, entrapment within the VRF membrane's structure, and sieving at the membrane surface. The protein and virus may interact with the membrane surface and deposit or adsorb to the membrane. Virus removal studies should be conducted in the process solutions identical to those used in manufacturing, and should contain the protein of interest in a typical concentration range. Viruses small enough to enter the membrane pore structure may become entrapped within the membrane pores. Viruses larger than the VRF membrane pore size will be separated from the product stream through a sieving mechanism at the membrane surface. The separation of virus from protein products by VRF devices is based on the size difference between product and virus. Clearly, as the size of the protein increases, the difference between the diameter of the virus and the effective diameter of the protein (the hydrodynamic radius) is reduced, and the separation becomes more difficult. This will result in the selection of a membrane with a larger pore size, thus decreasing the degree of virus removal, or give rise to significant retention of the product by the membrane, potentially decreasing the yield or increasing the processing time. Process development studies should determine the extent of product retention by the membrane selected, and define a process that will provide adequate passage of product despite large rejection coefficients. Some proteins are too large to pass through membranes which have adequate viral retention properties; a typical IgM would have a molecular weight of approximately 900 kDa, and a hydrodynamic radius of 22 nm, as large as some parvovirus (18 to 24 nm) [34].

For a separation mechanism based on the selective sieving of a virus by the membrane alone, it is expected that the log reduction value will correlate with virus size. In the case of filtration steps with significant polarization of the membrane surface by the product or other macromolecules, the log reduction value may not be a strong function of virus size, as the formation of a gel layer may serve to exclude all viruses independent of their size.

The two principal modes of filtration that have been used for virus retaining filters are tangential-flow filtration (TFF) and dead-end, or normal-flow filtration (NFF). The two filtration modes differ in their operation and performance characteristics.

1. Normal-Flow Filtration

During NFF, a pressure gradient normal to the membrane surface drives fluid flow through the membrane [35]. As permeate passes through the membrane, retained species build up on the membrane surface. If the membrane has a high porosity and the pores are large enough to admit solutes or virus particles, the filter acts as a depth filter, where solutes and solids collect in the membrane's structure. If the pore size is small or the porosity low, the solutes and solids build up on the membrane surface, and the filter acts as an absolute filter, where all test organisms are completely retained by the filter. NFF filters are more prone to polarization than TFF filters, due to this accumulation of solutes and solids at the membrane surface. However, these filters are quite simple to operate, the only requirement being a pressure source or a pump connected to a vessel holding the feedstream and the VRF filter.

2. Tangential-Flow Filtration

In TFF, the membrane surface is continuously swept clean by a fluid flow tangential to the membrane surface with the intent of eliminating or minimizing surface polarization. There are effectively two pressure gradients in these systems, one normal to the membrane surface to drive the fluid flow through the membrane, and one tangential to the membrane surface to drive the fluid flow past the membrane. The tangential liquid flow reduces the accumulation of solutes or solids at the membrane surface, reducing the polarization and gel layer formation observed during NFF. A TFF system minimally requires one pump, and equipment for pressure and flow rate monitoring. For many VRF systems, however, a diafiltration (constant-volume wash) of the retentate is required for maximal product recovery. Also, the permeate flow rate may have to be controlled to minimize the membrane polarization. Thus, three pumps are typically needed, which results in a more complex mechanical system than the NFF systems.

Both modes of filtration have demonstrated the ability to remove virus from various feedstreams. The development of processes that provide high product recovery and maximal virus retention for the TFF mode of operation will be described below.

B. Membranes, Vendors, Modules, and Areas

Table 1 provides a list of vendors who provide VRF filters and lists the mode of filtration, membrane areas available, membrane pore size, materials of construction, module configuration, and integrity test method. A large variety of filters have been used for viral filtration, and could be considered for evaluation, process development, and implementation.

Table 1 Available Virus Removal Filtration Products and Their Characteristics

VRF filter	Manufacturer	Membrane material	Separation basis	Device format	Mode of operation	Integrity test	Integrity test correlated to virus retention
Omega VR100	Pall Filtron	PES	Sieving	Cassette	2-pump TFF	Air diffusion	No
Omega VR200	Pall Filtron	PES	Sieving	Cassette	2-pump TFF	Air diffusion	No
Omega VR300	Pall Filtron	PES	Sieving	Cassette	2-pump TFF	Air diffusion	No
Planova 15N	Asahi	Cellulosic	Depth	Hollow fiber	Dead end	Gold particle retention	Yes
Planova 35N	Asahi	Cellulosic	Depth	Hollow fiber	Dead end	Gold particle retention	Yes
Ultipor(TM) DV20	Pall	PVDF	Depth	Cartridge	Dead end	Air diffusion	Yes
Ultipor(TM) DV50	Pall	PVDF	Depth	Cartridge	Dead end	Air diffusion	Yes
Viresolve(TM)70	Millipore	PVDF	Sieving	Open channel module	2-pump TFF	Liquid porosimetry	Yes
Viresolve(TM)180	Millipore	PVDF	Sieving	Open channel module	2-pump TFF	Liquid porosimetry	Yes
VIRA/GARD 300	AG Technology	PS	Sieving	Hollow fiber	2-pump TFF	Air diffusion	Yes
VIRA/GARD 500	AG Technology	PS	Sieving	Hollow fiber	2-pump TFF	Air diffusion	Yes
VIRA/GARD 750	AG Technology	PS	Sieving	Hollow Fiber	2-pump TFF	Air diffusion	Yes

C. Typical Performance: LRVs

Data on the performance of some of these VRF filters have been gathered in a recent review [32], which contains references to the original publications. Different viruses and feedstreams have been tested for each module. Figure 2 shows the LRV observed as a function of the average virus diameter for modules that have been tested for the removal of more than one virus. For certain filters, such as the Viresolve module, there is a clear increase in the LRV with virus diameter, as one would expect for a size-based separation. Because of limitations in either the concentration of virus which can be spiked into the feedstream during validation, or the limit of quantitation in the product pool [1], the LRVs have to be expressed as a lower limit, making the establishment of a "standard curve" of LRV as a function of virus diameter difficult. In some cases, a feedstream containing a high concentration of protein may result in elevated LRVs due to the polarization of the membrane by the protein [36,37]. However, in other cases the absence of protein and buffer salts may also lead to elevated LRVs possibly due to virus aggregation or adsorption of virus to the membrane [38]. While these data suggest that any of the VRF filters will provide some reduction in a viral load to the process, there is enough variability introduced by changes to the feedstream and operating conditions that a generic LRV cannot

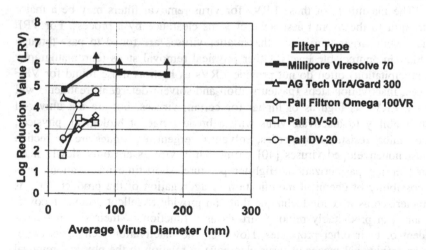

Figure 2   Relationship between the LRV of virus for several VRF devices as a function of average virus size (in nm) using the same protein product with a molecular weight of ~55 kDa. Typically, the viral retention of the VRF devices tested increased with the size of the virus. The upward pointing arrows signify the theoretical LRV based on the Poisson distribution when no virus was detected in the permeate pool.

be claimed. Instead, a validation study should be performed for each application of VRF.

In general, small viruses provided the greatest separations challenge for VRF devices. Parvovirus, being among the smallest commonly encountered virus family, could represent a worst-case validation study for biopharmaceutical products. In some cases, vendors test their VRF devices with a bacteriophage, which is slightly smaller than Parvoviridae.

The LRVs reported for many of these filters demonstrate a useful reduction of virus particles. For large viruses, some VRF filters provide >6 logs of clearance—a very large reduction, equivalent to or exceeding the best chromatographic steps. Generally, steps with >6 logs clearance or inactivation are regarded as excellent, >4 logs clearance or inactivation would be considered effective, and <1 log clearance or inactivation should not be counted. In addition, some regulatory reviewers [39] state that VRF is a robust process step; that is, normal variation in processing parameters such as solution properties (pH, protein concentration, conductivity, and excipient concentration), operating conditions (transmembrane pressure, flow rates, temperature), and membrane lot would have minimal influence on the assurance of virus removal. While some processing steps such as pasteurization and solvent-detergent inactivation are also considered robust, chromatographic separations were not automatically regarded as robust, given the complex nature of the potential interaction among virus, product, buffers, and resin.

The magnitude of these LRVs for virus removal filters may be a major contributor to the overall assurance of virus clearance by a process. For VRF devices with larger pore sizes, the smaller viruses are found to pass through the filter in appreciable levels. Other physical removal steps (chromatography and precipitation) often do not provide LRVs as high as those found for VRF devices. Inactivation steps (pasteurization and solvent-detergent treatment) will provide LRVs as high as or higher for certain viruses, but are not universal in their ability to inactivate virus with a broad range of biological, physical, and chemical resistance properties. Solvent-detergent techniques are ineffective against nonenveloped viruses [40]. Some hardy viruses are only slowly inactivated during pasteurization. High-temperature, short-time inactivation raises the possibility of chemical modification or aggregation of the product. VRF is characterized as a gentle technique that can provide excellent product recoveries, and can predictably remove viruses as a function of their size but independent of their other properties. However, the inactivation techniques represent an orthogonal means of virus clearance in relation to the physical removal provided by VRF. The combination of a large LRV from a robust removal step with an inactivation step should provide a very high level of assurance of the purification process to eliminate virus particles present in the source material.

IV. TFF DEVELOPMENT FOR VRF STEPS

This chapter focuses on the development of TFF steps for VRF in processes for the manufacture of therapeutic proteins. Other references describe the development of NFF steps for standard ultrafiltration and microfiltration membranes, with an emphasis on product recovery and not virus removal [41].

A. Operating Conditions

The typical TFF system used for VRF is shown in Figure 3. The system consists of a feed reservoir with volume or weight monitoring. The feed reservoir must be capable of holding the entire feed volume in a batch process and operating a minimum volume for diafiltration without creating vortexing or foaming. The feed tank must also be well mixed to prevent short-circuiting of the retentate

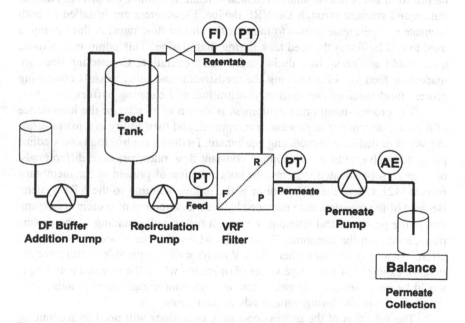

Figure 3 Schematic of typical system for TFF virus removal filtration devices. The system consists of a feed reservoir, recirculation pump, VRF module, retentate valve for creating backpressure, and pressure sensors on the feed, retentate, and permeate lines. The permeate flow rate is controlled with a positive displacement pump and it is useful to install a UV monitor in the permeate line to measure the protein concentration in the permeate during the trial. A third pump is used to add buffer to the feed reservoir during the diafiltration operation.

flow directly to the tank outlet and to assure efficient mixing of the diafiltration buffer. The feed pump should be selected to minimize the degradation of the protein product. Typically, the feed pump is a peristaltic pump for small systems and a rotary lobe pump for larger applications. Pressure sensors are placed in the feed, retentate, and permeate lines (close to the VRF module) to monitor the transmembrane pressure and retentate pressure drop of the VRF operations. The retentate stream is recirculated back to the feed reservoir. A valve is installed in the retentate line to create backpressure on the feed side of the VRF membrane. Typical operating pressures for the feed and retentate streams are <10 psig.

The permeate flow is controlled by use of a pump (typically peristaltic) to restrict the flow and control polarization at the membrane surface. The permeate pressure is monitored and the permeate flux is adjusted to maintain a stable transmembrane pressure and to keep the permeate pressure above atmospheric pressure to avoid off-gassing in the product stream. Ultraviolet (UV) monitors may be placed in the permeate and/or retentate stream to monitor the protein concentration and passage through the VRF device. Flowmeters are installed in both retentate and permeate stream to monitor the stream flow rates. A third pump is used to add buffer to the feed tank during diafiltration. This pump may also be used to add additional load during a fed-batch operation. Connecting lines are made to a feed vessel containing the feedstream, and other vessels containing process fluids such as sanitization, diafiltration, and cleaning buffers.

The process-monitoring equipment is shown to emphasize the importance of these measurements in process development, and their utility in a manufacturing step. For dedicated processing equipment, feedback control loops may adjust pump and valve settings to maintain constant flow rates, pressure differentials, or in more sophisticated systems, the concentration of protein at the membrane surface [42]. Often, the UV monitor is the unique addition to the VRF system. Because of the real-time data provided by the measurement of protein concentration in the permeate and retentate streams, a better understanding of the protein passage through the membrane is possible, which aids the process development effort. There may be cases where the UV monitor cannot provide useful information, for example, at very large scales of operation where the flowcell path length would be very large, at extremely low or high concentrations of protein, or in the presence of interfering compounds or excipients.

The selection of the process operating conditions will need to account for the effects of both fluid dynamics and protein biochemistry. Standard TFF principles and measurements must be employed to ensure control of the fluid flow, pressure gradients, and module operation. Product recovery, however, will be a strong function of the passage of protein through the membrane. Protein rejection can be a complex function of solution properties such as pH and conductivity [43], protein concentration, and membrane type. This may be due to either self-association of the protein, partitioning between the bulk solution and the pores, polarization, or other phenomena.

The principal measurement necessary for all process development activities is the rejection coefficient of the protein by the membrane. As defined in Eq. (3), the rejection coefficient is 1 minus the ratio of the concentration of the protein in the permeate to the concentration of protein in the bulk feed to the module.

$$R = 1 - \frac{C_{Perm}}{C_{Feed}} \qquad (3)$$

where R = the rejection coefficient, C_{Perm} = the protein concentration in the permeate, and C_{Feed} = the protein concentration in the feed.

The rejection coefficient influences many decisions during VRF development. A high rejection coefficient may suggest that the VRF module selected has a pore size that is too small for adequate passage of the product. Or, if the rejection coefficient is higher than would be expected from the known molecular weight of the protein and the membrane pore size, there may be some self-association of the protein, effectively raising its hydrodynamic radius and reducing its passage through the membrane.

The typical operations for a TFF step for VRF are listed in Table 2. Prior to introducing the product, the VRF system and module should be sanitized to

Table 2 Typical Steps Involved in a TFF VRF Process[a]

Clean check of system
Flush system hardware with purified water to remove system storage solution
VRF module installation
Preuse integrity test (optional)
Pre-use sanitization of system and VRF device
Flush with purified water to remove sanitization solution
Buffer equilibration
Sampling for endotoxin analysis prior to load addition
Load addition
Volume reduction
Diafiltration
Recovery of the product in the permeate pool
Cleaning of the system and VRF device
Flush with purified water to remove cleaning solution
Integrity test VRF Module
Discard VRF module following acceptable integrity test
Storage of system hardware in an appropriate solution

---

[a] In addition to the actual volume reduction and diafiltration operations that are typically thought of in developing a VRF step, there are several other important operations that must be considered.

inactivate any contaminating organisms (virus or otherwise) which may be contained in the module or the permeate fluid flowpath. This sanitization should employ solutions known to be bactericidal and viricidal [44–46], and must be compatible with the membrane chemistry and easily removed. Following sanitization, the module and fluid flowpath should be equilibrated in a buffer of similar pH and ionic strength to the load buffer; in the case of a VRF step which follows a chromatographic step, this may be the elution buffer of the previous column. After draining the system, the system is charged with the load containing the product. In the case where the load volume exceeds that of the retentate vessel volume, the load may be introduced into the feed tank in a fed-batch mode (as described below). The next step in the process is the volume reduction where the product passes through the VRF membrane and is collected in the permeate stream. Typically, a high degree of product passage through the membrane will allow the process step to first reduce the retentate volume to a minimum processing volume that still allows comfortable operation of the pump and monitoring systems. A volume limit may be reached prior to the minimum process volume if there is a critical protein concentration that should not be exceeded, and there is some retention of the protein [see Eq. (4) below]. It is during this volume reduction step that additional load may be added into the retentate vessel in a fed-batch manner to maintain a constant, maximum volume in the retentate vessel as permeate is removed from the system until the load vessel is emptied.

Following the volume reduction step, the remaining product in the retentate is washed through the membrane, often by a constant-volume diafiltration using a buffer similar or identical to the load buffer. After a sufficient volume of wash buffer has been processed during diafiltration, there should be essentially no product left in the retentate. The product pool containing the permeate from both the volume reduction and diafiltration operations can be removed for further processing. The module and system are then cleaned to remove residual protein and inactivate virus particles present. The VRF device is then integrity tested and discarded following an acceptable integrity test value. The system is stored in an appropriate storage solution.

The process operation must specify the extent of volume reduction and diafiltration volumes. Using a known rejection coefficient (and the assumption that this value remains constant over the process step), estimates can be made of the percentage of product in the permeate pool following volume reduction and diafiltration with Eq. (4), when a minimum process volume following volume reduction can be achieved:

$$\%\text{Recovery} = 100 \cdot [1 - (CF)^{R-1} \cdot e^{(R-1)DV}] \qquad (4)$$

where CF = the concentration factor of the volume reduction operation = V_o/V_{min}, V_o = the initial process volume prior to volume reduction, V_{min} = the system volume reached after volume reduction, R = the protein rejection coefficient, DV = the number of diavolumes = V_{DF}/V_{min}, and V_{DF} = the permeate volume from the diafiltration operation. Because of the significant product recovery during the volume reduction phase, a smaller number of diavolumes are typically performed than during an ultrafiltration buffer exchange operation.

Assuming a constant rejection coefficient, the maximum product concentration reached during the process operation is at the end of the volume reduction step, and is given by Eq. (5):

$$C_{max} = C_o(CF)^R = C_o\left(\frac{V_o}{V_{min}}\right)^R \tag{5}$$

where C_{max} = the maximum product concentration reached after volume reduction, C_o = the initial protein concentration in the feed at the start of the volume reduction, CF = the concentration factor of the volume reduction step = V_o/V_{min}, V_o = the initial process volume prior to volume reduction, V_{min} = the system volume reached after volume reduction, and R = the protein rejection coefficient.

If this maximum protein concentration exceeds a critical concentration established by process development studies, the volume reduction step should halt at the volume calculated using Eq. (5) after solving for the retentate volume at that critical protein concentration, and the diafiltration begun at that point. In the extreme case, there may be no volume reduction step, or the load may be diluted before processing. Note that slight changes in the initial protein concentration and volume will give rise to variable permeate pool volumes during the volume reduction operation; conservative estimates may allow more consistent operation in terms of permeate volumes, but will cause variation in the maximum protein concentration achieved. Often, the process is controlled by measurements of the retentate volume (retentate vessel level sensors, load cells, etc.) and permeate volume (permeate collection vessel level sensor, load cell, or flow rate totalizer).

For a given product recovery, an estimate can be made for the volume of diafiltration buffer required. Using Eq. (4), Table 3 was generated and lists the product recoveries that arise from various combinations of concentration factors and diafiltration volumes for products having various rejection coefficients. The diafiltration volumes were expressed as diavolumes, which is the diafiltration volume divided by the entire retentate volume. The diavolumes should be calculated based on the entire fluid volume of the retentate, including both the volume in the retentate vessel and the holdup volume of the lines, pump, monitors, and flowmeters. Failure to include this holdup volume will result in decreased product recovery.

Table 3 Simulation of Purification Process for a Protein Product Passing Through the Membrane and Being Recovered in the Permeate Pool Assuming a Constant 0%, 10%, or 30% Product Retention and 99.999% Virus Retention by the Membrane

Operation description	Product recovery @ 0% retention	Product recovery @ 10% retention	Product recovery @ 30% retention	Overall LRV @ 99.999% retention	Overall dilution factor
5× Volume reduction	80.0%	76.5%	67.6%	4.8	0.80
1× Diafiltration	92.6%	90.5%	83.9%	4.6	1.00
2× Diafiltration	97.3%	96.1%	92.0%	4.4	1.20
3× Diafiltration	99.0%	98.4%	96.0%	4.3	1.40
4× Diafiltration	99.6%	99.4%	98.0%	4.3	1.60
5× Diafiltration	99.9%	99.7%	99.0%	4.2	1.80
10× Volume reduction	90.0%	87.4%	80.0%	4.6	0.90
1× Diafiltration	96.3%	94.9%	90.1%	4.5	1.00
2× Diafiltration	98.6%	97.9%	95.1%	4.4	1.10
3× Diafiltration	99.5%	99.2%	97.6%	4.3	1.20
4× Diafiltration	99.8%	99.7%	98.8%	4.2	1.30
5× Diafiltration	99.9%	99.9%	99.4%	4.1	1.40
20× Volume reduction	95.0%	93.3%	87.7%	4.5	0.95
1× Diafiltration	98.2%	97.3%	93.9%	4.4	1.00
2× Diafiltration	99.3%	98.9%	97.0%	4.3	1.05
3× Diafiltration	99.8%	99.5%	98.5%	4.2	1.10
4× Diafiltration	99.9%	99.8%	99.3%	4.2	1.15
5× Diafiltration	100.0%	99.9%	99.6%	4.1	1.20

In all cases, the combination of volume reduction and diafiltration will dilute the product; the dilution factor is given by Eq. (6). This product dilution may have an influence on downstream steps, as described below.

$$DF = \frac{V_{final}}{V_o} = \frac{(V_o - V_{min}) + DV(V_{min})}{V_o} = 1 + \frac{DV - 1}{CF} \quad (6)$$

where DF = the overall product dilution factor for the entire step including volume reduction and diafiltration, V_{final} = the final process volume following the entire processing step, V_o = the initial process volume prior to volume reduction, V_{min} = the system volume reached after volume reduction, CF = the concentration factor of the volume reduction step = V_o/V_{min}, R = the protein rejection coefficient, DV = the number of diavolumes = V_{DF}/V_{min}, and V_{DF} = the permeate volume from the diafiltration operation.

The product recovery calculation provided by Eq. (4) can also be applied to the LRV estimated for viruses having a known or estimated rejection coefficient. The effect of process variables such as the concentration factor and diafiltration volume on the LRV can be estimated. In Table 3, the LRV for a virus that has an average rejection coefficient of 99.999% provides an estimate of the magnitude of the effect of these variables on virus removal. The LRV is found to be relatively insensitive to these modest changes in operating conditions. Because of the much lower rejection coefficient for product, the product recovery is more sensitive to these parameters than the virus removal.

B. Fluid Dynamics

The fluid flow dynamics within the module are set by the fluid properties and the pressure gradients resulting from the pumps' operation, valve settings, and on occasion, the polarization of the membrane by solute or solids. The transmembrane pressure (TMP), given in Eq. (7), is the average driving force for liquid to pass through the membrane. The tangential, or feed side pressure drop, is abbreviated ΔP, and is given in Eq. (8):

$$TMP = \frac{P_i + P_o}{2} - P_{Perm} \quad (7)$$

where TMP = the transmembrane pressure, P_i = the feed side inlet pressure, P_o = the feed side outlet pressure, and P_{Perm} = the permeate pressure.

$$\Delta P = P_i - P_o \quad (8)$$

where ΔP = the feed side pressure drop. Generally, the fluid flow through the pores is laminar, and therefore proportional to the TMP. The tangential flow rate usually follows a power-law dependence on pressure drop, with an exponent between 1.0 (laminar) and 0.6 (fully turbulent), depending on the Reynolds number of the retentate channel flow, and the absence or presence of turbulence-promoting screens which disrupt steady fluid flow [47]. The tangential-flow rate contributes to sweeping the membrane clean by reducing the boundary layer adjacent to the membrane. Correlations of the mass transfer coefficient to the cross-flow fluid velocity have been measured, and provide insight into the mechanisms of solute transport [23]. The conversion is defined by Eq. (9), and is the fraction of the feed which is drawn off in the permeate:

$$\%\text{Conversion} = 100 \left(\frac{Q_{\text{Perm}}}{Q_{\text{Feed}}} \right) \tag{9}$$

where Q_{Perm} = the permeate flow rate and Q_{Feed} = the feed flow rate.

The permeate pump acts as a valve to restrain the permeate flow rate, causing a positive pressure relative to the atmosphere on the permeate side of the module (in most TFF ultrafiltration operations, the permeate flow is not restricted, and is thus often close to atmospheric pressure, after accounting for fluid flow pressure drops and pressure head differences). This flow restriction controls the permeate flux through the membrane, and reduces the tendency of the membrane to polarize during unrestricted operation. In general, a positive permeate pressure should be maintained, to prevent off-gassing of the permeate stream, and the introduction of a relatively large area of gas-liquid interface, which may denature some proteins [48]. It has also been shown that a restriction in permeate flow, as well as cocurrent permeate flow, can improve the separation of similarly sized solutes [49], and could contribute to the maintaining acceptable product retention by the membrane.

The fluid dynamics of any module and hardware system can be evaluated with protein-free solutions to allow sizing of the hardware (pump, valves, and monitors). Mixing studies can be performed to evaluate the efficiency of the mixing of the diafiltration and retentate streams during the diafiltration step [50]. When protein is introduced into the system, differences in pressures and flows may be observed. Some differences will be expected in the TMP upon introduction of product, given the colloidal nature of the protein and the resulting colligative effects of different product concentrations in the feed and retentate [51], but this should be minor for VRF membranes that are highly permeable to the product. Should polarization take place, TMP values will rise. The high concentration of protein on the retentate side of the membrane may cause protein multimerization or aggregation, which should be avoided [52–54].

Polarization has been shown to have a slight influence on the LRVs measured for bacteriophage removal [37], but is not the principal component controlling the magnitude of the LRV. The minimum performance of the VRF system as measured by the LRV is set by the characteristics of the membrane, which under most situations is only mildly influenced by small changes in fluid flow conditions.

C. Biochemistry and Solution Properties of Proteins

Protein products may be prone to self-association, multimerization, and ultimately, aggregation [55]. These interactions are protein specific, and will depend upon solution conditions. Multimerization and aggregation may lead to increased TMP, product loss, and process variability. The effects of buffer conditions such as pH, ionic strength, temperature, and excipient levels may all influence product passage by affecting the protein self-association and therefore the product rejection coefficient. For certain mixtures, the difference in the molecules' size and/or multimerization state can lead to size fractionation by differential passage through the membrane. Such control of protein passage has been demonstrated for IgG and BSA separation using high-performance TFF (HPTFF), which employs permeate flux control as well as careful attention to the solution pH and composition [49].

Certain buffer excipients can affect protein passage, some in ways not fully understood. Modest concentrations of sodium chloride can increase protein passage [56–61], perhaps by decreasing the Debye double layer and effectively reducing the protein's effective hydrodynamic radius. Sucrose may also increase protein passage, but does not have a universal effect on all proteins (data not shown). The addition of detergents such as Triton X-100 or Tween 80 will increase passage for some proteins, probably by reducing self-association of proteins based on hydrophobic interactions [62]. Figure 4 shows the rejection coefficient of a protein through a 180-kDa VRF membrane in a buffer containing various excipients (sodium chloride, sucrose, Tween). Protein concentration was also decreased by a factor of 10, to test whether a reduction in protein concentration would affect product passage. In this case, only the addition of Tween increased product passage, and was found to be effective at concentrations as low as 0.01% (v/v). The effects of these variables on the hydrodynamic radius of the protein can be evaluated by various biophysical methods, including gel filtration [63], analytical ultracentrifugation [64], and quasi-elastic light scattering [65].

In some cases multimers will form, and cause an increase in hydrodynamic radius. This will increase the rejection coefficient of the complex, and the retentate will be enriched in multimer. In some cases, the only protein that remains in the retentate will be multimer or aggregate; the failure of the retentate's UV absorbance to decrease to zero during extended diafiltration is an indication that the protein is present as a high-molecular-weight species which is being retained by the membrane.

High concentrations of protein may drive self-association, which follows higher-order reaction kinetics. Thus, modest increases in protein concentration

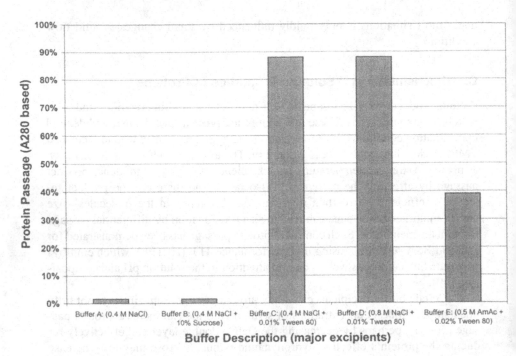

Figure 4   Passage of a ~170-kDa protein through a Viresolve/180 VRF device in various buffer conditions. Using buffer A, which contained 0.4 M NaCl, very little protein passed through the membrane. The addition of 10% sucrose in buffer B did not affect the passage. However, the addition of 0.01% Tween (buffer C) dramatically increased the protein passage to ~90%. The addition of more NaCl to 0.8 M in buffer D did not enhance the protein passage more than the addition of Tween 80. When the NaCl was changed to ammonium acetate with 0.02% Tween 80 (buffer E), the protein passage through the membrane decreased to 40%.

may have a dramatic effect on the protein aggregation state [66]. The equations above should be used for the design and analysis when the rejection coefficient does not change with protein concentration. These equations must be modified for the cases where the retention varies with protein concentration, or due to buffer pH or composition changes during diafiltration.

D. Experimental Evaluation and Data Acquisition

Process development experiments should gather information to be used for the design of the process step, according to the principles described above. If possible, the experimental TFF apparatus should include the monitoring and control equipment described in Figure 1. Process development studies should evaluate the effects of the variables shown in Table 4. This data set will include all relevant pressures, flow rates, and real-time estimates of the concentration factor and diavolumes. The rejection coefficient can be estimated by concurrent UV monitoring of the permeate and feedstreams, or by offline analysis of samples taken during the operation.

Initial studies should be conducted to test the rejection of the product for various solution conditions. Measuring hydrodynamic radius with a rapid experimental method such as quasi-elastic light scattering can accelerate this process. Figure 5 shows the hydrodynamic radius of an antibody as a function of pH, and clearly demonstrates that the operating range for the VRF should be above a pH of 4. There was also an increase in the molecule's effective hydrodynamic radius near its isoelectric point of ~7. Similar studies can be performed quickly using gel filtration, and changes in the retention time of the product as a function of solution conditions correlate to hydrodynamic radius. Special care should be to ensure that inaccurate results are not obtained due to interaction with the gel filtration resin [67].

Subsequent studies should examine the effects of permeate flux and retentate flow rate on the rejection coefficient. An efficient means of performing these studies is to operate the TFF system in total recycle mode, where the permeate and retentate are both returned to the retentate vessel. In this manner, there will be no change in the protein concentration of the feed. Measurements of the product concentration in the permeate allow estimates of the rejection coefficient for the different operating conditions. Prior to initiating the permeate flow, the absorption of the protein to the membrane and/or module can be tested by measuring the product concentration during total recycle. This is possible if the load concentration is relatively low and the available test methods are accurate enough to detect small changes in concentration. Permeate flow should then be initiated, with a slow and measured increase in permeate flow rate.

An example of the UV signal from the permeate stream is shown in Figure 6. In this case, the UV monitor also provides information about the changes in protein concentration following a change in permeate flux, and the duration of these transients before a steady state is reached.

The rejection coefficient and TMP as a function of permeate flux for another protein are shown in Figure 7. The rejection coefficient decreased with permeate flux, which would not be expected by simple polarization theories. Rather, the increase in permeate product concentration with permeate flux was likely the result of an increase in the concentration of protein at the membrane surface [68]. The actual rejection coefficient of the membrane is constant, but the effective (measured) rejection coefficient is changed by virtue of the increase in concentration gradient at the membrane surface. The TMP increased linearly with permeate flux.

Table 4 Sample Data Sheet Used for Collection of Data During Process Development Activities

Product: ___ Product MW: ___ Module Type: ___ Date: ___
Batch/Lot #: ___ Desired VCF 1: ___ Membrane: ___ Membrane Catalogue #: ___ Page: ___
Initial Volume: ___ Desired DF 1: ___ Element Size: ___ Membrane Lot #: ___ Operator: ___
Product Conc.: ___ Other VCF/DF: ___ Total Area Installed Membrane Serial #: ___ System: ___
GOALS: Run flux excursion in total recycle to determine effect of operating conditions on the product retention. Integrity Spec @ ___ psi:

Comment	RUN TIME	TIME	TEMP (°C)	Pfeed (psi)	Pret (psi)	Pperm (psi)	TMP (psi)	dP (psi)	Q ret (mL.pm)	Qperm (mL.pm)	Q total (mL.pm)	Flux (LMH)	V perm VCF (L)	VCF (X)	V perm DF (L)	# WV (N)	[feed]	[perm]	Retention %
Assemble system with one 0.33 ft2 Viresolve/180 module.																			
Add 2 L of buffer to tank. Start system and direct retentate and permeate streams to drain until stream conductivities and pH's are within specs. Add more buffer to tank as needed.																			
Buffer Permeability		1050	23	10.3	10	7.9	2.25	0.3	590	60	650	116.1							
Drain system and add ~2 L of the Load solution to the process tank and set up system in total recycle.																			
Flux excursion		1100	23	10	9.7	9	0.85	0.3	650	13.6	664	26.3					0.359	0.266	25.9%
		1105	23	10	9.7	9.1	0.75	0.3	650	13.6	664	26.3							
		1111	23	10	9.7	9	0.85	0.3	650	19.8	670	38.3					0.358	0.306	14.5%
		1117	23	10	9.7	9	0.85	0.3	650	19.8	670	38.3							
		1118	23	10	9.7	8.3	1.55	0.3	650	25.8	676	49.9					0.356	0.324	9.0%
		1124	24	10	9.7	8.3	1.55	0.3	650	25.8	676	49.9							
		1128	24	10	9.7	7.7	2.15	0.3	650	32	682	61.9					0.356	0.332	6.7%
		1134	24	10	9.7	7.7	2.15	0.3	650	32	682	61.9							
		1135	24	10	9.7	7.2	2.65	0.3	650	38.2	688	73.9					0.354	0.336	5.1%
		1140	24	10	9.7	7.2	2.65	0.3	650	38.2	688	73.9							
		1145	24	10	9.7	6.8	3.05	0.3	650	44	694	85.2					0.354	0.338	4.5%
		1152	24	10	9.7	6.8	3.05	0.3	640	44	684	85.2							
Decrease flux and feed flow rate. Repeat flux excursion.																			
Flux excursion		1155	24	10	9.7	9	0.85	0.3	440	13.5	454	26.1					0.358	0.246	31.3%
		1205	24	10	9.7	9	0.85	0.3	440	13.6	454	26.3					0.356	0.246	30.9%
		1210	24	10	9.7	8.9	0.85	0.3	440	13.5	454	26.1							
		1212	24	10	9.7	8.9	0.95	0.3	440	20	460	38.7					0.356	0.299	16.0%
		1222	24	10	9.7	8.4	0.95	0.3	440	20	460	38.7							
		1223	24	10	9.7	8.4	1.45	0.3	440	25.8	466	49.9					0.361	0.323	10.5%
		1233	25	10	9.7	7.8	1.45	0.3	440	25.8	466	49.9							
		1235	25	10	9.7	7.8	2.05	0.3	440	32.2	472	62.3					0.353	0.333	5.7%
		1245	25	10	9.7	7	2.05	0.3	440	32.2	472	62.3							
		1246	25	10	9.7	7	2.85	0.3	440	38.1	478	73.7					0.357	0.34	4.8%
		1256	25	10	9.7	6.5	2.85	0.3	440	38.1	478	73.7							
		1258	25	10	9.7	6.5	3.35	0.3	440	43.9	484	85.0					0.355	0.342	3.7%
		108	25	10	9.7	6.5	3.35	0.3	440	43.9	484	85.0							

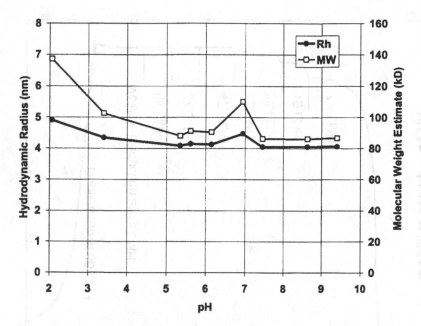

Figure 5 The hydrodynamic radius (Rh) and estimated molecular weight for an antibody as a function of the solution pH. The hydrodynamic radius of the molecule increased
as the pH was brought to <4. It is interesting to note that there is an increase in the
molecule's size near its isoelectric point of ~7.

During permeate flux excursion experiments, one may encounter rapid increases in transmembrane pressure during operation. This is an indication of the
buildup of protein on the membrane surface, resulting in polarization of the membrane. In some cases, this effect is reversible if the permeate flux is reduced
quickly, while in others, the effect may be permanent and the membrane must
be replaced or cleaned prior to continuation of the experiment. The maximum
permeate flux should be chosen which can still pass significant quantities of product while remaining comfortably below the polarization limit.

When a system is set up for total recycle, the protein concentration of the
feed will remain constant. The effect of protein concentration on the rejection
coefficient can be tested by either increasing or decreasing the protein concentration while in total recycle. Dilution with buffer will decrease the protein concentration; to increase the protein concentration, a second ultrafiltration membrane
system can be used to draw liquid from the retentate vessel, return the retentate,
and withdraw the UF permeate, thus increasing the protein concentration in the
VRF retentate vessel. In one such study, the rejection coefficient of another protein was found to be dependent on the concentration of protein, as shown in

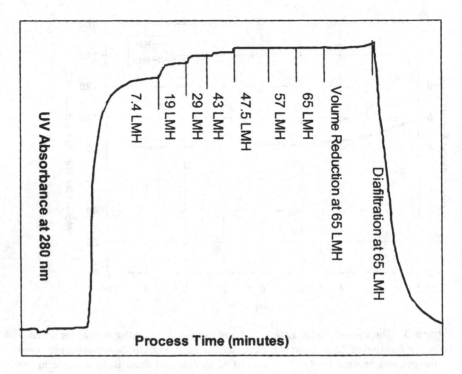

Figure 6 Typical flux excursion experiment. The UV absorbance (A280) of the permeate stream during a flux excursion and process simulation experiment with a ~32-kDa
protein using a Viresolve/70 VRF device is shown. This flux excursion experiment was
conducted in total recycle and therefore the feed concentration of protein remained constant. As the flux is increased step wise, the permeate concentration also increases in a
stepwise manner. The permeate passage approaches 100% as the flux is increased from
7.4 LMH to 65 LMH. Following the flux excursion, a process simulation was run at 65
LMH consisting of a volume reduction followed by a diafiltration operation.

Figure 8. For this case, it appears that one can operate safely up to approximately
7 mg/mL protein concentration before rejection increases to ~30%. In many
cases, protein concentrations <1 to 2 mg/mL show little dependence of the rejection coefficient on protein concentration.

There may be a critical protein concentration above which the membrane
will become polarized and the protein rejection increases dramatically. In such
a case there will be a need to limit the concentration factor during the initial
volume reduction step due to an increasing protein concentration in the feed. The
dilution factor will therefore increase due to the need to conduct the diafiltration
operation at a larger retentate volume.

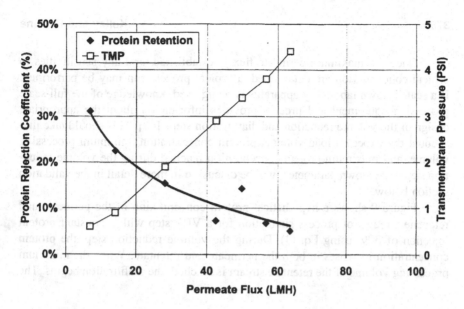

Figure 7 Protein rejection and TMP by a Viresolve/70 VRF device as a function of the permeate flux for a protein product with a molecular weight of ~55 kDa. The rejection coefficient of the protein decreased to ~5% as the flux was increased 60 LMH. The TMP increased linearly with increased permeate flux.

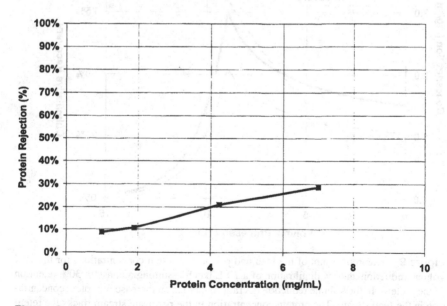

Figure 8 Protein rejection by a Viresolve/70 VRF device as a function of the feed concentration. The retention of protein at a given flux increases as the feed concentration is increased. Processing at high protein retention levels will result in elevating the feed concentration during the volume reduction operation. There will also be a need to increase the amount of diafiltration to achieve high step recovery.

Once the maximum permeate flux is established, and any sensitivities to protein concentration are determined, a typical process run may be performed. If a scaled-down laboratory apparatus is being used, knowledge of the full-scale processing equipment and process provides information about the appropriate design of the volume reduction and diafiltration steps [69]. This should take into account the expected load volume, protein concentration, minimum processing volume, and maximum protein concentration reached during the volume reduction step. Scaledown parameters will be discussed in more detail in the validation section below.

Figure 9 shows a hypothetical protein concentration of the permeate and retentate streams of process simulation for a VRF step with a constant protein rejection of 30% using Eq. (4). During the volume reduction step, the protein concentration increases in both the permeate and retentate. When the minimum processing volume of the retentate stream is reached, the diafiltration begins. The

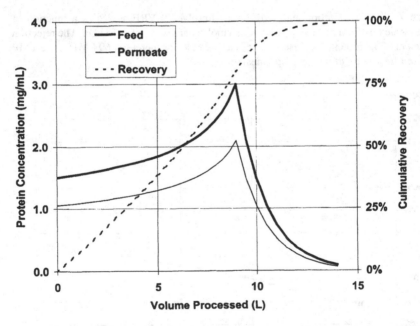

Figure 9 Theoretical plot of the feed and permeate protein concentrations for the 10X volume reduction and 5X diafiltration of a 10-L batch assuming a constant 30% retention of the protein. In the volume reduction operation, there is an increase in protein concentration in the feed stream. The protein concentration in the permeate stream tracks the retentate concentration with a similar but lower profile. During the diafiltration operation, both the feed and permeate protein concentrations asymptotically approach zero. The cumulative protein recovery in the permeate pool is shown on the secondary y-axis. The theoretical protein recovery in the combined permeate pool for this process is ~99.4%.

product washout during the diafiltration step shows the expected exponential decrease in concentration of both retentate and permeate streams. A semilogarithmic plot of protein concentration versus time should provide a linear washout profile, with the slope of the line proportional to the product of the one minus the rejection coefficient times the number of diafiltration wash volumes. As shown in Table 3, five diavolumes are needed for this case to ensure ~99.4% product recovery. The process development effort should strive for a maximum product recovery, but strike the necessary compromise considering aspects of product dilution, process time, and membrane area. With sufficient diafiltration volumes, the retentate holdup volume and membrane area will not adversely affect product recovery.

The operating limits of retentate flow rate, permeate flux, protein concentration, and maximum concentration factor should be tested during development to ensure product recovery is not affected. The effect of some of these variables on virus reduction can be estimated using Eq. (4). At this stage, the process optimization and robustness usually focuses on product recovery and the prevention of polarization, and not on virus reduction.

E. Placement in Purification Train

The placement of a VRF step in purification train balances several aspects of TFF and process design. Considerations should include protein concentration and purity, buffer conditions and their effects on hydrodynamic radius, total process volume, and compatibility with subsequent steps. Figure 10 shows a hypothetical process flow sheet with process volumes, protein concentrations, and purity, and comprises several chromatographic steps. The following placements could be considered.

1. Early Placement

Early in the process following the capture step, the protein is often in dilute concentration, which may improve the protein passage through the membrane. This would necessitate the design of a fairly large TFF system to allow the processing of a large volume of permeate in a reasonable time period. If the protein is not very pure, however, complications may arise from high-molecular-weight contaminants that would be retained by the membrane and possibly cause polarization. Depending on the subsequent processing step, there may be flexibility in adding certain buffer excipients to aid in protein passage, provided they do not affect subsequent steps and can be removed in the chromatography steps downstream.

2. Middle Placement

Once the protein is pure and fairly concentrated midway through the purification, the VRF step could be introduced prior to the final polishing steps. A smaller system could be used, and the complications of contaminating proteins elimi-

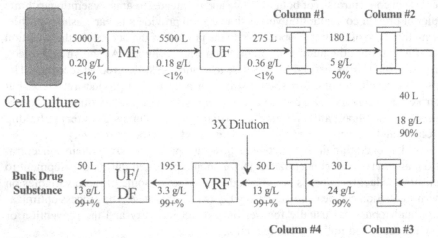

Figure 10 Hypothetical process flow sheet of a purification process for a recombinant product showing the volume, concentration, and purity of the protein. Placement of the VRF step later in the process and before an ultrafiltration step provides a relatively pure stream to process and the ability to concentrate the protein following the dilution from the VRF step.

nated. Care should be taken to understand the effects of the inevitable dilution of the product stream by the VRF step, as some chromatographic steps are sensitive to the volume of the load, especially size exclusion chromatography [63].

3. Late Placement

Many processes end with an ultrafiltration/diafiltration (UF/DF) step to adjust the concentration of the protein and buffer exchange it into the formulation buffer. The placement of the VRF step after the last column step and immediately before the final UF/DF is often a natural fit. The dilution of the protein is easily corrected by operating the UF portion of the UF/DF step for a longer period to compensate for the additional UF/DF load volume. If buffer excipients are added to the VRF step to increase passage, these could be removed by the DF step. Be certain that the excipients are removed to the necessary limit when conducting the final DF step; if high concentrations of sodium chloride are needed to increase product passage, the VRF permeate will require significant diafiltration to remove the salt to sufficiently low levels. Caution should be exercised if surfactants are added to the VRF step, as they will form micelles above their critical micelle concentration [70,71] and the partitioning of the detergent through the VRF step and subsequent removal by the UF/DF can be complex.

It is possible that the VRF step could be conducted with a diafiltration buffer that is not equivalent to the load buffer. If the subsequent step is ion exchange chromatography, a low ionic strength may be required for protein binding. By diafiltering with a solution of low ionic strength, the VRF permeate may be adjusted to the correct conductivity prior to the chromatography step. Conversely, if the subsequent step is a hydrophobic interaction step, the diafiltration buffer could contain a high concentration of the necessary salt to cause adsorption to the resin. Because the extent of dilution is a function of the initial load volume to the VRF step and the number of diavolumes, this strategy of changing the composition of the diafiltration buffer would give rise to variability in the permeate pool composition if not carefully controlled for these variables.

V. IMPACT ON PRODUCT QUALITY

The VRF step should not affect the product quality, although there may be modifications that arise from the processing. Some proteins are shear sensitive, and the TFF operation could result in multimerization and aggregation. Loss of activity could be associated with changes to the tertiary structure of the protein. Chemical modifications may occur, and could alter the protein isoform distribution.

A. Aggregation, Multimerization, and Denaturation

Protein aggregation and multimerization may occur for proteins sensitive to self-association. There are many examples of TFF steps causing protein aggregation [42,52,53]. The aggregation may result from elevated liquid shear generated by the pump or valving, the introduction of air-liquid interfaces that provide a hydrophobic surface resulting in denaturation [48], or uncontrolled temperature rises in the process fluid. While small-scale laboratory systems can be designed to test the effects of these variables on the product, it is difficult to translate such data to full-scale operation, as the maximum shear rate and total air-liquid interface may be difficult to predict for the process equipment. Still, these experiments can provide an early warning of potential scale-related problems which may assist in equipment design and process definition.

These modifications to the protein structure may be measured by several methods. Size exclusion HPLC gives a reliable, quantitative determination of the level of aggregation, and may also distinguish multimers from the product. UV absorbance at the near UV wavelengths such as 320 nm will detect aggregation even at low levels. While less quantitative than SEC-HPLC, this method is very rapid and sufficiently sensitive.

The consequences of the formation of multimers and aggregates are sometimes reflected in the process monitoring equipment. A rise in the TMP while maintaining a constant permeate flux could indicate polarization and the start of an aggregation process. However, not every aggregation event will cause an increase in TMP. The retentate may contain more product than would be predicted by Eq. (2), and this retained product should be subjected to SEC-HPLC to determine its multimeric state. A UV monitor on the retentate stream would register an anomalous UV trace if significant aggregate is formed, and light in the near UV range is also scattered by the aggregate.

B. Chemical Modifications

The VRF step requires the product to be present in solution for a defined period during processing, including the time required to set up the system prior to operation. If the protein is sensitive to chemical modifications in the buffer conditions used for the VRF step, changes to the product quality may result. Examples of chemical modifications that may occur include deamidation, pyroglutamic acid formation from N-terminal glutamine residues, oxidation, or disulfide bond scrambling at high pH [72]. The tendency for the protein to react under the buffer conditions of the VRF step can be tested by hold control samples which are not subjected to the VRF step, but are stored at the appropriate temperature and tested at intervals up to and beyond the time period required for VRF processing.

C. Product-Specific Characterization Assays

During the process development of a VRF step, all aspects of product quality should be tested for samples subjected to VRF processing. These should include product-specific release and characterization tests. Often, the steps subsequent to VRF step must be performed to prepare the protein for testing, and exchange it into the appropriate buffer matrix for the assays. Such tests could include an activity assay, measurements of secondary or tertiary structure like circular dichroism and analytical ultracentrifugation, SDS-PAGE gels, etc. Usually, physical separation methods like VRF are unlikely to cause major product changes.

VI. IMPLEMENTATION AND MANUFACTURING

The transfer of technology between process development laboratories and the manufacturing environment is a crucial element in the implementation of any processing step for biologics manufacturing. Specific issues that often arise relate to the scaleup of the process step, communication during technology transfer, data acquisition at manufacturing scale, confirmation of predicted performance, and ultimately, implementation of process improvements.

A. Scaleup Issues

Successful and efficient translation of a VRF process from a laboratory scale to full-scale manufacturing is aided by a data package from laboratory experiments which defines operating ranges, establishes critical limits such as maximum protein concentrations, and evaluates the robustness of the process to changes in feedstreams and raw materials. Perhaps the most important attribute of the development program is the confirmation that the small-scale experimental apparatus is an appropriate scaledown of the full-size manufacturing process [69]. This will ensure that the process limits and performance projections are accurate, and are also critical for the process validation effort described below.

Various approaches to the scaleup of TFF systems have been described [22,73], including descriptions of appropriate system flowpath design; preferences for specific types of instrumentation, pump, valve, and monitor; and details of process control loops and data acquisition. TFF equipment and processes for VRF have no unique requirements which complicate the design of a TFF system, other than the introduction of a permeate pump and the utility of a UV monitor on the permeate stream.

For very large systems comprising many individual membrane modules, the integrity testing becomes problematic. If tested all at once, a single module's failure may not be detected by some of the integrity test methods described above. That is, the specification limit set for the integrity test value is not low enough to detect the failure of one module in an assembly of 20 identical modules when compared to the specification limit based on the loss of integrity of the entire membrane area. Rather, preuse testing may be used to confirm the integrity of each module prior to assembly, and then an integrity test measurement could be used to set a performance limit that is unique to that combination of modules. Postuse integrity testing could then compare the integrity test value to the preuse value, and an individual specification set for the system under evaluation. The inability to integrity test a combination of VRF modules will require the individual testing of each module used in the VRF step. If even one of the VRF modules fails the integrity test, the combination of filters used for the process step must be deemed as failing to perform the step's intended purpose and the feed material must be reprocessed.

B. Technology Transfer and Data Acquisition

The transfer of a process developed in a laboratory to the manufacturing facility requires the coordination of a number of groups (process development, manufacturing, validation, engineering, etc.), and the communication and recording of a great deal of information. A complete list of system specifications, control ranges, and critical process-monitoring variables should be generated at the conclusion of the process development effort. These are used for the design of the TFF system, and the resulting process flow diagrams and piping and instrumentation diagrams should be reviewed by the process development group for accuracy. The system operations performed in the development lab must then be translated in standard operating procedures and manufacturing batch records to be used for the manufacturing system. The validation effort will then confirm that the system was fabricated as intended, perform the necessary operations and control sequences required for processing, and finally perform the actual process correctly.

It may be useful during the early trial runs of a VRF process to perform additional process monitoring and gather information which characterize the system performance to a degree beyond that needed for typical process control. This data set may include noncritical measurements or trend analyses of protein rejection at various points in the process, the effect of TMP on protein concentration, and characterization of any product, productrelated species, or contaminants which remain in the retentate following operation. In these early runs, it is informative to test for changes in the feedstream properties or quality by running development-scale systems concurrently, and checking the process performance against expectations. This study may serve also to identify any scale-related differences between the full-scale and scaledown systems that must be corrected before validation studies are performed.

C. Improvements for Manufacturing Use

Following implementation of a VRF step into a manufacturing process, there may be a desire to modify or improve the operations to reduce costs, labor requirements, or improve process consistency. The reuse of TFF membranes used for VRF is not recommended by any vendors supplying these membranes. However, the cost of the processing step would of course be reduced through reuse of VRF membranes much as standard ultrafiltration membranes are reused. The critical issue that prevents casual reuse of the membranes is the concern over a loss of membrane integrity for a device that has undergone multiple cleaning and sanitization cycles. However, it could be expected that a validation package could be assembled which addresses the performance of membranes that have been reused for multiple cycles. Special attention should be paid to the conditions of the regeneration cycle to ensure that the pore size distribution of the membrane or its separation capabilities are not significantly altered. The ability to effectively clean the VRF device and system to prevent carryover or cross-contamination between batches must also be validated. In combination with nondestructive post-use integrity testing, this should allow sufficient assurance that the membranes are still suitable for use after one or more cycles. The ability to reuse a VRF device may be technically possible, but it is not clear that this would be a viable business strategy in the biopharmaceutical industry.

The automation of the integrity test is also a useful operation that is not part of a typical manufacturing system, at least not at small scale. The consistency of the integrity test could be improved by automation, however, and for tests based on porosimetry methods, a significant labor reduction can be realized. While many VRF membranes are integrity tested by the vendor prior to shipment, an automated integrity test on the assembled system modules could be used to ensure that the modules were not damaged during shipment or assembly.

VII. VALIDATION OF THE VIRUS RETENTION

Validation of the VRF system to remove viruses from a feedstream is required for each process developed using a VRF device. In addition to certain general principles of validation of filtration systems [74–76], specific studies must be executed to determine the capacity of the VRF step to remove viruses from the process stream. While there may be reason to claim a minimum reduction of virus for a VRF module whose performance has been well characterized by multiple validation studies, changes in the buffers and the presence of the product may influence the actual removal values achieved. These validation studies should be designed and executed with an understanding of the expectations of the regulatory agencies governing the countries where the product is expected to be marketed. The depth and breadth of the validation package may reflect the stage at which the product is being evaluated, whether for an early-phase clinical study of an experimental product candidate, or a license application for a new product.

A. Design and Qualification of the Scaledown System

Except for the smallest of process scales, the VRF validation studies will employ a scaledown processing system [77]. The scaledown system used for the validation studies must provide an accurate replication of all the full-scale process operations, and reflect an understanding of the effects of scale on TFF operation.

Vendors have worked to provide a wide range of module areas to allow for small-scale development and validation studies to be performed (see Table 1). However, very small-scale systems (<0.01 m^2) are not commonly produced. As an example, the Viresolve modules manufactured by Millipore are currently available in three modules of very different design, containing the same VRF membrane. The process modules are 0.7 and 1.4 m^2, the intermediate scale modules are 0.3 and 0.1 m^2 and are used for pilot studies or small-scale manufacturing, and 0.001-m^2 modules are available for feasibility testing. The vendor does not claim or provide data to support the contention that these three different module designs are readily scalable and hence suitable for scaleup studies or validations. Asahi does provide the Planova series of VRF filters in a wide range of membrane areas (0.001, 0.01, 0.12, 0.3 and 1 m^2), but the path length of the hollow fiber varies between ~8.5 and ~21 cm.

The restrictions of validation studies, however, are such that often scaledown systems must be used. The virus titer stocks are expensive and may be difficult to produce in large quantities, which argues for the use of small-scale systems. Also, the consumption of product required for the validation studies can be very large, especially at an early stage of product manufacturing, where every gram of product may be allocated for various development activities. For these reasons, the smallest scale of operation should be evaluated for the viral validation studies, provided the scaledown system can be qualified as an appropriate model for the full-scale manufacturing process.

The scaledown parameters that should be maintained for TFF process operations include the ratio of volume to surface area, the volume reduction factor, the maximum number of diafiltration wash volumes allowed in the process, and the permeate flux (LMH), which should result in the same process time. The crossflow velocity is a difficult parameter to set for scaledown systems that employ a different module design that may change the retentate pathlength, the channel height, or the presence or absence of turbulence-promoting screens. No strict recommendations can be made here; each vendor's family of modules must be evaluated separately based on testing of the appropriate process performance based on fundamental studies of mass transfer coefficients, process performance including protein retention, and polarization observations.

The qualification of the scaledown system must assess critical performance parameters and compare them to those of the full-scale manufacturing system. Typically, one to three small-scale runs are performed, and compared to a larger manufacturing data set. The concentration factor during the volume reduction and the number of diafiltration wash volumes must match those of the manufacturing process. Worst-case conditions should be tested (maximum concentration factors and maximum number of diafiltration volumes). The UV monitoring of the permeate stream should show the expected profile during both the volume reduction and diafiltration operations (recall that the diafiltration profile should be linear on a semilogarithmic graph). The protein rejection coefficient should be measured and compared to the rejection coefficient measured in the full-scale system. Overall product recovery should be assessed, and a mass balance performed by measuring both the permeate and retentate product concentrations. Statistical methods such as a Student's t-test can be used to compare the mean values of critical performance parameters such as product recovery of the small-scale system to the manufacturing process [78].

If at all possible, the load material used for the qualification runs should be representative of the full-scale manufacturing process and, ideally, taken from a full-scale run. The expense incurred by diverting material that could be used for commercial sale or by extending a clinical manufacturing campaign can be significant, depending upon the scaledown factor of the laboratory system. For this reason, material from startup runs or shakedown runs for the manufacturing process can be used, provided the material is not significantly different from the full-scale cGMP manufacturing material.

B. Choice of Viruses

The panel of viruses used for the validation study should be chosen carefully. Table 5 lists commonly used viruses, and properties relevant to their selection as part of a virus validation package. Because the physical principle of the removal of virus by a VRF step is based on the size difference between the virus and the product, the size of the viruses used is an important criteria for selection. Certainly a small virus should be chosen as one of the viruses tested (a parvovirus or picornavirus, for instance) to provide a worst-case estimate of the virus removal provided by the VRF step, given the correlation of LRV to virus size. It is reasonable to expect that larger viruses would be removed to a greater degree, provided there are no complications with the particular virus tested and any potential it may have to aggregate and therefore be more easily removed. Validating the removal of a large virus would be expected to demonstrate the full capability of the VRF step by testing the removal of a virus at the upper end of the size range. The selection of a range of virus sizes, then, could provide an estimate of the maximum and minimum clearance that can be claimed for a VRF step.

The other physical and biological properties of the viruses (presence of an envelope, genome type, shape, etc.) should be considered so that a wide spectrum of virus types is used. With a broad representation of virus types, the panel would be likely to contain a model virus which is representative of unknown viruses present in the feedstream. This argument is held very strongly for the validation of inactivation steps, or removal steps such as chromatography or precipitation, but may not be as relevant for the validation of a VRF step which has predictable performance based on the size of the virus (see Figure 2). However, the virological safety profile of the source material used to generate the product will affect the selection of the viruses used for validation studies. The spectrum of viruses which may be present in human serum is different from the spectrum of viruses which have been shown to infect well-characterized cell lines such as CHO cells, a popular host cell line for recombinant protein production. The viruses selected for removal validation studies should reflect this difference, and viruses typically chosen for plasma products (HIV, HAV, HBV, polio) are different from those

Table 5 Properties of Commonly Used Viruses for Validation Studies

Virus	Size (nm)	Natural host	Envelope	Shape	Genome	Family	Physicochemical resistance
Canine parvovirus	18–24	Canine	No	Icosahedral	DNA	Parvo	Very high
Porcine parvovirus	18–24	Porcine	No	Icosahedral	DNA	Parvo	Very high
Minute virus of mice	18–26	Mouse	No	Icosahedral	DNA	Parvo	Very high
Poliovirus Sabin type 1	25–30	Human	No	Icosahedral	RNA	Picorna	Medium
Encephalomyocarditis virus	25–30	Mouse	No	Icosahedral	RNA	Picorna	Medium
Simian virus 40	40–50	Monkey	No	Icosahedral	DNA	Papova	Very high
Bovine viral diarrhea virus	50–70	Bovine	Yes	Pleospherical	RNA	Flavi	Low
Sindbis virus	50–70	Human	Yes	Spherical	RNA	Toga	Low
Reovirus-3	60–80	Mammals	No	Icosahedral	DNA	Reo	Medium
Adenovirus	70–90	Mammals	No	Icosahedral	DNA	Adeno	Medium
Vesicular stomatitis virus	70 × 175	Bovine	Yes	Bullet	RNA	Rhabdo	Low
Xenotropic murine leukemia virus	80–110	Mammals	Yes	Spherical	RNA	Retro	Low
Human immunodeficiency virus	80–120	Mammals	Yes	Spherical	RNA	Retro	Low
Herpes simplex virus	120–150	Mammals	Yes	Spherical	DNA	Herpes	Medium
Pseudorabies virus	120–200	Swine	Yes	Spherical	DNA	Herpes	Medium
Parainfluenza virus	150–300	Mammals	Yes	Pleospherical	RNA	Paramyxo	Low

that serve as models for adventitious viruses which can infect CHO cells (MVM, PI3, MuLV, etc.).

When conducting virus removal filtration validation studies with actual virus, high-titer virus stocks (at $\sim 10^7$ to 10^8 virus particles/mL) are used to spike the load material. The spike volume can vary from 1% to 10% of the load volume depending on the desired volume reduction factor of the process. Using a high virus spike-to-load ratio may result in a validation run with excessive polarization caused by the virus and its associated media when the process has a high volume reduction factor (10–20X). The spike ratio is also dependent on the size of the load pool being used for the study. Using large amounts of virus stocks can be expensive and it also may be difficult to supply the desired amount of virus.

C. Study Design

The validation studies should be designed to meet current regulatory expectations. Replicate runs using the same virus are recommended (ICH) to demonstrate process consistency. A validation package containing data on the removal of multiple viruses provides a strong foundation on which to claim consistent and significant virus reduction. Recommendations have been made for the appropriate statistical analyses of the virus titration data [1–4]. Controls must be performed to test for interference of the load solution on the viral titration assay, either through cytotoxicity of the indicator cells, or interference with the infection cycle. Hold controls should be taken to determine if the feedstream containing the product inactivates the virus during the duration of the process, and held at the process temperature.

All liquid streams should be sampled for virus titration, including the spiked load, the permeate pool, and the final retentate pool. This sampling plan allows a determination of the mass balance of the virus spiked into the load, and in combination with the hold control, will indicate if significant inactivation of the virus occurred during the process. More detailed sampling could include samples of the permeate which may be taken at various points during the volume reduction or diafiltration operations. These time point samples would show whether the virus retention is consistent during the process, and are often analyzed for information only.

The sampling plan for the validation study should be carefully considered. A disposable prefilter with a pore size of 0.2 μm or 0.45 μm is often used to remove potential virus aggregates from the spiked load solution. If a prefilter is used, samples should be taken after the filtration to verify the actual virus titer of the solution being used to challenge the VRF device. Measuring the virus titer before and after the prefiltration will provide information on the degree of aggregation of the virus. The virus titer for the post-VRF product is measured from the combined permeate pool from the volume reduction and diafiltration operations.

The large virus reduction values provided by some VRF processes introduce a complication in the determination of the extent of removal. In many cases, there is no detectable virus in the permeate pool. The LRV reported is a minimum LRV based on the known number of viruses present in the load, and maximum number of viruses which could be present in the permeate. This estimate of the number of viruses in the permeate is the limit of detection of the assay, and is based on the statistical evaluation of the viral titration performed [79]. The minimum concentration of virus that may be detected by the viral titration assay is a function of the volume of sample which is tested, reflecting the Poisson distribution of a dilute solution of virus, as well as the level of interference of the test article on viral infectivity or cytotoxicity, which will require a minimum dilution of the sample.

One means of increasing the level of virus removal that can be estimated is to employ very high virus challenges to the VRF step in the load. Both the viral titer of the stock and the volume of the virus spike can be increased to achieve this. While high titer virus stocks provided by concentration of virus cultures may provide high-titer stocks, there is concern over the potential for virus aggregation. Still, the ICH guidelines on viral removal validations recommend the use of high-titer virus stocks for validation studies. The benefit provided by an increase in the volume of the virus spike for a removal validation study is modest at best, as the typical range of virus spike is between 1% and 5% (vol/vol) and only one-half of a log of challenge is gained by increasing the spike volume 3.2-fold. Larger spike volumes result in both a dilution of the product and buffer in the load material, but also may introduce additional solutes present in the virus stock, such as serum, bovine serum albumin, or inactive virus or virus fragments. For large spike volumes, it is recommended that a mock run with the virus spike buffer added in place of the virus stock be performed prior to the validation runs to ensure that no process upsets will occur. The virus spike buffer may contain culture media and serum proteins that could affect the VRF operation. When it is expected that there will be complete retention of the virus by the VRF device, a larger permeate sample volume should be assayed to decrease the detection limit.

The studies should be documented to a degree consistent with the intention of filing the results with regulatory agencies. If the studies are performed at a contract lab, it is important to ensure that the lab is managed under good laboratory practices. An audit of the laboratory and all aspects of its quality and process control should be performed.

D. Execution of Studies

The execution of the virus removal studies provide the data necessary to establish the claim that the VRF step is a consistent and significant element contributing to the virological safety profile of the product. Attention must be paid to the operation of the scaledown system during the validation studies to ensure the data accurately reflect the VRF step performance. Following the completion of the process step, each VRF module must pass the appropriate postuse integrity test, based on the test used for the process modules.

The introduction of high-titer virus spikes and any proteins present in media may affect the operation of the module. In some cases, the membrane may become polarized, and an increase in TMP may occur. If the TMP rises to levels which threaten the integrity of the feed or permeate tubing, fittings, or pump, the permeate flow rate should be decreased to prevent catastrophic failure of the system. If the permeate flow rate is reduced, it is still important to complete the volume reduction operation with the required concentration factor, and for the diafiltration operation to continue for the required number of wash volumes. The reduction in permeate flow rate, then, will increase the process time. However, the removal of virus by the VRF step should be more representative for a process which has processed the required volume than for a step which decreased the amount of permeate volume to maintain the same process time.

Other aspects of the study execution should reflect the sensitivity of the assays to detect small levels of contamination of the permeate stream by minute contamination from the virus spike or the spiked load. Care must be taken to ensure no cross-contamination could arise from sampling, liquid collection, or mislabeling. For viruses that are known to be infectious in humans, appropriate precautions should be taken to prevent transmission by liquid contact, aerosols, or residual deposits on hardware. Disposable labware should be used when at all possible.

E. Published Results

Several published case studies allow comparison of some of the virus removal filters currently available [80–84].

VIII. CONCLUSIONS

Virus removal filtration can improve the virological safety profile of therapeutic products derived from human plasma or recombinant animal cell culture. VRF steps provide an orthogonal method of virus clearance to inactivation methods, or purification steps such as chromatography or precipitation. Other advantages of VRF steps are that they are robust to changes in operating parameters or feedstream composition, they provide large clearance factors for most viruses, and they are gentle to protein products and typically provide excellent product recovery. The principle of a VRF step is based on the removal of viruses from the product by virtue of their size difference, which allows the product to pass through an ultrafiltration membrane while the virus particles are retained. Processing flexibility is provided by selection of either tangential- or normal-flow filtration modules. In addition to the standard equipment used for tangential-flow ultrafiltration applications, a tangential-flow VRF system often requires the use of a permeate pump, and will benefit from a UV monitor on the permeate stream to provide information about the product passage. A tangential flow VRF process typically employs a volume reduction step followed by a diafiltration step to provide maximal product recovery. The process development for VRF steps should measure the product retention by the membrane for different buffer conditions and product concentrations, and determine if a critical product concentration or permeate flux must be avoided to prevent membrane polarization. Product recovery can be accurately predicted by equations that account for the retention of product by the membrane during filtration. Finally, the VRF step must be validated for the removal of viruses, and the validation studies must be designed in concert with the guidance documents of the appropriate regulatory agencies.

ACKNOWLEDGMENTS

The authors would like to thank coworkers and colleagues at Genetics Institute for valuable insight and assistance in developing VRF steps, including Jon Coffman, Bob Costigan, Jeff Deetz, Barry Foster, Andrea Knight, Scott Orlando, Kim Sterl, Steve Vicik, and Suresh Vunnum. Lastly, the vast experience and practical advice from George Oulundsen and Herb Lutz of Millipore Corporation are most greatly appreciated.

REFERENCES

1. ICH. Viral safety document. Viral safety evaluation of biotechnology products derived from cell lines of human or animal origin. Fed Reg 1996; 21:881–891.
2. ICH. Quality of biotechnology products: derivatization and characterization of cell substrates used for production of biotechnology/biological products. 1996.
3. FDA. Points to consider in the characterization of cell lines used to produce biologicals. 1993.
4. Center for Biologics Evaluation and Research, FDA. Draft points to consider in the manufacture and testing of monoclonal antibody products for human use. 1994.
5. Nissen E, Konig P, Feinstone SM, Pauli G. Inactivation of hepatitis A and other enteroviruses during heat treatment (pasteurization). Biologicals 1996; 24:339–341.
6. Horowitz B, Wiebe ME, Lippin A, Stryker MH. Inactivation of viruses in labile blood derivatives. I. Disruption of lipid-enveloped viruses by tri(n-butyl)phosphate detergent combinations. Transfusion 1985; 2:516–522.
7. Horowitz B, Prince AM, Hamman J, Watklevicz C. Viral safety of solvent/detergent-treated blood products. Blood Coag Fibrin 1994; 5:S21–S28.

8. Charm SL, Landau S, Williams B, Horowitz B, Prince AM, Pascual D. High-temperature short-time heat inactivation of HIV and other viruses in human blood plasma. Vox Sang 1992; 62:12–20.

9. Bos OJM, Sunye DGJ, Nieuweboer CEF, Van Engelenburg FAC, Schuitemaker H, Over J. Virus validation of pH-treated human immunoglobulin products produced by the Cohn fractionation process. Biologicals 1998; 26:267–276.

10. Corash L. Inactivation of viruses in human cellular blood components. Vox Sang 1994:211–216.

11. Ben-Hur E, Horowitz B. Virus inactivation in blood. AIDS 1996; 10:1183–1190.

12. Brown TT. Laboratory evaluation of selected disinfectants as viricidal agents against porcine parvovirus, pseudorabies virus, and transmissible gastroenteritis virus. Am J Vet Res 1981; 42:1033–1036.

13. DiLeo AJ, Allegrezza AE, Builder SE. High resolution removal of virus from protein solutions using a membrane of unique structure. Biotechnology 1992; 10:182–188.

14. Sekiguchi S, Ito K, Kobayashi M, Kosuda M, Kwon KW, Ikeda H. Preparation of virus-free pyridoxylated hemoglobin from the blood of HTLV-1 healthy carriers. Biomater Art Cells Art Org 1988; 16(1–3):113–121.

15. Aranha-Creado H, Oshima K, Jafari S, Howard G Jr, Brandwein H. Virus retention by a hydrophilic triple-layer PVDF microporous membrane filter. PDA J Pharm Sci Technol 1997; 51:119–124.

16. Adamson S, Bonam D, Brodeur S, Charlebois T, Clancy B, Costigan R, Drapeau D, Hamilton M, Hanley K, Kelley B, Knight A, Leonard M, McCarthy M, Oakes P, Sterl K, Switzer M, Walsh R, Foster W, Harrison S. The manufacturing process for recombinant Factor IX. Sem Hematol 1998; 35:4–10.

17. Adamson S, Charlebois T, O'Connell B, Foster W. Viral safety of rFIX. Sem Hematol 1998; 35:22–27.

18. Petrone J, Knight A, Kelley BD, Foster B, Costigan B, Vunnum S. Virus removal filtration for recombinant Factor IX. Presented at North American Membrane Society National Meeting, Cleveland, OH, May 19, 1998.

19. Lieber MM, Benveniste RE, Livingston DM, Todaro GJ. Mammalian cells in cell culture frequently release type C viruses. Science 1973; 182:56–59.

20. Levy J, Lee H, Kawahata R, Spitler L. Purification of monoclonal antibodies from mouse ascites eliminates contaminating infectious mouse type C viruses and nucleic acids. Clin Exp Immunol 1984; 56:114–120.

21. Darling AJ, Spaltro JJ. Process validation for virus removal: considerations for design of process studies and viral assays. BioPharm 1996; 9(Oct):42–50.

22. Lydersen BK, D'Elia NA, Nelson K, Rudolph EA, MacDonald JH. Tangential Flow Filtration Systems for Clarification and Concentration. Bioprocess Engineering: Systems, Equipment and Facilities. New York: Wiley-Interscience, 1994.

23. Cheryan M. Ultrafiltration Handbook. Lancaster, PA: Technomic Publishing, 1986.

24. Meltzer TH, Jornitz MW. Filtration in the Biopharmaceutical Industry. New York: Marcel Dekker, 1998.

25. Zeman LJ, Zydney AL. Microfiltration and Ultrafiltration: Principles and Applications. New York: Marcel Dekker, 1996.

26. Hirasaki T, Noda T, Nakano H, Ishizaki Y, Manabe S, Yamamoto N. Mechanism of removing monodisperse gold particles using cuprammonium regenerated cellulose

hollow fiber (iBMM or BMM) from aqueous solution containing protein. Polymer J 1994; 26:1244–1256.

27. Tsurumi T, Osawa N, Hirasaki T, Yamaguchi K, Manabe S, Yamashiki T. Mechanism of removing Japanese encephalitis virus (JEV) and gold particles from a suspension using cuprammonium regenerated cellulose hollow fiber (BMM hollow fiber). Polymer J 1990; 22:304–311.

28. Reti AR. An assessment of test criteria in evaluating the performance of sterilizing filters. Bull Parenter Drug Assoc 1977; 31(4):187–194.

29. Emory S. Principles of integrity-testing hydrophilic microporous membrane filters, Part 1. Pharm Technol 1989; 13(9):68–77.

30. Emory S. Principles of integrity-testing hydrophilic microporous membrane filters, Part 2. Pharm Technol 1989; 13(10):36–46.

31. Meltzer TH. A critical review of filter integrity testing. Part 1. The bubble-point method; assessing filter compatibility; initial and final testing. Ultrapure Water 1989; 6(4):40–51.

32. Levy RV, Phillips MW, Lutz H. Filtration and removal of viruses from biopharmaceuticals. In: Meltzer TH, Jornitz MW, eds. Filtration in the Biopharmaceutical Industry. New York: Marcel Dekker, 1998:619–646.

33. Phillips M, DiLeo AJ. A validatible porosimetric technique for verifying the integrity of virus-retentive membranes. Biologicals 1996; 24:243–253.

34. Wilhelm P, Pilz I, Palm W, Bauer K. Small-angle X-ray studies of a human immunoglobulin M. Eur J Biochem 1978; 84:457–463.

35. Goel V, Accomazzo MA, DiLeo AJ, Meier P, Pitt A, Pluskal M, Kaiser R. Dead-end microfiltration: applications, design, and cost. In: Ho WSW, Sirkar KK, eds. Membrane Handbook. New York: Van Nostrand Reinhold, 1992.

36. DiLeo AJ, Vacante DA, Deane EF. Size exclusion removal of model mammalian viruses using a unique membrane system, Part I. Membrane qualification. Biologicals 1993; 21:275–286.

37. DiLeo AJ, Vacante DA, Deane EF. Size exclusion removal of model mammalian viruses using a unique membrane system, Part II. Module qualification and process simulation. Biologicals 1993; 21:287–296.

38. Pall Corporation Scientific and Technical Report STR-PUF 24. Virus Retention by Ultipor VF Grade DV50 Membrane Filters. 1995.

39. Levy RV, Phillips MW, Lutz H. Filtration and removal of viruses from Biopharmaceuticals. In: Meltzer TH, Jornitz MW, eds. Filtration in the Biopharmaceutical Industry. New York: Marcel Dekker 1998:622.

40. Horowitz B, Wiebe ME, Lippin A, Stryker MH. Inactivation of viruses in labile blood derivatives. Transfusion 1985; 25:516–522.

41. Troccoli NM, McIver J, Losikoff A, Poiley J. Removal of viruses from human intravenous immune globulin by 35 nm nanofiltration. Biologicals 1998; 26:321–329.

42. Van Reis R, Goodrich EM, Yson CL, Frautschy LN, Whiteley R, Zydney AL. Constant C_{wall} ultrafiltration process control. J Membrane Sci 1997; 130:123–140.

43. Saksena S, Zydney AL. Effect of solution pH and ionic strength on the separation of albumin from immunoglobulins by selective filtration. Biotechnol Bioeng 1994; 43:960–968.

44. Lydersen BK, D'Elia NA, Nelson KL, Thompson PW. Cleaning of process equipment: design and practice. In: Bioprocess Engineering: Systems, Equipment and Facilities. New York: Wiley-Interscience, 1994:471–497.

45. Sofer GK, Hagel L. Handbook of Process Chromatography: A Guide to Optimization, Scale Up, and Validation. Orlando, FL: Academic Press, 1997.

46. Klein M, Deforest A. Antiviral action of germacides. Soap Chem Specialties 1963; 39:70–72,95.

47. Tutunjian RS. Ultrafiltration processes in biotechnology. In: Cooney CL, Humphrey AE, eds. Comprehensive Biotechnology. Elmsford, NY: Pergamon Press, 1985:411–437.

48. Maa YF, Hsu CC. Protein denaturation by combined effect of shear and air-liquid interface. Biotechnol Bioeng 1997; 54:503–512.

49. Van Reis R, Gadam S, Frautschy LN, Orlando SE, Goodrich EM, Saksena S, Kuriyel R, Simpson CM, Pearl S, Zydney AL. High performance tangential flow filtration. Biotechnol Bioeng 1997; 56:71–82.

50. Tatterson GB. Scaleup and Design of Industrial Mixing Processes. New York: McGraw-Hill, 1994.

51. Zydney AL, Pujar NS. Protein transport through porous membranes: effects of colloidal interactions. Colloids Surf A Physicochem Eng Asp 1998; 138(2–3):133–143.

52. Meireles M, Aimar P, Sanchez V. Albumin denaturation during ultrafiltration: effects of operating conditions and consequences on membrane fouling. Biotechnol Bioeng 1991; 38:528–534.

53. Kim KJ, Chen V, Fane AG. Some factors determining protein aggregation during ultrafiltration. Biotechnol Bioeng 1993; 42:260–265.

54. Kelly ST, Zydney AL. Protein fouling during microfiltration: comparative behavior of different model proteins. Biotechnol Bioeng 1997; 55:91–100.

55. Manning MC, Patel K, Borchardt RT. Stability of protein pharmaceuticals. Pharm Res 1989; 6(11):903–918.

56. Van Eijndhoven RHCM, Saksena S, Zydney AL. Protein fractionation using electrostatic interactions in membrane filtration. Biotechnol Bioeng 1995; 48:406–414.

57. Smith FG III, Deen WM. Electrostatic double-layer interactions of spherical colloids in cylindrical pores. J Colloid Interface Sci 1980; 78:444–465.

58. Smith FG III, Deen WM. Electrostatic effects on the partitioning of spherical colloids between dilute bulk solution and cylindrical pores. J Colloid Interface Sci 1983; 91: 571–590.

59. Munch WD, Zestar LP, Anderson JL. Rejection of polyelectrolytes from microporous membranes. J Membr Sci 1979; 5:77–102.

60. Deen WM. Hindered transport of large molecules in liquid-filled ores. AIChE J 1987; 33:1409–1425.

61. Bil'dyukevich AV, Ostrovskii EG, Kaputskii FN. Ultrafiltration of model solutions of high-molecular-weight compounds. Influence of the ionic strength on the ultrafiltration of protein solutions. Colloid J USSR 1989; 51:300–303.

62. Van Holten RW, Quinton GJ, Oulundsen GE. Viral clearance process. U.S. patent 6,096,872, (2000).

63. Hagel L. Gel filtration. In: Janson JC, Ryden L, eds. Protein Purification: Principles, High Resolution Methods, and Applications, 2nd ed. New York: John Wiley & Sons, 1998:9–143.

64. Liu J, Shire SJ. Analytical ultracentrifugation in the pharmaceutical industry. J Pharm Sci 1999; 88:1237–1241.

65. Bowen WR, Hall NJ, Pan LC, Sharif AO, Williams PM. The relevance of particle size and zeta-potential in protein processing. Nature Biotechnol 1998; 16:785–786.

66. Kiefhaber T, Rudolph R, Kohler HH, Buchner J. Protein aggregation in vitro and in vivo: a quantitative model of the kinetic competition between folding and aggregation. Biotechnology 1991; 9:825–829.

67. Scopes RK. Protein Purification: Principles and Practice, 3rd ed. New York: Springer-Verlag, 1994.

68. Opong WS, Zydney AL. Diffusive and convective protein transport through asymmetric membranes. AIChE J 1991; 37:1497–1510.

69. Van Reis R, Goodrich EM, Yson CL, Frautschy LH, Dzengeleski S, Lutz H. Linear scale ultrafiltration. Biotechnol Bioeng 1997; 55:737–746.

70. Shaw DJ. Introduction to Colloid and Surface Chemistry. London: Butterworth, 1980.

71. Everett DH. Basic Principles of Colloid Science. London: Royal Soc Chem Pub, 1988.

72. Hejnaes K, Matthiesen F, Skriver L. Protein stability in downstream processing. In: Subramanian G, ed. Bioseparations and Bioprocessing, Vol. II. New York: Wiley-VCH, 1998:31–65.

73. Michaels SL, Antoniou C, Goel V, Keating P, Kuriyel R, Michaels AS, Pearl SR, de los Reyes G, Rudolph E, Siwak M. Tangential flow filtration. In: Olson WP, ed. Separations Technology: Pharmaceutical and Biotechnology Applications. Buffalo Grove, IL: Interpharm Press, 1995.

74. Biotechnology Task Force on Purification and Scale-up. Industrial perspective on validation of tangential flow filtration in biopharmaceutical applications. PDA J Parenter Sci Technol 1996; 46:S3-S13.

75. Millipore Corp. Validation of tangential flow filtration systems. Technical brief, 1991.

76. Michaels SL. Validation of tangential flow filtration systems. J Parenter Sci Technol 1991; 45:218–223.

77. Walter JK, Werz W, Berthold W. Process scale considerations in evaluation studies and scale-up. Dev Biol Stud 1996; 88:99–108.

78. Box GEP, Hunter WG, Hunter JS. Statistics for Experimenters. New York: John Wiley, 1978:21–56.

79. Darling AJ. Validation of the purification process for viral clearance evaluation. In: Sofer G, Zabriske DW, eds. Biopharmaceutical Process Validation. New York: Marcel Dekker, 2000.

80. O'Grady L, Losikoff A, Poiley J, Fickett D, Oliver C. Virus removal studies using nanofiltration membranes. Dev Biol Stud 1996; 88:319–326.

81. Maerz H, Hahn SO, Maassen A, Meisel H, Roggenbuck D, Sato T, Tanzmann H, Emmrich F, Marx U. Improved removal of virus-like particles from purified monoclonal antibody IgM preparation via virus filtration. Nature Biotechnol 1996; 14:651–652.

82. Manabe SI. Removal of virus through novel membrane filtration method. Dev Biol Stud 1996; 88:81–90.

83. Hughes B, Bradburne BA, Sheppar A, Young D. Evaluation of anti-viral filters. Dev Biol Stud 1996; 88:91–98.

84. Walter J, Allgaier H. Validation of downstream processes. In: Hauser H, Wagner R, eds. Mammalian Cell Biotechnology in Protein Production. Berlin, Germany: Walter de Gruyter, 1997:453–482.

Index

Printed in the United States
by Baker & Taylor Publisher Services